DIFFERENTIAL CALCULUS AND ITS APPLICATIONS

Michael J. Field
University of Houston

Dover Publications, Inc.
Mineola, New York

Bibliographical Note

This Dover edition, first published in 2012, is an unabridged republication of the work originally published in 1976 by the Van Nostrand Reinhold Company Limited, New York and London.

Library of Congress Cataloging-in-Publication Data

Field, Mike.
Differential calulus and its applications / Michael J. Field. — Dover ed.
p. cm.
Originally published: New York : Van Nostrand Reinhold, 1976.
Summary: "This text offers a synthesis of theory and application related to modern techniques of differentiation. Based on undergraduate courses in advanced calculus, the treatment covers a wide range of topics, from soft functional analysis and finite-dimensional linear algebra to differential equations on submanifolds of Euclidean space. 1976 edition"—Provided by publisher.
Includes bibliographical references and index.
ISBN-13: 978-0-486-49795-2 (pbk.)
ISBN-10: 0-486-49795-X (pbk.)
1. Differential calculus. I. Title.

QA304.F5 2012
515'.33—dc23

2012020148

Manufactured in the United States by Courier Corporation
49795X02 2014
www.doverpublications.com

Contents

Preface

My primary concern in writing this book about several-variable differential calculus has been to achieve a synthesis of theory and application in the framework of the modern approach to differentiation as expounded in Dieudonné's treatise on the Foundations of Modern Analysis. By 'application' I mean here not only a working knowledge and familiarity with the techniques of differentiation but also applications of the theory to applied mathematics and other areas of pure mathematics, notably differential topology and differential equations.

The text is based on undergraduate courses on advanced calculus that I have taught at the Universities of Minnesota and Warwick and covers a fairly wide range–from soft functional analysis and finite-dimensional linear algebra to differential equations on submanifolds of Euclidean space. I hope that the book will not only be suitable for the final two years of an undergraduate degree course in pure or applied mathematics but will also form a suitable basis for graduate level courses on advanced calculus and differential manifolds.

Chapter 0 contains a brief resumé of prerequisites. These essentially consist in a knowledge of elementary linear algebra (linear maps, matrices, bases, etc.) together with a little point set topology. For the latter, it is sufficient for most of the text to know only the rudiments of metric space theory (that is, the definitions of a metric space, metric space topology, continuous map and completeness). Apart from the material referred to in Chapter 0, the book is relatively self-contained.

Chapter 1 naturally splits into two parts. The first half is mainly soft functional analysis and includes the elementary theory of normed vector spaces, inner product and Hilbert spaces, spaces of continuous linear and multilinear maps and the Riesz representation and Hahn–Banach theorems. The second half is devoted to finite-dimensional theory and includes the spectral theorem for self-adjoint and orthogonal maps (though not the Jordan canonical form theorem) and the elementary theory of quadratic forms.

In Chapter 2 we give a systematic exposition of the elementary theory of differentiation on normed vector spaces. Stress is laid on examples of differentiable functions defined on open subsets of Euclidean space. There are sections on extrema, curves in $\mathbf{R}^n$, various forms of Taylor's theorem and higher derivatives.

More advanced topics in the theory of differentiation are covered in Chapter 3. Among these are the inverse-function theorem, implicit-function theorem and the rank theorem. In the final section of this chapter, coordinate transformations are discussed together with applications to the solution of partial differential equations.

The final chapter is intended both as a consolidation of the theory given in the preceding chapters and as an introduction to differential manifolds and differential equations. The key results of Chapters 2 and 3 are repeatedly used in this chapter. Critical point theory, differential structures, submanifolds of Euclidean space, orientation of hypersurfaces, vector fields, tangent spaces and bundles are among the topics discussed.

Finally, one or two remarks about notation. The derivative of a map $f: E \longrightarrow F$ at x will be denoted by Df_x (not $Df(x)$) and will be a linear map from E to F. The value of Df_x at $e \epsilon E$ will be denoted by $Df_x(e)$ (not $Df(x)(e)$). Arbitrary functions will be denoted by $f, g, h, \ldots$ but linear maps will usually be denoted by capitals $S, T, U, \ldots$ We write the composition of the functions g and f by gf but the composition of the linear maps S and T by $S \, . \, T$. This 'dot' notation is expanded considerably in the text; suffice it to say that its use helps simplify many computations and formulae. For typographical convenience and clarity of exposition I have adopted the notation '$f_{;j}(x)$' to mean the jth partial derivative of f at x. These, and other conventions, are explained more fully in the text.

September, 1975 M. J. FIELD

CHAPTER 0

Preliminaries

0.1 Vector spaces and linear maps

In this section we shall make a brief survey of that part of the elementary theory of vector spaces and linear maps that will be assumed in the rest of the text. For more detailed expositions we refer the reader to any of the many basic texts on linear algebra and finite-dimensional vector spaces (Halmos [9] is particularly recommended).

The symbol K will always be used to denote a field which, in practice, will either be **R** (the 'real numbers') or **C** (the 'complex numbers'). The identity element of K will be denoted by 1 and the zero by 0.

Definition 0.1.1

A vector space over the field K consists of a set E, with distinguished element 0_E, together with operations

$$E \times E \longrightarrow E; (e, f) \longmapsto e + f \quad \text{(vector space addition)}$$

$$K \times E \longrightarrow E; (k, e) \longmapsto ke \quad \text{(scalar multiplication)}$$

such that the following conditions are satisfied.

1. The set E forms an Abelian group under + with zero element 0_E
   (a) $e + f = f + e$ for all $e, f \in E$.
   (b) $e + (f + g) = (e + f) + g$ for all $e, f, g \in E$.
   (c) $e + 0_E = 0_E + e$ for all $e \in E$.
   (d) For every $e \in E$, there exists a unique element $-e \in E$ such that $e + (-e) = 0_E$.
2. $k(e + f) = ke + kf$ for all $e, f \in E$ and $k \in K$.
3. $(k + l)e = ke + le$ for all $k, l \in K$ and $e \in E$.
4. $k(le) = (kl)e$, for all $k, l \in K$ and $e \in E$.
5. $1e = e$ for all $e \in E$.

We shall denote a vector space by the symbol used for its underlying set. Thus, in the definition, the vector space would be denoted by E. In the sequel we shall usually denote the zero element of E by 0, as opposed to 0_E. Thus we shall write $0e = 0$ rather than $0e = 0_E$. We shall also often omit reference to the underlying field of scalars as it will generally remain fixed during the course of discussion.

We recall the definition of a vector subspace.

Definition 0.1.2

A subset F of the vector space E is said to be a vector subspace of E if the restriction of the vector space operations of E to F induces on F the structure of a vector space.

Remark. In order that F be a vector subspace of E it is clearly necessary and sufficient that

(a) F is closed under addition: $e + f \in F$ for all $e, f \in F$.
(b) F is closed under scalar multiplication: $ke \in F$ for all $k \in K$ and $e \in F$.

The basic objects of study in the theory of vector spaces are linear maps, and we recall their definition.

Definition 0.1.3

Let E and F be vector spaces over the field K and let $T:E \longrightarrow F$ be a function from E to F. We call T a linear map or linear homomorphism (strictly, a K-linear map or K-linear homomorphism) if the following condition is satisfied for all $e, f \in E$ and $k \in K$:

$$T(ke + f) = kT(e) + T(f)$$

Associated with a linear map $T:E \longrightarrow F$ we have the following vector spaces:

The kernel of T, Ker T, defined by Ker $T = \{e \in E : T(e) = 0\}$.
The image of T, Im T, defined by Im $T = \{f \in F : \exists\, e \in E \text{ with } T(e) = f\}$.
The cokernel of T, Coker T, defined by Coker $T = F/\text{Im}\, T$.

Ker T is a vector subspace of E and Im T is a vector subspace of F.

T is said to be an injection (or monomorphism, or 1 : 1) if Ker $T = \{0\}$.
T is said to be a surjection (or epimorphism or onto) if Im $T = F$.
T is said to be an isomorphism if T is injective and surjective.
If $T:E \longrightarrow E$ we say that T is an endomorphism of E.

Definition 0.1.4

Let E be a vector space and $\mathscr{E} = (e_i)_{i \in I}$ be a collection of elements of E indexed by the set I. We call $\mathscr{E}$ a (Hammel) basis for E if

(a) Every $e \in E$ can be written as a finite linear combination of elements of $\mathscr{E}$:

$$e = \sum_{j=1}^{m} k_j e_j, \qquad \text{where } k_j \in K, \quad e_j \in \mathscr{E}, \quad 1 \leqslant j \leqslant m.$$

(b) No element of $\mathscr{E}$ can be written as a finite linear combination of any of the remaining elements of $\mathscr{E}$.

A standard theorem of linear algebra asserts that every vector space has a basis.

We are particularly interested in the case when I is finite. (If I is infinite, Definition 0.1.4 does not give a very useful definition of basis.) We have the following fundamental theorem.

Theorem 0.1.5

Let E be a vector space over the field K and let $\mathscr{E} = \{e_1, \ldots, e_m\}$ be a basis for E containing m elements. Then every other basis for E has exactly m elements. m is called the *dimension* of E, $\dim_K E$. Moreover, every element $x \in E$ can be written uniquely in the form

$$x = \sum_{i=1}^{m} x_i e_i,$$

where $x_i \in K$, $1 \leqslant i \leqslant m$.

Proof. In any text on linear algebra. ■

Remarks

1. The m-tuple $(x_1, \ldots, x_m) \in K^m$ given by Theorem 0.1.5 is said to define the coordinates of x relative to the basis $\mathscr{E}$ of E.
2. It can be shown that if E has an infinite basis, then the cardinal of the basis is an invariant of the space. The coordinate representation given by the basis is, however, no longer useful.

Examples

1. Let $\mathscr{E} = \{e_1, \ldots, e_m\}$ be a basis for the vector space E. Then e_1 has coordinates $(1, 0, 0, \ldots, 0)$. Similarly for $e_j, j > 1$.
2. K^m has a standard or canonical basis defined by taking the set of m-tuples

$$\{(1, 0, 0, \ldots, 0), (0, 1, 0, \ldots, 0), \ldots, (0, \ldots, 0, 1)\}.$$

We have an obvious identification between the coordinates of a basis vector and the vector (compare the previous example). We always refer to this basis on K^m as the *standard basis* of K^m.

0.1.6 THE MATRIX OF A LINEAR MAP WITH RESPECT TO A BASIS

For computational purposes it is often useful to have a numerical description of a linear map in terms of elements of the underlying field K.

Let $A : E \longrightarrow F$ be a K-linear map between the vector spaces E and F. Suppose that E and F are finite dimensional with respective bases

$$\mathscr{E} = (e_i)_{i=1}^{m} \qquad \text{and} \qquad \mathscr{F} = (f_j)_{j=1}^{n}.$$

Define elements $a_{ij} \in K$ by the rule

$$A(e_j) = \sum_{i=1}^{n} a_{ij} f_i, \qquad j = 1, \ldots, m.$$

That is

$$a_{ij} = i\text{th coordinate of } A(e_j)$$

The $m \times n$ array of elements of K,

$$\begin{bmatrix} a_{11} & a_{12} & \cdots & a_{1m} \\ a_{21} & a_{22} & \cdots & a_{2m} \\ \vdots & \vdots & & \\ a_{n1} & a_{n2} & \cdots & a_{nm} \end{bmatrix}$$

is called the matrix of A with respect to the bases $\mathscr{E}$ and $\mathscr{F}$.

We usually denote the matrix of A in abbreviated form by

$$[A] \qquad \text{or} \qquad [a_{ij}].$$

If it is necessary to emphasize the dependence of the matrix of A on the choice of bases $\mathscr{E}$ and $\mathscr{F}$, we may also write

$$[A]_{\mathscr{E},\mathscr{F}} \qquad \text{or} \qquad [a_{ij}]_{\mathscr{E},\mathscr{F}}.$$

Conversely, suppose we are given bases $\mathscr{E}$ and $\mathscr{F}$ for E and F respectively and an $m \times n$ matrix $[a_{ij}]$ of elements of K. Using the fact that a linear map is uniquely defined by specifying its values on the basis elements of E, we may define a linear map $A:E \longrightarrow F$, associated with the matrix $[a_{ij}]$, by the rule

$$A(e_j) = \sum_{i=1}^{n} a_{ij}f_i, \qquad j = 1, \ldots, m.$$

Notice that, if $x \in E$ has coordinates $(x_1, \ldots, x_m)$ relative to the basis $\mathscr{E}$ of E, then $A(x)$ is given, relative to the coordinate system on F, by

$$\left(\sum_{j=1}^{m} a_{1j}x_j, \ldots, \sum_{j=1}^{m} a_{ij}x_j, \ldots, \sum_{j=1}^{m} a_{nj}x_j\right).$$

In other words, the coordinates of $A(x)$ are given by the usual rules of matrix multiplication.

Example. Let $\mathscr{E} = \{e_1, \ldots, e_m\}$ be a basis of E. We may define a linear map $A:E \longrightarrow K^m$ by

$$A(e_j) = (0, 0, \ldots, 0, \underset{\substack{\uparrow \\ j\text{th place}}}{1}, 0, \ldots, 0), \; j = 1, \ldots, m.$$

The matrix of A is the identity matrix

$$\begin{bmatrix} 1 & 0 & 0 & \ldots & 0 \\ 0 & 1 & 0 & \ldots & 0 \\ 0 & 0 & 1 & \ldots & 0 \\ \vdots & \vdots & \vdots & & \vdots \\ 0 & 0 & 0 & \ldots & 1 \end{bmatrix}.$$

Notice that A is not the identity map unless $E = K^m$ and $\mathscr{E}$ is the standard basis of K^m!

Finally, we note that the above description of matrices easily gives us the correct rule for matrix multiplication. More specifically, suppose that $A:E \longrightarrow F$ and $B:F \longrightarrow G$ are linear maps. We shall denote the composite of A and B by $B.A:E \longrightarrow G$. The reader may easily verify that $B.A$ is a linear map. Let $(e_i)_{i=1}^p$, $(f_j)_{j=1}^q$ and $(g_k)_{k=1}^r$ be bases for E, F and G respectively. We set

$$[A] = [a_{ij}], \qquad B = [b_{ij}], \qquad B.A = [c_{ij}];$$

$[A]$, $[B]$ and $[B.A]$ are $p \times q$, $q \times r$ and $p \times r$ matrices respectively. Then,

$$\begin{aligned} c_{ij} &= i\text{th coordinate of } B.A(e_j) \\ &= i\text{th coordinate of } B\left(\sum_{k=1}^{q} a_{kj} f_k\right) \\ &= i\text{th coordinate of } \sum_{k=1}^{q} a_{kj} B(f_k) \\ &= i\text{th coordinate of } \sum_{k=1}^{q} \sum_{l=1}^{r} a_{kj} b_{lk} g_l = \sum_{k=1}^{q} b_{ik} a_{kj}. \end{aligned}$$

Definition 0.1.7

Let F and G be vector subspaces of E. Suppose that $F \cap G = \{0\}$. The *direct sum* of F and G is the vector subspace $F \oplus G$ of E defined by

$$F \oplus G = \{f + g : f \in F, g \in G\}.$$

If $F \oplus G = E$, then F and G are said to be *complementary* subspaces of E.

Closely related to the notion of direct sum we have the definition of a product of vector spaces.

Definition 0.1.8

Let E and F be two vector spaces over the same field, The product of E and F is the vector space $E \times F$ defined by

$$E \times F = \{(e, f) : e \in E, f \in F\},$$

with addition and scalar multiplication defined coordinatewise in the obvious way.

Example. The relation between direct sum and product is shown by the following: Let F and G be complementary subspaces of E; then $F \oplus G$ is canonically isomorphic to $F \times G$.

Remark. The term 'canonical' used in the above example means that the isomorphism is defined independently of any choice of bases for E and F. Sometimes the more suggestive term 'natural' is used instead, but we shall regard these terms as interchangeable in the sequel.

Exercises

1. Let E and F be vector spaces and suppose that $T: E \longrightarrow F$ is linear. Show that the graph of T is a vector subspace of $E \times F$.
2. Let

$$\mathscr{E} = (e_i)_{i=1}^{m} \qquad \text{and} \qquad \mathscr{E}'(e_i')_{i=1}^{m}$$

be two bases for E and

$$\mathscr{F} = (f_j)_{j=1}^{n} \qquad \text{and} \qquad \mathscr{F}' = (f_j')_{j=1}^{n}$$

be two bases for F.

The change of coordinates map for the bases $\mathscr{E}$ and $\mathscr{E}'$ is the linear map $A: E \longrightarrow E$ defined by

$$A(e_i) = e_i', \qquad 1 \leqslant i \leqslant m.$$

The change of coordinates map $B: F \longrightarrow F$ for the bases $\mathscr{F}$ and $\mathscr{F}'$ is defined similarly. Show that if $[T]$ is the matrix of a linear map $T:E \longrightarrow F$, relative to the bases $\mathscr{E}$ and $\mathscr{F}$, then

$$[T]_{\mathscr{E}', \mathscr{F}'} = [B]^{-1}_{\mathscr{F}, \mathscr{F}} [T] [A]_{\mathscr{E}, \mathscr{E}}.$$

What is the corresponding result if we wish to find how the linear map varies if we fix the matrix and vary the bases?

4. Let E and F have respective bases $\mathscr{E}$ and $\mathscr{F}$ and dimensions m and n. Let $M(m, n; k)$ denote the set of all $m \times n$ matrices with coefficients in K and $L(E, F)$ denote the set of all linear maps from E to F. Prove that
   (a) $M(m, n; k)$ is a K-vector space of dimension mn under the usual operations of matrix addition and scalar multiplication.
   (b) The map $L(E, F) \longrightarrow M(m, n; k); T \longmapsto [T]$, is bijective. Find operations of addition and scalar multiplication which make $L(E, F)$ a K-vector space and the map of (b) a vector-space isomorphism. What is the inverse of the map given in (b)?

0.2 Topological and metric spaces

In this section we briefly list some of the basic definitions and properties of topological and metric spaces. For more details we refer the reader to [11], [15]. The material presented here on *metric* spaces will be of the greater relevance in the sequel.

First let us recall the definition of a topological space.

Definition 0.2.1

Let X be a set and $\mathscr{U}$ a collection of subsets of X. We say the pair $(X, \mathscr{U})$ is a *topological space* with open sets given by $\mathscr{U}$ if

1. $X \in \mathscr{U}$ and $\emptyset \in \mathscr{U}$ ($\emptyset$ denotes the 'empty set').
2. The intersection of any finite collection of elements of $\mathscr{U}$ lies in $\mathscr{U}$

$$\bigcap_{i=1}^{n} U_i \in \mathscr{U} \qquad \text{where } U_i \in \mathscr{U} \text{ for } i = 1, \ldots, n$$

3. The union of an arbitrary family of elements of $\mathscr{U}$ lies in $\mathscr{U}$

$$\bigcup_{i \in I} U_i \in \mathscr{U} \qquad \text{where } U_i \in \mathscr{U} \text{ for all } i \in I.$$

Elements of $\mathscr{U}$ are called *open subsets* of the topological space $(X, \mathscr{U})$.

In the sequel we shall often abbreviate $(X, \mathscr{U})$ to X, and refer instead to the topological space X.

Definition 0.2.2

Let $(X, \mathscr{U})$ be a topological space. A subset V of X is said to be *closed* if $X \backslash V \in \mathscr{U}$.

A large source of examples of topological spaces is given by metric spaces.

Definition 0.2.3

Let X be a set and $d: X \times X \longrightarrow \mathbf{R}$ be a function. Then the pair (X, d) is said to define a *metric space* if

1. $d(x, y) \geqslant 0$ for all pairs $(x, y) \in X \times X$.

2. $d(x, y) = 0$ if and only if $x = y$.
3. $d(x, y) + d(y, z) \geqslant d(x, z)$ for all $x, y, z \in X$ ('triangle inequality').
4. $d(x, y) = d(y, x)$ for all $(x, y) \in X \times X$.

d is called the *distance function* or *metric* on X.

In the sequel we shall often refer to (X, d) as the metric space X.

If (X, d) is a metric space then we can construct a topology on X, 'the metric topology of X', as follows:

First, for all $r > 0$ and $x \in X$, we set

$$B_r(x) = \{y \in X : d(x, y) < r\}$$

$$\bar{B}_r(x) = \{y \in X : d(x, y) \leqslant r\}$$

$B_r(x)$ is called the *open disc* centre x and radius r, and $\bar{B}_r(x)$ is called the *closed disc* centre x and radius r.

An open set for the metric topology of X is then defined to be any subset of X which can be expressed as a union of open discs (technically, the open discs form a *basis* for the metric topology of X). We denote the collection of such open sets by $\mathscr{U}_d$. We have the following well-known result.

Theorem 0.2.4

1. $(X, \mathscr{U}_d)$ is a topological space.
2. $U \in \mathscr{U}_d$ if and only if for every $x \in U$, there exists $r > 0$ such that $B_r(x) \subset U$.

Examples

Metric spaces form in many ways the most intuitive collection of topological spaces. For example, the surface of the earth or of a ball is a 2-sphere, denoted by S^2, and a metric may be defined by taking the shortest distance between two points measured along the surface of the sphere. Intuitively, we measure distance between two points X and Y on the earth's surface as the length of a piece of string stretched taut between the two points X and Y. That is, distance is measured along great circles.

In this context the triangle equality merely states that the distance from London to New York is less than or equal to the distance from London to New York via, say, Madrid.

Another example of a metric space is given by the Cartesian product of two circles, $S^1 \times S^1$. This space is called a *torus* (Fig. 0.1). In this case a metric can be defined by drawing a string taut between points on the surface subject to the constraint that the string is forced to lie on the surface.

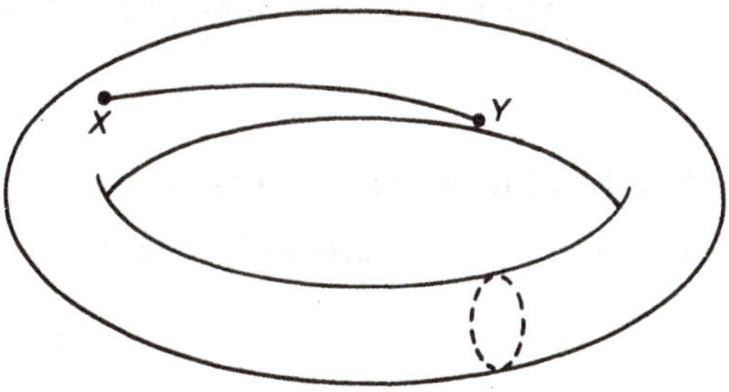

FIG. 0.1

Other examples of metric spaces are given by $\mathbf{R}^n$, $n \geq 1$, with the *Euclidean* distance defined by

$$d((x_1, \ldots, x_n), (y_1, \ldots, y_n)) = \sqrt{\left(\sum_{i=1}^{n} (x_i - y_i)^2\right)}$$

See also Section 1.1.

In practice many of the functions studied in, say, mathematical physics are defined on metric spaces like S^2 or $S^1 \times S^1$ rather than on open subsets of $\mathbf{R}^n$.

Next we wish to recall the definition and some of the elementary properties of continuous functions.

Definition 0.2.5

Let $(X, \mathscr{U})$ and $(Y, \mathscr{V})$ be topological spaces and $f: X \longrightarrow Y$. f is said to be *continuous* if

$$f^{-1}(V) \in \mathscr{U} \qquad \text{for all } V \in \mathscr{V}.$$

This definition may be given in many equivalent forms. In particular, f is continuous if and only if for every $x \in X$ and every open subset V of Y containing $f(x)$ there exists an open neighbourhood U of x such that $f(U) \subset V$.

If (X, d) and (Y, ρ) are metric spaces, then continuity may be characterized by any of the following equivalent conditions

1. $f^{-1}(B_r(y))$ is open for every open disc $B_r(y) \subset Y$.
2. $f^{-1}(\bar{B}_r(y))$ is closed for every closed disc $\bar{B}_r(y) \subset Y$.
3. Given any open disc of the form $B_r(f(x))$ contained in Y, there exists a disc $B_\epsilon(x) \subset X$ such that $f(B_\epsilon(x)) \subset B_r(f(x))$.
4. Let $x \in X$. Then for every $r > 0$, there exists $\epsilon > 0$ such that $d(f(x), f(x')) < r$ for all $x' \in X$ satisfying $d(x, x') < \epsilon$.

Let $x_1, \ldots, x_n, \ldots$ be a sequence of points of X. We shall write the sequence in abbreviated form as $(x_n)_{n=1}^{\infty}$ or, more simply, just as (x_n).

Definition 0.2.6

Let (x_n) be a sequence of points of the metric space (X, d). The sequence (x_n) is said to *converge* to the point $x \in X$ if $d(x_n, x) \longrightarrow 0$ as $n \longrightarrow \infty$.

Definition 0.2.7

The sequence (x_n) of points of the metric space (X, d) is said to be a *Cauchy sequence* if $d(x_n, x_m) \longrightarrow 0$ as $n, m \longrightarrow \infty$.

Remark. In general a Cauchy sequence need not converge. See the exercises at the end of this section.

Definition 0.2.8

The metric space (X, d) is said to be *complete* if every Cauchy sequence converges.

Returning to the study of topological spaces we have the following important definition.

Definition 0.2.9

X is said to be a *compact* topological space if, for every collection $\{U_i\}_{i \in I}$ of open subsets that cover X ($\bigcup_{i\in I} U_i = X$), it is possible to find a finite subcollection $\{U_i\}_{i\in I'}$ ($I' \subset I$ and I' finite) which still covers X.

Examples

1. Every closed and bounded subset of $\mathbf{R}^n$ is compact. For example, S^2 and $S^1 \times S^1$ are compact. Conversely every compact subset of $\mathbf{R}^n$ is closed and bounded.
2. $\mathbf{R}^n$ is not compact.

Definition 0.2.10

A topological space is said to be *locally compact* if every point has a neighbourhood with compact closure.

Examples

1. Every compact topological space is locally compact.
2. $\mathbf{R}^n$ is locally compact.

Exercises

1. Show that an open subset of $\mathbf{R}$ provides an example of an incomplete metric space.
2. Let (X, d) be a metric space. Define an equivalence relation on the set of all Cauchy sequences of points of X as follows: Say the two Cauchy sequences (x_n) and (y_n) are equivalent if $d(x_n, y_n) \longrightarrow 0$ as $n \longrightarrow \infty$. Show

   (a) If (X, d) is complete, then (x_n) and (y_n) are equivalent if and only if they are convergent to the same point of X.
   (b) The above defined relation on Cauchy sequences is an equivalence relation.
   (c) Letting $\hat{X}$ denote the set of equivalence classes, show that $\hat{X}$ naturally contains X and the metric d extends to a metric $\hat{d}$ on X. $(\hat{X}, \hat{d})$ is called the completion of (X, d).
   (d) $(\hat{X}, \hat{d})$ is complete.

3. Show that a closed subset of a compact topological space is compact. Show that a product of n closed bounded intervals is a compact subset of $\mathbf{R}^n$ (Assume the Bolzano-Weierstrass theorem). Hence show that every closed and bounded subset of $\mathbf{R}^n$ is compact.

CHAPTER 1

Linear algebra and normed vector spaces

1.1 Normed vector spaces

In this chapter K will always denote either the field of real numbers, $\mathbf{R}$, or the field of complex numbers, $\mathbf{C}$. $|\quad|$ will denote the absolute value or modulus of numbers in K.

Let E be a vector space over K of not necessarily finite dimension. As it stands, E is too general to admit of interesting study. Thus, without any topology on E, questions like 'Is vector space addition a continuous operation?' or 'Are linear maps continuous?' are meaningless. We therefore wish to start by studying vector spaces with some additional structure such as a topology or metric. One such structure that the reader will have already encountered in the study of the vector spaces $\mathbf{R}$, $\mathbf{R}^2$, $\mathbf{R}^3$ and $\mathbf{C}$ is the important notation of the *length* of a vector. The generalization of length to an arbitrary vector space is given by the following definition.

Definition 1.1.1

Let E be a vector space over the field K. A function

$$\|\quad\| : E \longrightarrow \mathbf{R};\ x \longmapsto \|x\|$$

is called a *norm* on E if

1. $\|x\| = 0$ if and only if $x = 0$
2. $\|x + y\| \leqslant \|x\| + \|y\|$ for all $x, y \in E$ ('triangle inequality')
3. $\|kx\| = |k|\,\|x\|$ for all $x \in E$ and $k \in K$.

The pair $(E, \|\quad\|)$ is then called a *normed vector space.*

A norm, then, is our generalization of length. The following proposition shows that the norm of a vector is, like length, always positive.

Proposition 1.1.2

Let $(E, \|\quad\|)$ be a normed vector space. Then $\|x\| \geqslant 0$ for all $x \in E$.

Proof. $0 = \|0\| = \|x + -x\| \leqslant \|x\| + \|-x\| = 2\,\|x\|$. ■

Associated with a norm on a vector space E we may define a translation-invariant metric on E.

Proposition 1.1.3

Let $(E, \| \ \|)$ be a normed vector space. Define $d(x, y) = \| x - y \|$, $x, y \in E$. Then d is a metric on E, 'the metric associated with $\| \ \|$', and d satisfies the following additional properties:

1. $d(x + z, y + z) = d(x, y)$ for all $x, y, z \in E$ (d is 'translation invariant').
2. $d(kx, ky) = | k | d(x, y)$ for all $x, y \in E$ and $k \in K$.

Proof. Left to the exercises. ∎

The translation invariance of d is shown by Fig. 1.1. All three unbroken lines have equal length $\| x - y \|$.

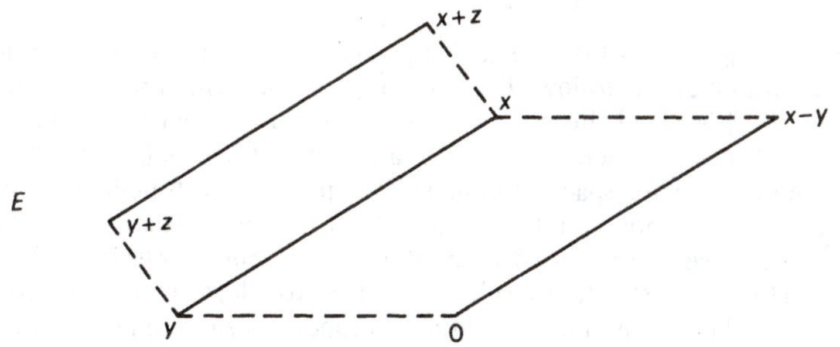

FIG. 1.1

Examples

1. K is a normed vector space with norm defined by taking absolute value. Thus, if $k \in K$, $\| k \| = | k |$. In the sequel we shall *always* take this norm on K.
2. Let $\mathbf{R}^n = \{(x_1, \ldots, x_n): x_1, \ldots, x_n \in \mathbf{R}\}$ (real n-space) and $\mathbf{C}^n = \{(z_1, \ldots, z_n): z_1, \ldots, z_n \in \mathbf{C}\}$ (complex n-space).

   We can define many different norms on $\mathbf{R}^n$ and $\mathbf{C}^n$.

   (a) $\| x \|_1 = \sum_{i=1}^n | x_i |$, $x = (x_1, \ldots, x_n) \in \mathbf{R}^n$.

   (b) $\| x \|_\infty = \max_i | x_i |$, $x = (x_1, \ldots, x_n) \in \mathbf{R}^n$.

   (c) More generally, for $p \geqslant 1$, we may define the *p-norm* on $\mathbf{R}^n$ by

$$\| x \|_p = \left(\sum_{i=1}^{n} | x_i |^p \right)^{1/p}, \qquad x = (x_1, \ldots, x_n) \in \mathbf{R}^n$$

   All the above norms may be defined on $\mathbf{C}^n$ as well. The reader may easily check that all of the above norms satisfy properties 1 and 3 of Definition 1.1.1. Property 2 is rather more difficult. For $\| \ \|_1$ and $\| \ \|_\infty$ it follows from the ordinary triangle inequality for real numbers. The triangle inequality for $\| \ \|_p$ is known as *Minkowski's inequality.* We leave the proof of this important inequality to the exercises at the end of this section. We have, therefore, constructed infinitely many different metrics on $\mathbf{R}^n$ (at least for $n \geqslant 2$). However, we shall show later that all these metrics have the same associated topology.

3. Let X denote an arbitrary set and $B(X)$ denote the set of all bounded real valued functions on X. That is, $f \in B(X)$ if and only if $\exists M \geqslant 0$ such that $|f(x)| \leqslant M$ for all $x \in X$. We can easily give $B(X)$ the structure of a real vector space if we define vector space addition and scalar multiplication respectively by

$$(f+g)(x) = f(x) + g(x) \qquad \text{for all } f, g \in B(X),$$

$$(kf)(x) = k(f(x)) \qquad \text{for all } f \in B(X) \text{ and } k \in K.$$

Define $\|f\| = \sup_{x \in X} |f(x)|$, $f \in B(X)$. Observe that

$$\begin{aligned}\|f+g\| &= \sup_X |f(x)+g(x)| \leqslant \sup_X (|f(x)| + |g(x)|) \\ &\leqslant \sup_X |f(x)| + \sup_X |g(x)| = \|f\| + \|g\|.\end{aligned}$$

Hence $\|\ \|$ satisfies the triangle inequality. The other axioms for a norm are easily verified and it follows that $(B(X), \|\ \|)$ is a (real) normed vector space. $\|\ \|$ is called the *sup norm* on $B(X)$ and we shall have occasion to refer to it in the sequel. We may similarly define $(B_C(X), \|\ \|)$, the (complex) normed vector space of bounded complex valued functions on X.

4. Injecting some topology into Example 2, let us suppose that X is a compact topological space. We let $C(X)$ denote the set of continuous real-valued functions on X. Then, by a standard result of point-set topology, $C(X)$ is a subset of $B(X)$. Standard properties of continuous functions show that $C(X)$ is a vector subspace of $B(X)$. Restricting the sup norm defined in Example 2 to $C(X)$, it follows that $(C(X), \|\ \|)$ is a normed vector subspace of $(B(X), \|\ \|)$.

   The metric associated with the sup norm on $C(X)$ is given by

$$d(f, g) = \sup_X |f(x) - g(x)|, \qquad f, g \in C(X).$$

We call the topology associated with this metric on $C(X)$, the topology of *uniform convergence* on $C(X)$. The reader should check, in the special case when X is the compact interval $[a, b]$, that if we are given a sequence of continuous functions (f_n) on X, then $f_n \longrightarrow f$ uniformly (classical definition) if and only if $f_n \longrightarrow f$ in the topology of uniform convergence on $C(X)$. See also the exercises at the end of this section. We may similarly define $(C_C(X), \|\ \|)$, the (complex) normed vector space of continuous complex-valued functions on X.

5. Many examples of normed vector spaces are given by *sequence* spaces. In what follows, $\mathbf{N}^+$ denotes the set of strictly positive integers.

   (a) Let m denote the set of all bounded sequences of real numbers. Thus $x = (x_i)_{i=1}^{\infty} \in m$ if and only if $\exists M \geqslant 0$ such that $|x_n| \leqslant M$ for all $n \in \mathbf{N}^+$. We may give m the structure of a real vector space by defining addition and scalar multiplication coordinatewise. For example, if $(x_i), (y_i) \in m$, then $(x_i) + (y_i)$ is the sequence

$$(x_1 + y_1, \ldots, x_n + y_n, \ldots) = (x_n + y_n)_{n=1}^{\infty}$$

We define a norm on m by setting

$$\|x\| = \sup_i |x_i|, \qquad x = (x_i) \in m.$$

We leave it as an exercise for the reader to verify that $(m, \| \ \|)$ is a real normed vector space. We may similarly define the space of bounded complex sequences which again is a normed vector space with obvious norm.

Let us look at some subspaces of m.

(b) Let c denote the set of convergent sequences of real numbers and c_0 denote the set of sequences convergent to 0. We leave it to the reader to verify that c and c_0 are normed vector subspaces of m.

(c) Let s denote the subset of m consisting of sequences with only finitely many terms non-zero. That is $(x_i)_{i=1}^{\infty} \in s$ if and only if $\exists\, p \in \mathbf{N}^+$, such that $x_n = 0$, $n \geqslant p$. It is easily seen that s is a normed vector subspace of m. It is also a subspace of c_0.

(d) We may define many other vector subspaces of m with, however, different norms from that induced by m. For example, let l^1 denote the space of real sequences such that $(x_n)_{n=1}^{\infty} \in l^1$ if and only if

$$\| (x_n)_{n=1}^{\infty} \|_1 = \sum_{n=1}^{\infty} |x_n| < \infty.$$

Since $(1, 1, \ldots, 1, \ldots) \notin l^1$, $l^1 \neq m$. It is not difficult to show that l^1 is a vector subspace of m and that $\| \ \|_1$ defines a norm on l^1 different from the norm induced by $\| \ \|$.

More generally, let l^p denote the set of sequences $(x_n)_{n=1}^{\infty}$ satisfying the condition

$$\| (x_n)_{n=1}^{\infty} \|_p = \left(\sum_{i=1}^{\infty} |x_i|^p \right)^{1/p} < \infty,$$

where p is greater than or equal to 1.

However, now we need Minkowski's inequality even to verify that l^p is a vector space. We refer the reader to the exercises for this verification and also the fact that $(l^p, \| \ \|_p)$ is a normed vector space. Unlike the case of $\mathbf{R}^n$, however, the metric topologies associated with the norms $\| \ \|_p$ are all different, and indeed, regarded as vector subspaces of m, $l^p = l^{p'}$ if and only if $p = p'$.

Exercises

1. Prove Proposition 1.1.3.
2. Given a norm $\| \ \|$ on $\mathbf{R}^2$, interpret geometrically the triangle inequality.
3. Find all norms on the real line $\mathbf{R}$ and the complex line $\mathbf{C}$ (regarded as a complex vector space).
4. In this exercise we indicate the main steps of proof of Minkowski's inequality.

   (a) Consider the function $f(x, y) = \alpha x + \beta y - x^{\alpha} y^{\beta}$, where $0 < \alpha, \beta < 1$, $\alpha + \beta = 1$ and x and y are assumed positive. By finding the minimum value of f for a fixed value of y, show that

   $$\alpha x + \beta y \geqslant x^{\alpha} y^{\beta} \qquad \text{for all } x, y \geqslant 0.$$

   (b) Let (a_n) and (b_n) be real (or complex) sequences consisting of N terms. Let $r, r' > 1$ and satisfy the condition

   $$1/r + 1/r' = 1.$$

   Define for $1 \leqslant m \leqslant N$

   $$A_m = a_m \Big/ \left(\sum_{i=1}^{N} |a_i|^r \right)^{1/r} \quad \text{and} \quad B_m = b_m \Big/ \left(\sum_{i=1}^{N} |b_i|^{r'} \right)^{1/r'}.$$

Thus we have

$$\sum_{i=1}^{N} |A_i|^r = 1 \qquad \text{and} \qquad \sum_{i=1}^{N} |B_i|^{r'} = 1.$$

From the inequality proved in step (a), it follows that

$$|A_m B_m| \leqslant |A_m|^r/r + |B_m|^{r'}/r'$$

and hence by summation over m that

$$\sum_{i=1}^{N} |A_i B_i| \leqslant 1 \leqslant \left(\sum_{i=1}^{N} |A_i|^r\right)^{1/r} \left(\sum_{i=1}^{N} |B_i|^{r'}\right)^{1/r'}.$$

Hence deduce Hölder's inequality:

$$\sum_{i=1}^{N} |a_i b_i| \leqslant \left(\sum_{i=1}^{N} |a_i|^r\right)^{1/r} \left(\sum_{i=1}^{N} |b_i|^{r'}\right)^{1/r'}.$$

(c) With the assumptions of step (b) show that

$$\sum_{i=1}^{N} |a_i + b_i|^r \leqslant \sum_{i=1}^{N} |a_i + b_i|^{r-1} |a_i| + \sum_{i=1}^{N} |a_i + b_i|^{r-1} |b_i|$$

and apply Hölder's inequality to the sums on the right to deduce Minkowski's inequality:

$$\left(\sum_{i=1}^{N} |a_i + b_i|^r\right)^{1/r} \leqslant \left(\sum_{i=1}^{N} |a_i|^r\right)^{1/r} + \left(\sum_{i=1}^{N} |b_i|^r\right)^{1/r}.$$

Use Minkowski's inequality to verify that the functions $\| \;\|_p$ defined on $\mathbf{R}^n$ in this section are indeed norms.

5. By extending Minkowski's inequality to infinite sequences, prove that $(l^p, \| \;\|_p)$ is a normed vector space.
6. What goes wrong if we try to define the p-norm $\| \;\|_p$ for p *less* than 1?
7. Consider the set of all continuous functions on the closed unit interval $[0, 1]$. Define

$$\|f\|_1 = \int_0^1 |f(t)|\, \mathrm{d}t, \qquad f \in C[0, 1].$$

Prove that $(C[0, 1], \| \;\|_1)$ is a normed vector space. How might one define $(C[0, 1], \| \;\|_p), p > 1$?

8. Consider the vector space $C[0, 1]$ together with the sup norm and let d denote the associated metric. Suppose that (f_n) is a Cauchy sequence in $C[0, 1]$. Prove (using any standard results about uniformly convergent sequences of functions you may need) that (f_n) converges to an element $f \in C[0, 1]$ and hence that $(C[0, 1], d)$ is a *complete* metric space.

1.2 Inner-product spaces

The metric on $\mathbf{R}^2$ most frequently met in applications is induced from the Euclidean norm $\| \;\|_2$ defined by

$$\|(x_1, x_2)\|_2 = \sqrt{(x_1^2 + x_2^2)}, \qquad (x_1, x_2) \in \mathbf{R}^2.$$

This norm and its associated metric are particularly well adapted to the study of geometry on $\mathbf{R}^2$. For example, $\| \quad \|_2$ is rotation invariant. With the notation of Fig. 1.2, if we rotate the line segment through the origin and the point X through an angle θ about the origin, then $\|X\|_2 = \|X'\|_2$. This property is certainly not true for any of the other norms $\| \quad \|_p$ on $\mathbf{R}^2$ (see the exercises). More generally, as we shall see in a later section, it is possible to define the angle between lines in $\mathbf{R}^2$ using this norm.

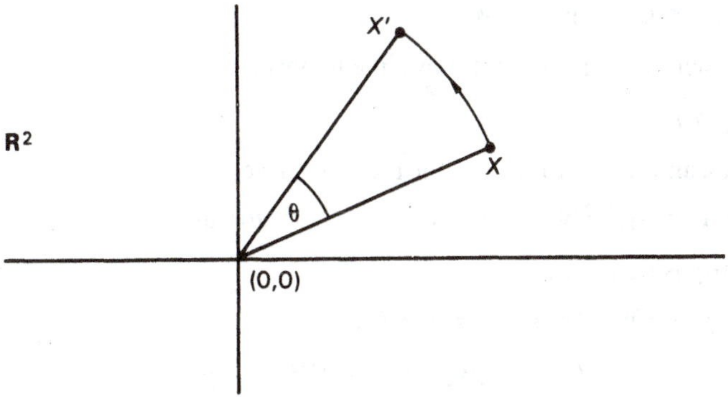

FIG. 1.2

Those norms which allow us to define angles and talk about rotations form a particularly important class of norms and, as we shall soon see, are induced from an *inner product* on the vector space.

Definition 1.2.1

Let E be a real vector space and $\langle\ ,\ \rangle: E \times E \longrightarrow \mathbf{R}; (x, y) \longmapsto \langle x, y\rangle$ be a real-valued function on $E \times E$. The pair $(E, \langle\ ,\ \rangle)$ is said to be a (real) *inner-product space* if the following conditions are satisfied.

1. $\langle a_1x_1 + a_2x_2, y\rangle = a_1\langle x_1, y\rangle + a_2\langle x_2, y\rangle$ for all $x, y_1, y_2 \in E$ and $a_1, a_2 \in \mathbf{R}$.
2. $\langle x, y\rangle = \langle y, x\rangle$ for all $x, y \in E$ ('symmetry').
3. $\langle x, x\rangle \geqslant 0$ for all $x \in E$. $\langle x, x\rangle = 0$ if and only if $x = 0$.

Remarks

1. Conditions 1 and 2 imply that we also have the relation $\langle x, a_1y_1 + a_2y_2\rangle = a_1\langle x, y_1\rangle + a_2\langle x, y_2\rangle$ for all $x, y_1, y_2 \in E$ and $a_1, a_2 \in \mathbf{R}$. This condition, together with condition 1, on $\langle\ ,\ \rangle$ is usually referred to as the *bilinearity* property of the inner product.
2. $\langle x, y\rangle$ is sometimes referred to as the *dot product* of x and y and is then written $x \,.\, y$.

The fundamental result about inner products is Schwartz' inequality.

Proposition 1.2.2. (Schwartz' inequality)

Let $(E, \langle\ ,\ \rangle)$ be a real inner-product space. Then for all $x, y \in E$ we have

$$\langle x, y\rangle^2 \leqslant \langle x, x\rangle\langle y, y\rangle.$$

Proof. Fix $x, y \in E$ and consider the expression $\langle ax + by, ax + by\rangle$, where $a, b \in \mathbf{R}$. By property 3 of the inner product, this expression is positive for all $a, b \in \mathbf{R}$. Now, using the bilinearity of $\langle\ ,\ \rangle$, we have

$$\langle ax + by, ax + by\rangle = a^2\langle x, x\rangle + 2ab\langle x, y\rangle + b^2\langle y, y\rangle,$$

and regarding this expression as a quadratic form in the variables a and b, we may use the well-known necessary and sufficient condition for the positivity of such a form to deduce

$$\langle x, y\rangle^2 \leqslant \langle x, x\rangle\langle y, y\rangle. \quad \blacksquare$$

From the Schwartz inequality follows the important

Corollary 1.2.2.1

With the assumptions of Proposition 1.2.2, we have

$$\langle x + y, x + y\rangle^{1/2} \leqslant \langle x, x\rangle^{1/2} + \langle y, y\rangle^{1/2} \qquad \text{for all } x, y \in E.$$

Proof. By Proposition 1.2.2,

$$\begin{aligned}\langle x + y, x + y\rangle &= \langle x, x\rangle + 2\langle x, y\rangle + \langle y, y\rangle \\ &\leqslant \langle x, x\rangle + 2\langle x, x\rangle^{1/2}\langle y, y\rangle^{1/2} + \langle y, y\rangle.\end{aligned}$$

But the latter expression is just $(\langle x, x\rangle^{1/2} + \langle y, y\rangle^{1/2})^2$ and so, taking square roots, the result follows. $\blacksquare$

Corollary 1.2.2.1 implies immediately the non-trivial part of :

Theorem 1.2.3

Let $(E, \langle\ ,\ \rangle)$ be a real inner-product space. If we define $\|x\| = \langle x, x\rangle^{1/2}$, $x \in E$, then $(E, \|\ \ \|)$ is a normed vector space.

Examples

1. $\mathbf{R}^n$ is an inner-product space with inner product defined by

$$\langle(x_1, \ldots, x_n), (y_1, \ldots, y_n)\rangle = \sum_{i=1}^{n} x_i y_i, \qquad (x_1, \ldots, x_n), (y_1, \ldots, y_n) \in \mathbf{R}^n.$$

The norm associated with $\langle\ ,\ \rangle$ is the norm $\|\ \ \|_2$ defined on $\mathbf{R}^n$ in Section 1.1.

2. The norm $\|\ \ \|_2$ on the space of sequences l^2 is associated with the inner product defined on l^2 by

$$\langle(x_i), (y_i)\rangle = \sum_{i=1}^{\infty} x_i y_i, \qquad (x_i), (y_i) \in l^2.$$

3. An inner product can be defined on the vector space of continuous functions on the unit interval $[0, 1]$ by the rule

$$\langle f, g\rangle = \int_0^1 f(t)g(t)\, \mathrm{d}t, \qquad f, g \in C[0, 1].$$

The associated norm is then given by

$$\|f\|^2 = \int_0^1 f(t)^2\, \mathrm{d}t, \qquad f \in C[0, 1].$$

A natural question to ask is whether it is possible to find conditions on a norm which will determine whether it is induced from an inner product. The following theorem gives an answer to this question.

Theorem 1.2.4 (Parallelogram law)

Let $(E, \|\ \ \|)$ denote a normed vector space. A necessary and sufficient condition for the norm to be associated with an inner product on E is that the 'parallelogram law' holds:

$$2(\|x\|^2 + \|y\|^2) = \|x - y\|^2 + \|x + y\|^2 \qquad \text{for all } x, y \in E.$$

Proof. Necessity. Suppose the norm $\|\ \ \|$ is given by the inner product $\langle\ ,\ \rangle$. Then

$$\|x - y\|^2 = \langle x - y, x - y\rangle = \|x\|^2 + \|y\|^2 - 2\langle x, y\rangle$$

and

$$\|x + y\|^2 = \langle x + y, x + y\rangle = \|x\|^2 + \|y\|^2 + 2\langle x, y\rangle.$$

Addition of these two identities proves the necessity of the condition.

The proof of sufficiency is rather less trivial and, since the result will not be used in the sequel, the reader can omit what follows if he so wishes.

First notice that if $\|\ \ \|$ is associated with the inner product $\langle\ ,\ \rangle$, then $\langle\ ,\ \rangle$ is uniquely determined by the norm. In fact we must have

$$\langle x, y\rangle = \tfrac{1}{2}(\|x + y\|^2 - \|x\|^2 - \|y\|^2) \qquad \text{for all } x, y \in E.$$

We therefore use the right-hand side of this expression to *define* our conjectured inner product on E.

It is clear that whether or not the parallelogram law holds we have $\langle x, x\rangle \geqslant 0$ for all $x \in E$ and $\langle x, x\rangle = 0$ if and only if $x = 0$. Symmetry is also trivial. The hard property to verify is the bilinearity. By definition of $\langle\ ,\ \rangle$ we have

$$\langle x + y, z\rangle = \tfrac{1}{2}(\|x + y + z\|^2 - \|x + y\|^2 - \|z\|^2).$$

By the parallelogram law,

$$\begin{aligned}\tfrac{1}{2}\|x + y + z\|^2 &= \tfrac{1}{2}\|(x + z) + y\|^2 = \|x + z\|^2 + \|y\|^2 - \tfrac{1}{2}\|x + z - y\|^2 \\ &= \tfrac{1}{2}\|(y + z) + x\|^2 = \|y + z\|^2 + \|x\|^2 - \tfrac{1}{2}\|y + z - x\|^2\end{aligned}$$

Substituting in the above expression for $\langle x + y, z\rangle$ yields

$$\begin{aligned}\langle x + y, z\rangle = \tfrac{1}{2}(&\|x + z\|^2 + \|y + z\|^2 + \|y\|^2 + \|x\|^2 \\ &- 2\|z\|^2 - (\|x - y\|^2 + \|x + y\|^2)).\end{aligned}$$

Again by the parallelogram law,

$$\|x - y\|^2 + \|x + y\|^2 = 2(\|x\|^2 + \|y\|^2),$$

and substitution gives

$$\begin{aligned}\langle x + y, z\rangle &= \tfrac{1}{2}(\|x + z\|^2 - \|x\|^2 + \|y + z\|^2 - \|y\|^2 - 2\|z\|^2) \\ &= \langle x, z\rangle + \langle y, z\rangle.\end{aligned}$$

From this result it follows inductively that

$$\langle nx, z\rangle = n\langle x, z\rangle \qquad \text{for all positive or negative integers } n.$$

In particular, for a non-zero integer n, we have

$$n\langle x/n, y\rangle = \langle n(x/n), y\rangle = \langle x, y\rangle.$$

Hence $\langle x/n, y\rangle = 1/n\langle x, y\rangle$. It follows easily that for any *rational* number p/q we have

$$\langle (p/q)x, y\rangle = p/q\langle x, y\rangle.$$

Thus far we have shown that the bilinearity property holds for the subfield of **R** of *rational* numbers. To complete the proof we use a continuity argument to extend the bilinearity to the field of real numbers.

First we remark that, for fixed x and y, the map defined on **R** by

$$\lambda \longmapsto \langle \lambda x, y\rangle$$

is continuous. Indeed, since

$$\begin{aligned}\langle \lambda x, y\rangle &= \tfrac{1}{2}(\| \lambda x + y\|^2 - \|\lambda x\|^2 - \|y\|^2)\\ &= \tfrac{1}{2}(\|\lambda x + y\|^2 - \lambda^2\|x\|^2 - \|y\|^2)\end{aligned}$$

this amounts to showing that the map $\lambda \longmapsto \|\lambda x + y\|$ is continuous. But from the triangle inequality it follows that, for $\lambda, \lambda' \in \mathbf{R}$, we have

$$\begin{aligned}|\|\lambda x + y\| - \|\lambda' x + y\|| &\leqslant \|(\lambda - \lambda')x\|\\ &= |\lambda - \lambda'|\|x\|\end{aligned}$$

and the continuity follows immediately since x and y are assumed fixed.

Now consider any real number k. We can approximate k by a sequence of rational numbers k_n

$$k_n \longrightarrow k, \qquad n \longrightarrow \infty$$

By what we have already proved $\langle k_n x, y\rangle = k_n\langle x, y\rangle$. Taking limits we therefore have

$$\begin{aligned}k\langle x, y\rangle = \lim_{n\to\infty} k_n\langle x, y\rangle &= \lim_{n\to\infty} \langle k_n x, y\rangle\\ &= \langle kx, y\rangle, \qquad \text{by the continuity proved above.}\end{aligned}$$

The bilinearity of $\langle\ ,\ \rangle$ now follows immediately. ∎

Remark. The parallelogram law

$$2(\|x\|^2 + \|y\|^2) = \|x - y\|^2 + \|x + y\|^2$$

is so called because of the theorem of plane Euclidean geometry which states that the sum of the squares of the lengths of the diagonals of a parallelogram is equal to the sum of the squares of the lengths of the sides (Fig. 1.3).

In practice, the parallelogram law provides a very effective means of checking whether or not a given norm is induced from an inner product—see Exercise 4 at the end of the section.

So far we have defined inner products only on real vector spaces. We have to modify our definitions somewhat to define an inner product on a complex vector space for which Theorem 1.2.3 still holds. We give the definition here and leave the proof of properties of the inner product to the exercises.

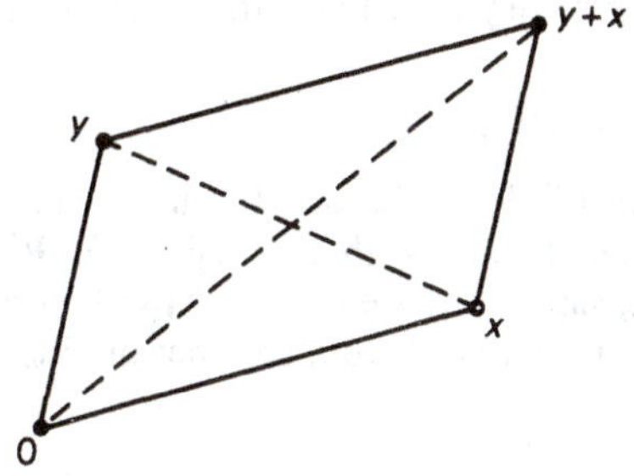

FIG. 1.3

Definition 1.2.5

Let E be a complex vector space and $\langle\,,\rangle: E \times E \longrightarrow \mathbf{C}$ be a complex valued function on $E \times E$. The pair $(E, \langle\,,\rangle)$ is said to be a *Hermitian inner-product space* or *complex inner-product space* if the following conditions are satisfied:

1. $\langle a_1x_1 + a_2x_2, y\rangle = a_1\langle x_1, y\rangle + a_2\langle x_2, y\rangle$ for all $x_1, x_2, y \in E$ and all $a_1, a_2 \in \mathbf{C}$.

1′. $\langle x, a_1y_1 + a_2y_2\rangle = \bar{a}_1\langle x, y_1\rangle + \bar{a}_2\langle x, y_2\rangle$ for all $x, y_1, y_2 \in E$ and all $a_1, a_2 \in \mathbf{C}$.

2. $\langle x, y\rangle = \overline{\langle y, x\rangle}$ for all $x, y \in E$ (Hermitian symmetry).
3. $\langle x, x\rangle \geqslant 0$, with equality if and only if $x = 0$.

Remark. Conditions 1 and 1′ are sometimes called the *sesquilinearity* of the Hermitian inner product.

Examples

1. $\mathbf{C}^n$ with complex inner product defined by

$$\langle (z_1, \ldots, z_n), (y_1, \ldots, y_n)\rangle = \sum_{i=1}^{n} z_i\bar{y}_i.$$

2. If we let $C[0, 1]_{\mathbf{C}}$ denote the complex vector space of continuous *complex*-valued functions on $[0, 1]$, then we may define a complex inner product on $C[0, 1]_{\mathbf{C}}$ by

$$\langle f, g\rangle = \int_0^1 f(t)\overline{g(t)}\,\mathrm{d}t, \qquad f, g \in C[0, 1]_{\mathbf{C}}.$$

3. If we let $l^2_{\mathbf{C}}$ denote the complex vector space of infinite sequences (z_n) of complex numbers satisfying

$$\sum_{i=1}^{\infty} |z_n|^2 < \infty,$$

then a Hermitian inner product can be defined on $l^2_{\mathbf{C}}$ by the formula

$$\langle (z_n), (y_n)\rangle = \sum_{i=1}^{\infty} z_n\bar{y}_n, \qquad (z_n), (y_n) \in l^2_{\mathbf{C}}.$$

Exercises

1. Verify that the norm $\|\quad\|_1$ on $\mathbf{R}^2$ is *not* invariant by rotations.
2. Check that the inner product defined on $C[0, 1]$ does indeed satisfy all the conditions for an inner product.

3. Show that the Schwartz inequality holds for a Hermitian inner product in the form:

$$|\langle x, y\rangle|^2 \leqslant \langle x, x\rangle\langle y, y\rangle$$

Deduce that Theorem 1.2.3 is valid for a Hermitian inner-product space.

4. Using the parallelogram law, show that $(\mathbf{R}^n, \| \quad \|_1)$, $(\mathbf{R}^n, \| \quad \|_\infty)$, $(s, \| \quad \|)$, $(m, \| \quad \|)$ and $(C(X)$, sup norm) are not inner-product spaces (that is, the given norm on the space is not induced from an inner product).

1.3 Topology on normed vector spaces

Let $(E, \| \quad \|)$ be a normed vector space. From Proposition 1.1.3 it follows that $d(x, y) = \|x - y\|$ defines a translation-invariant metric on E. In this section we consider the topology associated with this metric.

We first define a neighbourhood base for the origin. Let $B_r(0)$ denote the open disc centre 0, radius r:

$$B_r(0) = \{x : \|x\| < r\}.$$

Then $\{B_r(0)\}_{r>0}$ defines a base of open neighbourhoods for the origin. By translation we obtain a base of open neighbourhoods at $x \in E$:

$$\{x + B_r(0)\}_{r>0} = \{x + y : y \in B_r(0)\}_{r>0}.$$

The resulting collection of open sets is a base for the metric topology of E. We let

$$B_r(x) = \{y : \|x - y\| < r\} = x + B_r(0)$$

and

$$\bar{B}_r(x) = \{y : \|x - y\| \leqslant r\} = x + \bar{B}_r(0)$$

respectively denote the open and closed discs centre x and radius r.

The first indication of the association between the topology on a normed vector space and the underlying vector-space structure is given by

Proposition 1.3.1

Let $(E, \| \quad \|)$ be a normed vector space and F be a linear subspace of E. Then the closure of F, $\bar{F}$, is a linear subspace of E.

Proof. Let $x, y \in \bar{F}$. We show that $x + y \in \bar{F}$. Given $\epsilon > 0$, $\bar{B}_{\epsilon/2}(x) \cap F \neq \emptyset$. Therefore $\exists x_0 \in F$ such that $\|x - x_0\| \leqslant \epsilon/2$. Similarly $\exists y_0 \in F$ such that $\|y - y_0\| \leqslant \epsilon/2$. Hence,

$$\|x + y - (x_0 + y_0)\| \leqslant \epsilon.$$

Since $x_0 + y_0 \in F$, it follows that

$$\bar{B}_\epsilon(x + y) \cap F \neq \emptyset \qquad \text{for all } \epsilon > 0.$$

and so $x + y \in \bar{F}$.

The proof that $kx \in \bar{F}$ if $k \in K$ and $x \in \bar{F}$ is accomplished similarly and is left as an exercise. ■

Remark. Not every vector subspace of a normed vector space need be closed. See, for example, Exercise 3 at the end of this section.

Finally, we note the trivial

Proposition 1.3.2

Let $(E, \| \ \|)$ be a normed vector space. Then $\| \ \|: E \longrightarrow \mathbf{R}$ is a continuous function on E.

Exercises

1. Draw the unit discs in $\mathbf{R}^2$ for the norms $\| \ \|_1$, $\| \ \|_2$ and $\| \ \|_\infty$.
2. If A and B are subsets of the normed vector space E, we let

$$A + B = \{a + b : a \in A, b \in B\}.$$

Prove that

(a) A open implies $A + B$ open.
(b) A and B compact implies $A + B$ compact
(c) A closed and B compact implies $A + B$ closed.
Show, by producing an example, that A and B closed does not imply $A + B$ closed. (*Hint:* Construct a suitable pair of subsets of $\mathbf{R}^2$ which become asymptotically close at infinity.)

3. Show that the normed vector subspace s of m defined in Section 1.1 provides an example of a non-closed linear subspace of a normed vector subspace. What is $\bar{s}$?
4. Are the following subsets of l^2 closed? compact?

(a) $S = \{x : \|x\| = 1\}$ (Construct a sequence with no convergent subsequence).

(b) $E_1 = \left\{(x_i) \in l^2 : \sum_{i=1}^{\infty} i^2 x_i^2 \leqslant 1\right\}$.

(c) $E_2 = \left\{(x_i) \in l^2 : \sum_{i=1}^{\infty} i^2 x_i^2 = 1\right\}$.

5. Let $(E, \| \ \|)$ be a normed vector space. By finding a homeomorphism between $[0, 1)$ and $\{x \in \mathbf{R}: x \geqslant 0\}$ show that the open disc $B_r(0)$ of E is homeomorphic to E.

1.4 Products of normed vector spaces

In this and the next few sections we wish to consider methods of constructing new normed spaces which depend, in a natural way, on a given set of normed vector spaces. The product construction given in this section provides the first (and easiest!) example of this process. First, however, let us review some material on products of vector spaces.

Let $E_1, \ldots, E_n$ be a finite set of vector spaces. The product

$$E = E_1 \times \cdots \times E_n$$

may be given the structure of a vector space if we define vector space addition and multiplication coordinatewise. Thus, if

$$e = (e_1, \ldots, e_n), f = (f_1, \ldots, f_n) \in E_1 \times \cdots \times E_n,$$

we define

$$e + f = (e_1 + f_1, \ldots, e_n + f_n)$$

Associated with the vector space $E_1 \times \cdots \times E_n$ we have the *projection* maps

$$p_i: E_1 \times \cdots \times E_n \longrightarrow E_i, \qquad 1 \leqslant i \leqslant n,$$

and the *inclusion* maps

$$I_i: E_i \longrightarrow E_1 \times \cdots \times E_n, \qquad 1 \leqslant i \leqslant n$$

defined respectively by

$$p_i(e_1, \ldots, e_n) = e_i \qquad \text{for all } (e_1, \ldots, e_n) \in E_1 \times \cdots \times E_n, \quad 1 \leqslant i \leqslant n,$$

$$I_i(e_i) = (0, \ldots, 0, \underset{\substack{\uparrow \\ i\text{th place}}}{e_i}, 0, \ldots, 0) \qquad \text{for all } e_i \in E_i, \quad 1 \leqslant i \leqslant n.$$

Let I_F denote the identity map of the vector space F. Then, the reader may verify that p_i and I_i are linear maps, $1 \leqslant i \leqslant n$, and that they satisfy the conditions

1. $p_i . I_i = I_{E_i}, \qquad 1 \leqslant i \leqslant n.$
2. Every $x \in E_1 \times \cdots \times E_n$ may be uniquely written as the sum

$$x = \sum_{i=1}^{n} I_i . p_i(x)$$

Remarks

1. We often use the abbreviated notation $\mathsf{X}_{i=1}^{n} E_i$ for the product vector space $E_1 \times \cdots \times E_n$.
2. The product vector space $E_1 \times \cdots \times E_n$ is sometimes referred to as the *direct sum* of the vector spaces $E_1, \ldots, E_n$ and is then written $E_1 \oplus \cdots \oplus E_n$ or $\bigoplus_{i=1}^{n} E_i$. The direct sum $E_1 \oplus \cdots \oplus E_n$ is always considered as a vector space, unlike $E_1 \times \cdots \times E_n$, which can also be regarded as just the set-theoretic product of the vector spaces $E_1, \ldots, E_n$. The context, however, usually makes it clear whether we are considering $E_1 \times \cdots \times E_n$ as a vector space or merely a product of vector spaces, and we shall reserve the direct-sum notation for subspaces of a fixed vector space (as in Definition 0.1.7).

Proposition 1.4.1

Let $(E_1, \| \enspace \|_1), \ldots, (E_n, \| \enspace \|_n)$ be a finite set of normed vector spaces. Then the product vector space $E = E_1 \times \cdots \times E_n$ may be given the structure of a normed vector space, with norm defined by

$$\| e \| = \max_i \| p_i(e) \|_i \qquad \text{for all } e \in E$$

$$= \max_i \| e_i \|_i, \qquad \text{if } e = (e_1, \ldots, e_n) \in E_1 \times \cdots \times E_n.$$

Further, the disc centre 0, radius r, in E is given as the product

$$B_r(0_E) = B_r(0_{E_1}) \times \cdots \times B_r(0_{E_n}),$$

where $B_r(0_{E_i})$ denotes the disc radius r, centre 0, in E_i, $1 \leqslant i \leqslant n$.

Proof. Left to the exercises. ∎

Exercises

1. Prove Proposition 1.4.1.
2. By analogy with the definition of norms on $\mathbf{R}^n$, we may attempt to define a norm on the product $E = E_1 \times \cdots \times E_n$ by setting

$$\|e\|_p = \left(\sum_{i=1}^{n} \|p_i(e)\|_i^p\right)^{1/p} \qquad \text{for all } e \in E,$$

where $p \geqslant 1$ and the notations and assumptions of Proposition 1.4.1 apply. Show that $\| \quad \|_p$ does indeed define a norm on E. (In the sequel we will, however, always use the norm on E given by Proposition 1.4.1.)

1.5 Linear and multilinear maps

In this section we begin the study of continuous linear maps between normed vector spaces. First, however, we shall make the following general definition.

Definition 1.5.1

Let $E, E_1, E_2, \ldots, E_n, F$ denote vector spaces over the field K.

1. A map $T: E \longrightarrow F$ is called *linear* if

$$T(kx + y) = kT(x) + T(y) \qquad \text{for all } x, y \in E \text{ and } k \in K.$$

2. A map $T: E_1 \times E_2 \longrightarrow F$ is called *bilinear* if it is linear in each variable separately. That is, if for all $k \in K, x, x_1, x_2 \in E_1$ and $y, y_1, y_2 \in E_2$, we have

$$T(kx_1 + x_2, y) = kT(x_1, y) + T(x_2, y)$$

$$T(x, ky_1 + y_2) = kT(x, y_1) + T(x, y_2).$$

And, more generally,

3. A map $T: E_1 \times \cdots \times E_n \longrightarrow F$ is called *multilinear* or *n-linear* if it is linear in each variable separately.

Remark. In the above definition we do not make any use of the product vector-space structure on the Cartesian product $E_1 \times \cdots \times E_n$ (cf. Section 1.4). The only linear structure that is being used is that on the factors E_i, $1 \leqslant i \leqslant n$. Thus, if $E_1 \times \cdots \times E_n$ is given the product vector space structure, then a multilinear map $T: E_1 \times \cdots \times E_n \longrightarrow F$ will *not* be linear (unless $n = 1$). This fact will become clearer in the following examples.

Examples

1. Let E and F be finite dimensional real vector spaces and let $T: E \times F \longrightarrow \mathbf{R}$ be a bilinear map. Suppose that E and F have bases

$$\mathscr{E} = (e_i)_{i=1}^n \qquad \text{and} \qquad \mathscr{F} = (f_j)_{j=1}^m$$

respectively. We claim that T is uniquely determined by its values on the bases $\mathscr{E}$ and $\mathscr{F}$. Indeed, let $x \in E$ have coordinates $(x_1, \ldots, x_n)$ relative to $\mathscr{E}$ and

$y \in F$ have coordinates $(y_1, \ldots, y_m)$ relative to $\mathscr{F}$. We have

$$T(x, y) = T\left(\sum_{i=1}^{n} x_i e_i, \sum_{j=1}^{m} y_j f_j\right)$$
$$= \sum_{i=1}^{n} \sum_{j=1}^{m} x_i y_j T(e_i, f_j), \qquad \text{using the bilinearity of } T.$$

Let us set $T(e_i, f_j) = a_{ij} \in \mathbf{R}$. We then have the following formula for T in the coordinates determined by the bases $\mathscr{E}$ and $\mathscr{F}$:

$$T((x_1, \ldots, x_n), (y_1, \ldots, y_m)) = \sum_{i,j} a_{ij} x_i y_j. \tag{1.5.2}$$

Conversely, given bases $\mathscr{E}$ and $\mathscr{F}$ on E and F respectively, an $n \times m$ array of real numbers $[a_{ij}]$ determines a real-valued bilinear form on $E \times F$ according to formula (1.5.2).

More generally, if T is a G-valued bilinear form on $E \times F$, T may be written, relative to the bases $\mathscr{E}$ and $\mathscr{F}$, in the unique form

$$T((x_1, \ldots, x_n), (y_1, \ldots, y_m)) = \sum_{i,j} x_i y_j a_{ij},$$

where $T(e_i, f_j) = a_{ij} \in G$.

2. Let $T: E \times \cdots \times E \longrightarrow K$ be a K-valued m-linear map defined on the product of m copies of E. Suppose that E is finite dimensional with basis $\mathscr{E} = (e_i)_{i=1}^n$. Then T is uniquely determined by its values on the basis $\mathscr{E}$. In fact, if we set

$$T(e_{i_1}, \ldots, e_{i_m}) = a_{i_1 \cdots i_m}, \qquad 1 \leqslant i_1, \ldots, i_m \leqslant n,$$

we have, relative to the coordinate system on E given by $\mathscr{E}$,

$$T(x^1, \ldots, x^m) = \sum_{i_1, \ldots, i_m = 1}^{n} a_{i_1 \cdots i_m} x_{i_1}^1 \ldots x_{i_m}^m \tag{1.5.3}$$

where the sum is over all m-tuples of integers $(i_1, \ldots, i_m)$ satisfying $1 \leqslant i_j \leqslant n$ and $(x_1^k, \ldots, x_n^k)$ are the coordinates of the point $x^k \in E$, $1 \leqslant k \leqslant m$. Conversely, given a basis $\mathscr{E}$ on E, formula (1.5.3) uniquely determines a K-valued m-linear form on the m-fold product of E. The generalization of formula (1.5.3) to G-valued multilinear forms on a product $E_1 \times \cdots \times E_m$ is obvious.

Nowhere in Definition 1.5.1 did we put any topology on the vector spaces or make any requirement as to the continuity of the maps defined. We now wish to do this and will start by making a study of linear maps. An indication that the study of continuous linear maps on a normed vector space is rather spccial is given by the following lemma.

Lemma 1.5.4 (Uniform continuity of continuous linear maps)

Let $(E, \| \ \|)$ and $(F, \| \ \|)$ be normed vector spaces and let $T: E \longrightarrow F$ be a linear map. Then T is continuous if and only if it is continuous at $0 \in E$.

Proof. Let T be continuous at 0. Then, given $r > 0$, $\exists \delta > 0$ such that

$$T(B_\delta(0)) \subset B_r(0).$$

But, since $B_\delta(x) = x + B_\delta(0)$ and $B_r(T(x)) = T(x) + B_r(0)$, it follows using the linearity of T that

$$T(B_\delta(x)) \subset B_r(T(x)).$$

Since r and x are arbitrary, it follows that T is continuous on E. The reverse implication is trivial. ∎

Remark on notation. In the proof of Lemma 1.5.4, we did not specifically indicate that $B_\delta(0)$ was a disc in E whilst $B_r(0)$ was a disc in F. The context, however, made this distinction clear. In the sequel we shall not use specific designations for discs or norms unless the context is such as to lead to confusion. Thus norms will be written as '$\| \quad \|$' and discs as '$B_r(x)$' with no mention of the underlying vector space.

The next theorem gives a very powerful characterization of continuity for linear maps between normed vector spaces. It essentially tells us that a linear map is continuous if and only if it is bounded on discs. In other words, a linear map $T:E \longrightarrow F$ is continuous if and only if $T(B)$ is a bounded subset of F (that is, of finite diameter) whenever B is a bounded subset of E.

Theorem 1.5.5

Let E and F be normed vector spaces and $T:E \longrightarrow F$ be a linear map. The following conditions are equivalent.

1. T is continuous.
2. $\sup_{\|e\| \leqslant 1} \|T(e)\| < \infty$ (T is bounded on the unit disc of E)
3. $\exists \alpha \geqslant 0$ such that $\|T(e)\| \leqslant \alpha \|e\|$ for all $e \in E$.

Proof. We divide the proof into three steps.

Step 1. (1) implies (2).

Since T is continuous at 0_E, $\exists \delta > 0$ such that

$$T(\bar{B}_\delta(0_E)) \subset \bar{B}_1(0_F)$$

If $e \in \bar{B}_1(0_E)$ then $\delta e \in \bar{B}_\delta(0_E)$ and so $\|T(\delta e)\| \leqslant 1$. But

$$\|T(e)\| = (1/\delta)\|T(\delta e)\| \leqslant 1/\delta.$$

Hence $\sup_{\|e\| \leqslant 1} \|T(e)\| \leqslant 1/\delta$, proving (2).

Step 2. (2) implies (3).

Let $\alpha = \sup_{\|e\| \leqslant 1} \|T(e)\|$.

(a) If $e = 0$, condition (3) is certainly satisfied.

(b) If $e \neq 0$, set $e' = e/\|e\|$. Then $\|e'\| = 1$ and so $\|T(e')\| \leqslant \alpha$. Hence

$$\|T(e)\| = \|T(\|e\|e')\| = \|e\| \, \|T(e')\| \leqslant \alpha \|e\|.$$

Step 3. (3) implies (1).

Let $\delta > 0$, then $T(\bar{B}_{\delta/\alpha}(0)) \subset \bar{B}_\delta(0)$. Hence T is continuous at 0. Therefore, by Lemma 1.5.4, T is continuous. ∎

As an application of Theorem 1.5.5, we have

Proposition 1.5.6

Let E be a normed vector space; then vector-space addition $E \times E \longrightarrow E$; $(x, y) \longmapsto x + y$ is a continuous linear map.

Proof. Taking the product vector-space structure on $E \times E$, the reader may easily verify the linearity of vector-space addition. We shall prove the continuity. We let $T: E \times E \longrightarrow E$ denote the linear (not bilinear!) map defined by vector space addition: $T(x, y) = x + y$. We have

$$\sup_{\|(x,y)\| \leqslant 1} \|T(x,y)\| = \sup_{\substack{\|x\| \leqslant 1 \\ \|y\| \leqslant 1}} \|x + y\| \qquad \text{(definition of product norm)}$$
$$\leqslant \sup_{\substack{\|x\| \leqslant 1 \\ \|y\| \leqslant 1}} (\|x\| + \|y\|) = 2.$$

Hence T is bounded on the unit disc of $E \times E$ and so is continuous by Theorem 1.5.5. ■

In the remainder of this section we wish to show how Theorem 1.5.5 generalizes to bilinear and multilinear maps.

Theorem 1.5.7

Let $E_1, \ldots, E_n, F$ be normed vector spaces and $T: E_1 \times \cdots \times E_n \longrightarrow F$ be an n-linear map. The following statements are equivalent.

1. T is continuous.
2. $\displaystyle\sup_{\|(e_1,\ldots,e_n)\| \leqslant 1} \|T(e_1, \ldots, e_n)\| < \infty$
3. $\exists \alpha \geqslant 0$ such that

$$\|T(e_1, e_2, \ldots, e_n)\| \leqslant \alpha \|e_1\| \|e_2\| \cdots \|e_n\|$$

for all

$$(e_1, e_2, \ldots, e_n) \in E_1 \times E_2 \times \cdots \times E_n.$$

Here $E_1 \times \cdots \times E_n$ is considered as a normed vector space with the product norm given by Proposition 1.4.1.

Proof. The proof is similar to that of Theorem 1.5.5. We shall prove the theorem for bilinear maps and leave the general case to the exercises.

Step 1. (1) implies (2).

Since T is continuous at $0_{E_1 \times E_2}$, $\exists \delta > 0$ such that

$$T(\bar{B}_\delta(0_{E_1 \times E_2})) \subset \bar{B}_1(0_F).$$

Now $(\delta x, \delta y) \in \bar{B}_\delta(0_{E_1 \times E_2})$ if $(x, y) \in \bar{B}_1(0_{E_1 \times E_2})$. Hence, since

$$\|(x,y)\| = \max(\|x\|, \|y\|),$$

$$\sup_{\|(x,y)\| \leqslant 1} \|T(x,y)\| = \sup_{\substack{\|x\| \leqslant 1 \\ \|y\| \leqslant 1}} \|T(x,y)\| = \sup_{\substack{\|x\| \leqslant 1 \\ \|y\| \leqslant 1}} (1/\delta^2) \|T(\delta x, \delta y)\| \leqslant 1/\delta^2.$$

Step 2. (2) implies (3).

Set

$$\alpha = \sup_{\|(x,y)\| \leqslant 1} \|T(x,y)\|.$$

If either $x = 0$ or $y = 0$ then $T(x, y) = 0$ and (3) is true. So suppose $x \neq 0$ and $y \neq 0$. Set $x' = x/\|x\|$ and $y' = y/\|y\|$. Then $(x', y') \in \bar{B}_1(0_{E_1 \times E_2})$ and $\|T(x', y')\| \leqslant \alpha$. Substituting for x' and y', we get (3).

Step 3. (3) implies (1).

Note that for $x, x' \in E_1$ and $y, y' \in E_2$ we have

$$T(x, y) - T(x', y') = T(x - x', y) - T(x', y - y').$$

Hence

$$\|T(x, y) - T(x', y')\| \leqslant \|T(x - x', y)\| + \|T(x', y - y')\|$$
$$\leqslant \alpha\|x - x'\| \|y\| + \alpha\|x'\| \|y - y'\|,$$

and this is sufficient to show that T is continuous. In fact, given $1 > \epsilon > 0$ and $(x, y) \in E_1 \times E_2$, set $M = \max(\|x\|, \|y\|)$. Then if $(x', y') \in B_r((x, y))$, where $r = \epsilon/(2\alpha(M + 1))$, it follows that

$$\|T(x, y) - T(x', y')\| < \epsilon. \quad \blacksquare$$

As an application of Theorem 1.5.7, we have

Proposition 1.5.8

Let E be a normed vector space. Then scalar multiplication

$$K \times E \longrightarrow E; (k, e) \longmapsto ke$$

is a continuous bilinear map.

Proof. Scalar multiplication certainly defines a bilinear map from $K \times E$ to E. We must prove this map is continuous. This follows immediately from Theorem 1.5.7, since

$$\sup_{\substack{|k| \leqslant 1 \\ \|x\| \leqslant 1}} \|kx\| = \sup_{\substack{|k| \leqslant 1 \\ \|x\| \leqslant 1}} |k| \|x\| = 1. \quad \blacksquare$$

The importance of Theorems 1.5.5 and 1.5.7 is that they enable us to reduce problems about continuity of multilinear maps to problems of boundedness. In the next section we shall see powerful applications of these two theorems. For the moment, however, the reader should note the remarkably simple proofs of the continuity of vector-space addition and scalar multiplication given by these two theorems.

Exercises

1. Not all linear maps between normed vector spaces need be continuous. Find an example of a linear map $T: s \longrightarrow \mathbf{R}$ which is not continuous. (*Hint:* Use Theorem 1.5.5.)
2. Let $T: s \times s \longrightarrow \mathbf{R}$ be defined by

$$T((x_i), (y_i)) = \sum_{i,j=1}^{\infty} x_i y_j/(i + j)^2$$

Prove that T is a bilinear map which is continuous in each variable *separately* but that T is not continuous.
3. Prove that if $(E, \langle\ ,\ \rangle)$ is a real inner-product space, then $\langle\ ,\ \rangle: E \times E \longrightarrow \mathbf{R}$ is a continuous bilinear map (use Theorem 1.5.7).

4. Using the standard bases for $\mathbf{R}^p$ and $\mathbf{R}^q$, let A be a linear map from $\mathbf{R}^p$ to $\mathbf{R}^q$ with matrix $[a_{ij}]$. Verify that A is continuous with respect to the norm $\| \ \|_\infty$ on $\mathbf{R}^p$ and $\mathbf{R}^q$ by showing that A is bounded. What is $\sup_{\|x\|_\infty \leq 1} \|Ax\|_\infty$?
5. Consider the linear map $I : C[0, 1] \longrightarrow C[0, 1]$ defined by

$$I(f)(t) = \int_0^t f(u)\, \mathrm{d}u, \qquad f \in C[0, 1], \quad t \in [0, 1].$$

Prove using Theorem 1.5.5 and directly that I is continuous, relative to the sup norm on $C[0, 1]$.
6. Let $C_B^\infty(0, 1)$ denote the vector space of functions on the open interval $(0, 1)$ which are infinitely differentiable with all derivatives continuous and bounded. Consider the linear map $D : C_B^\infty(0, 1) \longrightarrow C_B^\infty(0, 1)$ defined by

$$(Df)(t) = \mathrm{d}f/\mathrm{d}t, \qquad f \in C_B^\infty(0, 1).$$

Regarding $C_B^\infty(0, 1)$ as a subspace of $B(0, 1)$, and taking the sup norm on $C_B^\infty(0, 1)$, show that D is not a continuous linear map. Draw pictures.
7. Let E and F be real normed vector spaces and $T : E \longrightarrow F$ be a continuous map such that $T(x + y) = T(x) + T(y)$ for all $x, y \in E$. Prove that T is linear (*Hint:* see the proof of Theorem 1.2.4). What happens if E and F are complex vector spaces?
8. If $T : E \longrightarrow F$ is a linear map between the normed spaces E and F, prove the following are equivalent.

   (a) T is continuous.
   (b) $(T(x_n))$ is a bounded sequence in F whenever (x_n) is bounded in E.
   (c) $(T(x_n))$ is a bounded sequence in F whenever $x_n \longrightarrow 0$.

9. Give examples of a normed space E and a continuous linear map from E to E which is

   (a) Injective but nor surjective.
   (b) Surjective but not injective.
   (c) Injective and surjective but without a continuous inverse.

   (*Note:* (c) cannot happen if E is a *complete* normed space. This result is known as the *closed-graph theorem.*)
10. Prove Theorem 1.5.7 for the general case of an n-linear map.
11. Is there a version of Lemma 1.5.4 for multilinear maps?

1.6 Normed spaces of continuous linear and multilinear maps

In this section we show how spaces of *continuous* linear and multilinear maps between normed vector spaces naturally have the structure of normed vector spaces.

Definition 1.6.1

Let $E, E_1, \ldots, E_n, F$ denote normed vector spaces. We let

1. $L(E, F)$ denote the set of all continuous linear maps from E to F.
2. $L(E_1, \ldots, E_n; F)$ denote the set of all continuous n-linear maps from $E_1 \times \cdots \times E_n$ to F.

   In the special case $E_1 = \cdots = E_n = E$, we let $L^n(E; F)$ denote the set of all continuous n-linear maps from the n-fold product of E to F. An element of

$L^n(E; F)$ is called an n-linear form on E with values in F. We define $L^0(E; F)$ to be equal to F.

Remark. The space $L(E_1, \ldots, E_n; F)$ of n-linear maps should not be confused with the space $L(E_1 \times \cdots \times E_n, F)$ of *linear* maps from the product vector space $E_1 \times \cdots \times E_n$ to F.

Theorem 1.6.2

Let E and F denote normed vector spaces. Then $L(E, F)$ has the natural structure of a normed vector space with norm defined by

$$\|T\| = \sup_{\|x\| \leqslant 1} \|T(x)\|, \qquad T \in L(E, F)$$

In addition, for all $T \in L(E, F)$ we have the inequality

$$\|T(x)\| \leqslant \|T\| \, \|x\|, \qquad x \in E.$$

Proof. Vector-space addition and scalar multiplication on $L(E, F)$ are defined by

$$(T + S)(x) = T(x) + S(x) \qquad \text{for all } S, T \in L(E, F) \text{ and } x \in E,$$

$$(kS)(x) = kS(x) \qquad \text{for all } S \in L(E, F),\ k \in K \text{ and } x \in E.$$

It is trivial to show that these definitions give $L(E, F)$ the structure of a vector space once we have verified that $T + S$ and kS are continuous linear maps. We leave the easy verification of the linearity of $T + S$ and kS to the reader and prove the continuity. We have

$$\begin{aligned} \sup_{\|x\| \leqslant 1} \|(T + S)(x)\| &= \sup_{\|x\| \leqslant 1} \|T(x) + S(x)\| \\ &\leqslant \sup_{\|x\| \leqslant 1} (\|T(x)\| + \|S(x)\|) \\ &\leqslant \sup_{\|x\| \leqslant 1} \|T(x)\| + \sup_{\|x\| \leqslant 1} \|S(x)\| < \infty. \end{aligned}$$

Hence, by Theorem 1.5.5, $T + S$ is continuous. Moreover, we have also shown that $\|T + S\| \leqslant \|T\| + \|S\|$.

Similarly, kS is continuous and $\|kS\| = |k| \, \|S\|$. Since $\|T\| = 0$ if and only if $T(x) = 0$ for all $x \in \bar{B}_1(0)$, it follows that $\|T\| = 0$ if and only if $T = 0$. Thus we have proved parts (1) and (2) of the theorem. It remains to prove (3).

If $x = 0$, then (3) is trivially satisfied. Suppose, therefore, that $x \neq 0$. Then, since $x/\|x\| \in \bar{B}_1(0)$, it follows that

$$\|T(x/\|x\|)\| \leqslant \|T\|.$$

Multiplying both sides of this inequality by $\|x\|$ the result follows. ∎

Example. Let us consider the vector spaces $\mathbf{R}^m$ and $\mathbf{R}^n$ with the norm $\| \ \|_\infty$ (see Section 1.1 for the definition of $\| \ \|_\infty$). Now any linear map from $\mathbf{R}^m$ to $\mathbf{R}^n$ is continuous with respect to the topology given by $\| \ \|_\infty$. To see this let $A: \mathbf{R}^m \longrightarrow \mathbf{R}^n$ be linear with matrix $[a_{ij}]$ relative to the standard bases on $\mathbf{R}^m$ and $\mathbf{R}^n$. Then the reader may easily verify that

$$\|A(x)\|_\infty \leqslant \left(\max_i \left(\sum_{j=1}^m |a_{ij}| \right) \right) \|x\|_\infty, \qquad x \in \mathbf{R}^m.$$

and, indeed, that $\|A\| = \max_i (\Sigma_{j=1}^m |a_{ij}|)$.

Now $L(\mathbf{R}^m, \mathbf{R}^n)$ is linearly isomorphic to $M(m, n; \mathbf{R})$, the vector space of real $m \times n$ matrices (see Exercise 4, Section 0.1). A natural basis for $M(m, n; \mathbf{R})$ is given by the set of matrices $\{E^{\alpha\beta} : 1 \leqslant \alpha \leqslant m, 1 \leqslant \beta \leqslant n\}$, where $E^{\alpha\beta}$ is the $m \times n$ matrix all of whose entries are zero except the $\alpha\beta$ entry, whose value is 1. Hence $M(m, n; \mathbf{R})$ is isomorphic to $\mathbf{R}^{mn}$, where the isomorphism is defined by, say, mapping $E^{\alpha\beta}$ to the pth basis element of the standard basis for $\mathbf{R}^{mn}$, where $p = n(\alpha - 1) + \beta$.

The function γ on $M(m, n; \mathbf{R})$ defined by

$$\gamma([a_{ij}]) = \max_i \left(\sum_{j=1}^{m} |a_{ij}| \right)$$

is then clearly induced, via this isomorphism, from the norm $\| \ \|^*$ on $\mathbf{R}^{mn}$, where $\| \ \|^*$ is defined as the n-fold product of the norm $\| \ \|_1$ on $\mathbf{R}^m$.

The reader should carefully note that the norm we have constructed on $L(\mathbf{R}^m, \mathbf{R}^n)$ depends on our choice of norms on $\mathbf{R}^m$ and $\mathbf{R}^n$. Even in this case the computation of the norm on $L(\mathbf{R}^m, \mathbf{R}^n)$ is not completely straightforward. In the exercises at the end of this section it is shown how the choice of different norms on $\mathbf{R}^m$ and $\mathbf{R}^n$ results in a different norm on $L(\mathbf{R}^m, \mathbf{R}^n)$.

We now wish to generalize Theorem 1.6.2 to spaces of multilinear maps.

Theorem 1.6.3

Let $E_1, \ldots, E_n, F$ be normed vector spaces. Then $L(E_1, \ldots, E_n; F)$ has the natural structure of a normed vector space with norm defined by

$$\|T\| = \sup_{\|(x_1, \ldots, x_n)\| \leqslant 1} \|T(x_1, \ldots, x_n)\|, \qquad T \in L(E_1, \ldots, E_n; F).$$

Furthermore for all $T \in L(E_1, \ldots, E_n; F)$ we have the inequality

$$\|T(x_1, x_2, \ldots, x_n)\| \leqslant \|T\| \, \|x_1\| \, \|x_2\| \cdots \|x_n\|,$$

where $(x_1, x_2, \ldots, x_n) \in E_1 \times E_2 \times \cdots \times E_n$.

Proof. Vector-space addition and scalar multiplication on $L(E_1, \ldots, E_n; F)$ are defined in the obvious way. Thus for all $S, T \in L(E_1, \ldots, E_n; F)$ and $k \in K$ we set

$$(T + S)(x_1, \ldots, x_n) = T(x_1, \ldots, x_n) + S(x_1, \ldots, x_n),$$
$$(x_1, \ldots, x_n) \in E_1 \times \cdots \times E_n$$

$$(kS)(x_1, \ldots, x_n) = kS(x_1, \ldots, x_n), \ (x_1, \ldots, x_n) \in E_1 \times \cdots \times E_n.$$

Just as in the proof of Theorem 1.6.2, it is easy to show that these definitions give $L(E_1, \ldots, E_n; F)$ the structure of a vector space. That it has the structure of a normed vector space follows from Theorem 1.5.7 in the same way that Theorem 1.6.2 followed from Theorem 1.5.5. We leave details to the exercises. ∎

Examples

1. Consider the vector space $\mathbf{R}^n$ with the norm $\| \ \|_\infty$ and the bilinear map $T : \mathbf{R}^n \times \mathbf{R}^n \longrightarrow \mathbf{R}^n$ defined by

$$T((x_1, \ldots, x_n), (y_1, \ldots, y_n)) = (x_1 y_1, \ldots, x_n y_n).$$

We leave it to the reader to verify that $\|T\| = 1$ and that $\|T(x, y)\|_\infty \leqslant \|x\|_\infty \|y\|_\infty$ for all $x, y \in \mathbf{R}^n$.

2. Let $\langle\ ,\ \rangle$ be the standard inner product on $\mathbf{R}^n$ associated with the norm $\|\ \ \|_2$. Then $\langle\ ,\ \rangle \in L^2(\mathbf{R}^n; \mathbf{R})$ and, as a bilinear map, $\langle\ ,\ \rangle$ has norm 1.
The latter fact follows immediately from the Schwartz lemma.

Proposition 1.6.4

Let E, F, G be normed vector spaces. Then if $T \in L(E, F)$ and $S \in L(F, G)$, the composition $S \,.\, T \in L(E, G)$. Furthermore, we have

1. The map $L(F, G) \times L(E, F) \longrightarrow L(E, G); (S, T) \longmapsto S \,.\, T$ is a continuous bilinear map.
2. If $S \in L(F, G)$ and $T \in L(E, F)$, then $\|S \,.\, T\| \leqslant \|S\| \, \|T\|$.

Proof. Let $T \in L(E, F)$ and $S \in L(F, G)$; then the composition $S \,.\, T$ is trivially shown to be linear and the continuity of $S \,.\, T$ follows from general results about continuous maps. The map defined in (1) is clearly bilinear, and the continuity follows, via Theorem 1.5.7, once we have shown that (2) is true. By definition,

$$\|S \,.\, T\| = \sup_{\|x\| \leqslant 1} \|S(T(x))\| \leqslant \sup_{\|x\| \leqslant 1} \|S\| \, \|T(x)\| = \|S\| \, \|T\|,$$

proving (2). ∎

The following theorem will be very useful when we come to study higher derivatives in Chapter 2.

Theorem 1.6.5

Let E, F, G be normed vector spaces. Then $L(E, L(F, G))$ is *naturally* isomorphic to $L(E, F; G)$ in a norm-preserving way:

$$L(E, L(F, G)) \approx L(E, F; G).$$

Proof. We shall define an element $\tilde{T} \in L(E, F; G)$ associated with each point $T \in L(E, L(F, G))$. Given our limited assumptions about the spaces E, F and G, the only way we can hope to do this is by defining

$$\tilde{T}(e, f) = [T(e)](f), \qquad e, f \in E \times F.$$

The reader may easily verify that $\tilde{T}$ is a bilinear map from $E \times F$ to G. However, we have many other details to check.

1. $\tilde{T}$ is continuous and $\|\tilde{T}\| = \|T\|$. In fact we have

$$\sup_{\substack{\|e\| \leqslant 1 \\ \|f\| \leqslant 1}} \|\tilde{T}(e, f)\| = \sup_{\substack{\|e\| \leqslant 1 \\ \|f\| \leqslant 1}} \|[T(e)](f)\| = \sup_{\|e\| \leqslant 1} \|T(e)\| = \|T\|.$$

2. The map $L(E, L(F, G)) \longrightarrow L(E, F; G); T \longmapsto \tilde{T}$ is linear, continuous, norm preserving and an isomorphism. Linearity is trivial. Continuity follows from (1) and Theorem 1.5.5. We must prove the map is an isomorphism. Suppose, therefore, that $S \in L(E, F; G)$. Define $S' \in L(E, L(F, G))$ by

$$[S'(e)](f) = S(e, f), \qquad (e, f) \in E \times F.$$

We leave it to the reader to check that this formula defines a continuous linear inverse to the map $T \longmapsto \tilde{T}$. ∎

Corollary 1.6.5.1

Let $E_1, \ldots, E_n, F$ be normed vector spaces. Then we have a natural norm-preserving isomorphism

$$L(E_1, \ldots, E_n; F) \approx L(E_1, L(E_2, L(\ldots, L(E_n, F)\ldots))).$$

Proof. Exercise. ∎

We conclude this section by considering a number of results about products of normed vector spaces and spaces of linear maps between products of vector spaces.

Proposition 1.6.6

Let $(E_1, \|\ \|_1), \ldots, (E_n, \|\ \|_n)$ be a finite set of normed vector spaces. Then the projection map p_i,

$$p_i : E_1 \times \cdots \times E_n \longrightarrow E_i,$$

is a continuous linear map of norm 1, $1 \leqslant i \leqslant n$. The inclusion map I_i,

$$I_i : E_i \longrightarrow E_1 \times \cdots \times E_n,$$

is similarly a continuous linear map of norm 1, $1 \leqslant i \leqslant n$.

Proof. Both p_i and I_i are linear maps (see Section 1.4). We shall show that $\| p_i \| = 1$, $1 \leqslant i \leqslant n$. Indeed,

$$\begin{aligned} \| p_i \| &= \sup_{\|(e_1, \ldots, e_n)\| \leqslant 1} \| p_i(e_1, \ldots, e_n) \|_i \\ &= \sup_{\|(e_1, \ldots, e_n)\| \leqslant 1} \| e_i \|_i \\ &= 1, \qquad \text{by definition of the product norm.} \end{aligned}$$

That $\| I_i \| = 1$ is proved similarly. ∎

Proposition 1.6.7

Let $E_1, \ldots, E_p, F_1, \ldots, F_q$ denote normed vector spaces. Then there exists a canonical linear homeomorphism

$$L\left(\mathop{\times}_{i=1}^{p} E_i, \mathop{\times}_{j=1}^{q} F_j \right) \approx \mathop{\times}_{i=1}^{p} \mathop{\times}_{j=1}^{q} L(E_i, F_j).$$

This isomorphism is norm preserving if $p = 1$.

Proof. Let $A \in L(\times_{i=1}^{p} E_i, \times_{j=1}^{p} F_j)$. For $1 \leqslant m \leqslant p$ and $1 \leqslant n \leqslant q$, we may define $A_{nm} \in L(E_m, F_n)$ by the formula

$$A_{nm} = p_n \,.\, A \,.\, I_m.$$

This clearly defines for us a linear map

$$\gamma : L\left(\mathop{\times}_{i=1}^{p} E_i, \mathop{\times}_{j=1}^{q} F_j \right) \longrightarrow \mathop{\times}_{i=1}^{p} \mathop{\times}_{j=1}^{q} L(E_i, F_j).$$

Now $\gamma(A) = 0$ if and only if $A_{nm} = 0$, $1 \leqslant m \leqslant p$, $1 \leqslant n \leqslant q$, which clearly can happen if and only if $A = 0$. Hence γ is injective. On the other hand suppose we are given

$A_{nm} \in L(E_m, F_n)$, $1 \leqslant m \leqslant p$, $1 \leqslant n \leqslant q$. Then we may define a unique element $A \in L(\times_{i=1}^{p} E_i, \times_{j=1}^{q} F_j)$ by the formula

$$A(x_1, \ldots, x_p) = \sum_{m,n} I_n . A_{nm}(x_m) \qquad \text{for all } (x_1, \ldots, x_p) \in E_1 \times \cdots \times E_p$$

and this clearly defines a linear inverse to the map γ. Hence γ is a linear isomorphism. We must verify that γ is continuous. First note that, with the above notation, $\|A_{nm}\| \leqslant \|A\|$, $1 \leqslant m \leqslant p$, $1 \leqslant n \leqslant q$. Indeed this is an immediate consequence of Proposition 1.6.4, applied twice, and Proposition 1.6.6. But

$$\begin{aligned} \|\gamma(A)\| &= \max_{m,n} \|A_{nm}\| \\ &\leqslant \|A\|, \qquad \text{by the above remark.} \end{aligned}$$

Hence γ is continuous and $\|\gamma\| \leqslant 1$. We may similarly verify that γ^{-1} is continuous and so γ is a linear homeomorphism.

Finally, we leave it to the exercises for the reader to verify that if $p = 1$, γ is norm preserving. ■

Example. Suppose we consider $\mathbf{R}^n = \mathbf{R}^{n_1} \times \mathbf{R}^{n_2}$ and $\mathbf{R}^m = \mathbf{R}^{m_1} \times \mathbf{R}^{m_2}$, where $n = n_1 + n_2$ and $m = m_1 + m_2$ respectively. Then if $A \in L(\mathbf{R}^n, \mathbf{R}^m)$, we have maps $A_{ij} \in L(\mathbf{R}^{n_i}, \mathbf{R}^{m_j})$, $1 \leqslant i, j \leqslant 2$, defined by

$$A_{ij} = p_i . A . I_j.$$

Let us choose bases $\mathscr{E} = (e_i)_{i=1}^{n}$ and $\mathscr{F} = (f_j)_{j=1}^{m}$ for $\mathbf{R}^n$ and $\mathbf{R}^m$ respectively. Suppose that $\mathscr{E}_1 = (e_i)_{i=1}^{n_1}$ and $\mathscr{E}_2 = (e_i)_{i=n_1+1}^{n}$ are bases for $\mathbf{R}^{n_1}$ and $\mathbf{R}^{n_2}$ respectively and that $\mathscr{F}_1 = (f_j)_{j=1}^{m_1}$ and $\mathscr{F}_2 = (f_j)_{j=m_1+1}^{m}$ are bases for $\mathbf{R}^{m_1}$ and $\mathbf{R}^{m_2}$ respectively. The matrix of A relative to the bases $\mathscr{E}$ and $\mathscr{F}$ may then be written

$$[A] = [a_{ij}] = \left[\begin{array}{ccc|ccc} a_{11} & \cdots & a_{1n_1} & a_{1,n_1+1} & \cdots & a_{1n} \\ \vdots & & \vdots & \vdots & & \vdots \\ a_{m_1 1} & \cdots & a_{m_1 n_1} & a_{m_1,n_1+1} & \cdots & a_{m_1 n} \\ \hline a_{m_1+1,1} & \cdots & a_{m_1+1,n_1} & a_{m_1+1,n_1+1} & \cdots & a_{m_1+1,n} \\ \vdots & & \vdots & \vdots & & \vdots \\ a_{m1} & \cdots & a_{mn_1} & a_{m,n_1+1} & \cdots & a_{mn} \end{array}\right] = \begin{bmatrix} [A_{11}] & [A_{12}] \\ [A_{21}] & [A_{22}] \end{bmatrix}$$

Here the submatrices $[A_{ij}]$ of A correspond to the matrices of the maps A_{ij} relative to the bases $\mathscr{F}_i$ and $\mathscr{E}_j$.

Exercises

1. Consider the vector spaces $\mathbf{R}^n$ and $\mathbf{R}^m$ with the norm $\| \quad \|_2$. Let $A : \mathbf{R}^n \longrightarrow \mathbf{R}^m$

be linear with matrix $[a_{ij}]$ relative to the standard bases of $\mathbf{R}^n$ and $\mathbf{R}^m$. Show that

$$\|A\|^2 \leqslant \sum_{i,j} |a_{ij}|^2.$$

Need we have equality?

2. Complete the details of the proof of Theorem 1.6.3.
3. Prove Corollary 1.6.5.1 (use induction).
4. Complete the proof of Proposition 1.6.7.
5. Let $A \in L(\mathbf{R}^n, \mathbf{R}^m)$. Regarding $\mathbf{R}^n$ and $\mathbf{R}^m$ as the respective n- and m-fold products of $\mathbf{R}$, find the maps A_{ij} given by Proposition 1.6.7. Interpret, using the canonical isomorphism $L(\mathbf{R}, \mathbf{R}) \approx \mathbf{R}$, defined by mapping $A \in L(\mathbf{R}, \mathbf{R})$ to $A(1) \in \mathbf{R}$.

1.7 Some special spaces of linear and multilinear maps

Definition 1.7.1

Let E be a normed vector space over the field K. The normed vector space $L(E, K)$ is called the *dual space* of E or just the *dual* of E. We shall denote the dual of E by the symbol E^*.

Remarks

1. It should be carefully noted that E^* comprises the set of *continuous* linear K-valued maps on E. Not every linear K-valued map on E need be continuous—see Exercise 1 of Section 1.5.
2. Elements of E^* are often referred to as linear *functionals* or linear *forms*.
3. The norm on E^* is given, according to Theorem 1.6.2, by

$$\|f\| = \sup_{\|x\| \leqslant 1} |f(x)| \qquad \text{for all } f \in E^*.$$

It follows from Theorem 1.5.5 that if $f \in E^*$ we have

$$|f(x)| \leqslant \|f\| \, \|x\| \qquad \text{for all } x \in E.$$

4. Let E be a finite-dimensional vector space with basis $\mathscr{E} = (e_i)_{i=1}^n$. Associated with $\mathscr{E}$ we may define a basis $\mathscr{E}^* = (e_i^*)_{i=1}^n$ for E^*. Thus for $1 \leqslant i \leqslant n$ we define $e_i^* : E \longrightarrow K$ by

$$e_i^*(e_j) = \begin{cases} 0, & i \neq j \\ 1, & i = j. \end{cases}$$

The reader may verify that if $f \in E^*$, f can be written uniquely in the form

$$f = \sum_{j=1}^{n} f(e_j) e_j^*.$$

Hence $\mathscr{E}^*$ is a basis for E^*. We usually refer to $\mathscr{E}^*$ as the *dual basis* of $\mathscr{E}$. It follows from the above construction of $\mathscr{E}^*$ that $\dim_K E = \dim_K E^*$. Furthermore the basis $\mathscr{E}$ determines a linear isomorphism between E and E^* which is characterized by mapping e_i to e_i^*, $1 \leqslant i \leqslant n$. However, this isomorphism is not 'natural' (it depends on the choice of basis $\mathscr{E}$).

5. Given a normed vector space E, we have the *dual pairing* of E and E^*

$$E \times E^* \longrightarrow K$$

defined by mapping $(e, \phi) \in E \times E^*$ to $\phi(e) \in K$. This pairing is clearly bilinear. It is also continuous since $\| \phi(e) \| \leqslant \| \phi \| \| e \|$, and so we may apply Theorem 1.5.7. It is somewhat less trivial, however, to prove that the norm of the pairing map is actually *equal* to 1. We shall show later in this chapter that this is a corollary of the Hahn–Banach theorem.

Example. Consider the normed vector space ($C[0, 1]$, sup norm) and let g be any continuous function on $[0, 1]$ (or, more generally, Lebesgue-integrable function on $[0, 1]$). g defines the following linear map:

$$I_g : C[0, 1] \longrightarrow \mathbf{R}; f \longmapsto \int_0^1 g(x)f(x)\, dx.$$

We claim that $I_g \in E^*$. In fact the reader may verify that we have

$$\| I_g \| \leqslant \int_0^1 | g(u) |\, du.$$

(Actually we have equality.)

Remark. It can in fact be shown that every continuous linear form on $C[0, 1]$ can be represented as a suitable integral on $[0, 1]$. The proper enunciation of this result requires, however, measure theory and lies outside the scope of this text.

Notation. We let S_n denote the group of all permutations of the set $\{1, 2, \ldots, n\}$. If $\sigma \in S_n$, we let sign (σ) denote the signature of σ.

For $n \geqslant 1$, we shall let $\times^n E$ denote the n-fold Cartesian product of E.

Definition 1.7.2

Let E and F be normed vector spaces. An element $T \in L^n(E; F)$ is said to be a *symmetric* n-linear form if for all $(x_1, \ldots, x_n) \in \times^n E$ we have

$$T(x_1, \ldots, x_n) = T(x_{\sigma(1)}, \ldots, x_{\sigma(n)}) \qquad \text{for all } \sigma \in S_n.$$

We denote the set of symmetric n-linear forms from E to F by $L_s^n(E; F)$. We *define* $L_s^0(E; F)$ to be equal to F.

An element $T \in L^n(E; F)$ is said to be an *anti-symmetric* n-linear form if for all $(x_1, \ldots, x_n) \in \times^n E$ we have

$$T(x_1, \ldots, x_n) = \text{sign}\,(\sigma)\, T(x_{\sigma(1)}, \ldots, x_{\sigma(n)}) \qquad \text{for all } \sigma \in S_n.$$

We denote the set of anti-symmetric n-linear forms from E to F by $L_a^n(E; F)$. We *define* $L_a^0(E; F)$ to be equal to F.

Proposition 1.7.3

Let E and F be normed vector spaces. Then $L_s^n(E; F)$ and $L_a^n(E; F)$ are normed vector subspaces of $L^n(E; F)$.

Proof. Left to the exercises. ∎

Examples

1. Let $\langle\, , \rangle$ be an inner product on E. Then $\langle\, , \rangle \in L_s^2(E; \mathbf{R})$.
2. Let $T \in L^n(\mathbf{R}; \mathbf{R})$. We may write T in the form

$$T(x_1, \ldots, x_n) = k x_1 \cdots x_n, \qquad x_1, \ldots, x_n \in \mathbf{R},$$

where $k = T(1, 1, \ldots, 1)$. Since T is obviously symmetric, we have in this case

$$L^n(\mathbf{R}; \mathbf{R}) = L_s^n(\mathbf{R}; \mathbf{R}), \qquad n \geqslant 1.$$

Further the map $T \longmapsto T(1, 1, \ldots, 1) \in \mathbf{R}$ defines an isomorphism between $L^n(\mathbf{R}; \mathbf{R})$ and $\mathbf{R}$. Thus $L_s^n(\mathbf{R}; \mathbf{R}) \approx \mathbf{R}$, $n \geqslant 1$.

Suppose, on the other hand, that $T \in L_a^n(\mathbf{R}; \mathbf{R})$, $n > 1$. Then $T(1, 1, \ldots, 1) = 0$, since transposition of any two variables of T changes the sign of T. Consequently $T \equiv 0$. It follows that $L_a^n(\mathbf{R}; \mathbf{R}) = 0$, $n > 1$. Clearly when $n = 1$, $L_a^1(\mathbf{R}; \mathbf{R}) = L(\mathbf{R}, \mathbf{R})$.

3. If E and F are normed vector spaces then $L_s^1(E; F) = L_a^1(E; F) = L(E, F)$. This is immediate from the definition, since S_1 consists of only one element: the identity permutation.
4. Let $T \in L_a^n(E; F)$ and $(x_1, \ldots, x_n) \in \mathsf{X}^n E$. Then, if any two of $x_1, \ldots, x_n$ are equal, $T(x_1, \ldots, x_n) = 0$. This follows since transposition of two equal x's changes the sign of T.
5. Let $m > 1$ and consider $L^n(\mathbf{R}^m; \mathbf{R})$. If $n = 1$ it follows from 2 that

$$L_a^1(\mathbf{R}^m; \mathbf{R}) = L_s^1(\mathbf{R}^m; \mathbf{R}) = L(\mathbf{R}^m, \mathbf{R}) = (\mathbf{R}^m)^*.$$

Suppose that $n > 1$. Let $(e_i)_{i=1}^m$ denote a basis for $\mathbf{R}^m$ and set

$$T(e_{i_1}, \ldots, e_{i_n}) = a_{i_1 \cdots i_n}, \qquad 1 \leqslant i_1, \ldots, i_n \leqslant m.$$

If T is symmetric, we must have for all $1 \leqslant i_1, \ldots, i_n \leqslant m$

$$a_{i_1 \cdots i_n} = a_{\sigma(i_1) \cdots \sigma(i_n)}, \qquad \text{for all permutations } \sigma \text{ of } i_1, \ldots, i_n.$$

Conversely, if this condition is satisfied, T is symmetric. This follows from the expression given for an n-linear form, relative to a coordinate system, in Example 2 of Section 1.5.

Similarly it may be shown that a necessary and sufficient condition for an n-linear form $T \in L^n(\mathbf{R}^m; \mathbf{R})$ to be anti-symmetric is

$$a_{i_1 \cdots i_n} = \operatorname{sign}(\sigma) a_{\sigma(i_1) \cdots \sigma(i_n)}, \qquad \text{for all permutations } \sigma \text{ of } i_1, \ldots, i_n.$$

Let us conclude this example by considering two bilinear forms on $\mathbf{R}^2$.

$$S : \mathbf{R}^2 \times \mathbf{R}^2 \longrightarrow \mathbf{R}; ((x_1, y_1), (x_2, y_2)) \longmapsto x_1 x_2 + y_1 y_2$$

gives an example of a symmetric bilinear form on $\mathbf{R}^2$. On the other hand

$$T : \mathbf{R}^2 \times \mathbf{R}^2 \longrightarrow \mathbf{R}; ((x_1, y_1), (x_2, y_2)) \longmapsto x_1 y_2 - x_2 y_1$$

defines an anti-symmetric bilinear form on $\mathbf{R}^2$.

Definition 1.7.4.

A continuous function $P : E \longrightarrow F$ is said to be a *homogeneous polynomial of degree n* if there exists $T \in L_s^n(E; F)$ such that

$$P(x) = T(x, x, \ldots, x) \qquad \text{for all } x \in E.$$

We denote the set of all homogeneous polynomials from E to F of degree n by $H^n(E; F)$. If $n = 0$, we define $H^0(E; F) = F$.

Notation. We let $\Delta : E \longrightarrow X^n E$

denote the *diagonal map* defined

by $\Delta(x) = (x \ldots x)$, for all $x \in E$.

We shall often write $T(\Delta(x))$ in the abbreviated (and suggestive) form $T(x^n)$. Thus, if $T \in L_s^n(E; F)$, we may write the associated homogeneous polynomial in the form $p(x) = T(x^n)$.

Remark. It should be noticed that in Definition 1.7.4 we have not asserted that the symmetric map T is *unique.* This is in fact true but the proof is not completely straightforward. We shall return to this point later in this section.

Proposition 1.7.5

Let E and F be normed vector spaces and $P \in H^n(E; F)$. Then

$$P(\lambda x) = \lambda^n P(x) \qquad \text{for all } x \in E \text{ and } \lambda \in K.$$

Proof. By definition of P, $\exists T \in L_s^n(E; F)$ such that $P(x) = T(x, x, \ldots, x)$. Hence

$$P(\lambda x) = T(\lambda x, \ldots, \lambda x) = \lambda^n T(x, x, \ldots, x) = \lambda^n P(x). \quad \blacksquare$$

Proposition 1.7.6

Let E and F be normed vector spaces. Then the set of homogeneous polynomials from E to F of degree n has the natural structure of a normed vector space.

Proof. Let $T, S \in L_s^n(E; F)$ define the polynomials $P, Q \in H^n(E; F)$ respectively. Then we define $P + Q$ by

$$(P + Q)(x) = (T + S)(x^n).$$

Since $T + S \in L_s^n(E; F)$ (Proposition 1.7.3), it follows that $P + Q \in H^n(E; F)$.

Scalar multiplication is defined in the obvious way and the axioms for a vector space are readily verified.

Let $P \in H^n(E; F)$. We define

$$\|P\| = \sup_{\|x\| \leq 1} \|P(x)\|.$$

We leave it as an exercise to the reader to check that this definition gives $H^n(E; F)$ the structure of a normed vector space. $\blacksquare$

Remarks

1. In the sequel we shall not require the normed space structure on $H^n(E; F)$. The reader may like to verify, however, that we have a generalization of Theorem 1.5.7 to $H^n(E; F)$; namely,

$$\|P(x)\| \leq \|P\| \, \|x\|^n \qquad \text{for all } P \in H^n(E; F) \text{ and } x \in E.$$

2. The definition of norm in Proposition 1.7.6 gives us the possibility of defining an alternative norm on $L_s^n(E; F)$. Thus, if $T \in L_s^n(E; F)$, define

$$\|T\|' = \sup_{\|x\| \leq 1} \|T(x, x, \ldots, x)\|.$$

As yet we have not proved enough even to show that $\| \quad \|'$ is a norm on $L_s^n(E; F)$. (Which condition could fail?) It does, however, turn out that $\| \quad \|'$ defines a new norm on $L_s^n(E; F)$. Though this is not equal to the original norm on $L_s^n(E; F)$, both norms do, in fact, define the same topology on $L_s^n(E; F)$.

Example. Let $p \in H^n(\mathbf{R}; \mathbf{R})$, $n \geqslant 1$. Then $p(x) = kx^n$, where $k = p(1) \in \mathbf{R}$. The map $p \longmapsto p(1)$ defines a vector-space isomorphism between $H^n(\mathbf{R}; \mathbf{R})$ and $\mathbf{R}$.

Homogeneous polynomials are used in the generalization of Taylor's theorem to functions of several variables. We wish to give an expression in coordinates for a homogeneous polynomial evaluated at a point of a finite-dimensional vector space. First, however, we shall develop some more efficient notation.

1.7.7 MULTI-INDEX NOTATION

We let $\mathbf{N}$ denote the set of non-negative integers; $\mathbf{N} = \{0, 1, \ldots\}$. Let n denote a fixed positive integer. We shall write a point $(m_1, \ldots, m_n) \in \mathbf{N}^n$ in abbreviated form as m (m is called a *multi-index*). Choose some fixed basis and associated coordinate system on $\mathbf{R}^n$. Let $x = (x_1, \ldots, x_n) \in \mathbf{R}^n$ and $m \in \mathbf{N}^n$, then x^m is defined to be the monomial

$$x_1^{m_1} \cdots x_n^{m_n}.$$

Similarly we abbreviate the coefficient $a_{m_1 \cdots m_n} \in \mathbf{R}$ to a_m. We define $|m|$ to be $m_1 + \cdots + m_n$ and $m!$ to be $m_1! \cdots m_n!$ With these conventions, the sum

$$\sum_{|m|=p} \frac{p!}{m!} a_m x^m$$

is the multi-index notation for the finite sum

$$\sum_{m_1 + \cdots + m_n = p} \frac{p!}{m_1! \cdots m_n!} a_{m_1 \cdots m_n} x_1^{m_1} \cdots x_n^{m_n}.$$

As a first step towards obtaining a representation in coordinates of a homogeneous polynomial we shall prove the following lemma.

Lemma 1.7.8

Let E and F be normed vector spaces and let $T \in L_s^n(E; F)$. Suppose we are given scalars $t_i \in K$, $1 \leqslant i \leqslant p$, and vectors $e_i \in E$, $1 \leqslant i \leqslant p$, then we have the following identity:

$$T((t_1 e_1 + \cdots + t_p e_p)^n) = \sum_{|m|=n} \frac{n!}{m!} t^m T(e^m),$$

where by $T(e^m)$ is meant

$$T(\underbrace{e_1, \ldots, e_1}_{m_1}, e_2, \ldots, \underbrace{e_p, \ldots, e_p}_{m_p}).$$

Proof. Using the n-linearity property of T repeatedly yields

$$T((t_1 e_1 + \cdots + t_p e_p)^n) = \sum_{i_1, \ldots, i_n = 1}^{p} t_{i_1} \cdots t_{i_n} T(e_{i_1}, \ldots, e_{i_n}).$$

Since T is symmetric, many of the terms in this sum are equal. In fact, from the symmetry of T, any term $T(e_{i_1}, \ldots, e_{i_n})$ is equal to $T(e_1^{m_1}, \ldots, e_p^{m_p})$, where

$m_1, \ldots, m_p$ are unique positive integers satisfying $m_1 + \cdots + m_p = n$ and m_k is equal to the number of i_j subscripts equal to k, $1 \leqslant k \leqslant p$. From the elementary theory of permutations and combinations it follows easily that exactly $n!/(m_1! \cdots m_p!)$ terms equal to $T(e_1^{m_1}, \ldots, e_p^{m_p})$ occur in the above sum for each multi-index m satisfying $|m| = n$. ∎

As a corollary of this lemma we have

Proposition 1.7.9

Let $T \in L_s^n(\mathbf{R}^p; \mathbf{R})$ and let $(e_i)_{i=1}^p$ be a basis for $\mathbf{R}^p$. Then, if $x = (x_1, \ldots, x_p) \in \mathbf{R}^p$,

$$T(x^n) = \sum_{|m|=n} \frac{n!}{m!} a_m x^m,$$

where $a_m = T(e^m)$.

Proof. Immediate from Lemma 1.7.8, since $T(x^n) = T((x_1 e_1 + \cdots + x_p e_p)^n)$. ∎

Corollary 1.7.9.1

Let $P \in H^n(\mathbf{R}^p; \mathbf{R})$ be equal to $T.\,\Delta$, where $T \in L_s^n(\mathbf{R}^p; \mathbf{R})$. Suppose that $x \in \mathbf{R}^p$ has coordinates $(x_1, \ldots, x_p)$ relative to some basis $(e_i)_{i=1}^p$ of $\mathbf{R}^p$. Then

$$P(x) = \sum_{|m|=n} \frac{n!}{m!} a_m x^m,$$

where the coefficients a_m are given by $a_m = T(e^m)$, $|m| = n$.

Conversely, any sum of this type defines a homogeneous polynomial $P \in H^n(\mathbf{R}^p; \mathbf{R})$.

Proof. The first part of the corollary is immediate from Proposition 1.7.9. For the converse, it is sufficient to show that any monomial of the form $p_m(x) = x^m$, $|m| = n$, is an nth-order homogeneous polynomial, for we have already shown that $H^n(\mathbf{R}^p; \mathbf{R})$ is a vector space (Proposition 1.7.6). That is, given p_m we must find an element $T_m \in L_s^n(\mathbf{R}^p; \mathbf{R})$ such that $p_m(x) = T_m(x^m)$. Now it is easy to find a point $T \in L^n(\mathbf{R}^p; \mathbf{R})$ such that

$$T(x, x, \ldots, x) = x_1^{m_1} \cdots x_p^{m_p}, \qquad (m_1, \ldots, m_p) \in \mathbf{N}^p.$$

In fact let $y_1, \ldots, y_n \in \mathbf{R}^p$ and denote the coordinates of y_j, $1 \leqslant j \leqslant n$, by $(y_{j1}, \ldots, y_{ji}, \ldots, y_{jp})$. If we define

$$T(y_1, \ldots, y_n) = y_{11} y_{21} \cdots y_{m_1 1} y_{(m_1+1)2} \cdots y_{np},$$

then $T \in L^n(\mathbf{R}^p; \mathbf{R})$ and clearly satisfies the required condition. However, T is not generally symmetric. We may obtain a symmetric form from T by a process known as *symmetrization.* Thus, let $y_1, \ldots, y_n \in \mathbf{R}^p$ and define $T_m \in L^n(\mathbf{R}^p; \mathbf{R})$ by the formula

$$T_m(y_1, \ldots, y_n) = 1/n! \sum_{\sigma \in S_n} T(y_{\sigma(1)}, \ldots, y_{\sigma(n)}).$$

T_m is clearly symmetric. In addition, we also have

$$T_m(x, x, \ldots, x) = T(x, x, \ldots, x) \qquad \text{for all } x \in \mathbf{R}^p.$$

Hence $p_m(x) = T_m(x^n)$. ∎

Suppose we are given a homogeneous polynomial. How do we recover the symmetric multilinear map with which the polynomial is associated? The following important lemma provides the answer to this question.

Lemma 1.7.10 (Polarization lemma)

Let $T \in L_s^n(E;F)$ and $(x_1, \ldots, x_n) \in \times^n E$. Then

$$T(x_1, \ldots, x_n) = \frac{1}{2^n n!}\left(\sum_{\epsilon_i=\pm 1, 1\leqslant i\leqslant n} \epsilon_1 \cdots \epsilon_n T((\epsilon_1 x_1 + \cdots + \epsilon_n x_n)^n)\right).$$

Proof. If we expand $T((\epsilon_1 x_1 + \cdots + \epsilon_n x_n)^n)$ according to Lemma 1.7.8, we find that we have to consider the sum over $\epsilon_1, \ldots, \epsilon_n$ of terms of the form

$$\sum_{|m|=n} \frac{n!}{m_1! \cdots m_n!} \epsilon_1^{m_1+1} \cdots \epsilon_n^{m_n+1} T(x_1, \ldots, x_n).$$

Now suppose $(m_1, \ldots, m_n) \neq (1, \ldots, 1)$. Then it is easy to see that at least two of the indices $m_j + 1$, $1 \leqslant j \leqslant n$, must be odd. Without loss of generality, suppose $m_1 + 1$ and $m_2 + 1$ are odd. Sum over ϵ_1 and ϵ_2, keeping the remaining ϵ_i fixed. We obtain a contribution from ϵ_1 and ϵ_2 of

$$(+1)^{m_1+1}(+1)^{m_2+1} + (-1)^{m_1+1}(+1)^{m_2+1} + (+1)^{m_1+1}(-1)^{m_2+1} + (-1)^{m_1+1}(-1)^{m_2+1} = 0,$$

since $m_1 + 1$ and $m_2 + 1$ are odd. Therefore, if we fix $(m_1, \ldots, m_n)$ and sum over $\epsilon_1, \ldots, \epsilon_n$ we obtain zero. Hence all terms for which $(m_1, \ldots, m_n) \neq (1, 1, \ldots, 1)$ cancel on summation.

On the other hand, if $(m_1, \ldots, m_n) = (1, \ldots, 1)$, then

$$\epsilon_1^{m_1+1} \cdots \epsilon_n^{m_n+1} = 1.$$

Hence, summing over ϵ_i, we obtain the coefficient of $T(x_1, \ldots, x_n)$ to be $2^n n!$, since

$$\frac{n!}{1!, \ldots, 1!} = n!$$

and $(\epsilon_1, \ldots, \epsilon_n)$ takes precisely 2^n different values in the summation. ∎

As an application of the polarization lemma we have

Proposition 1.7.11

Let E and F be normed vector spaces. The map

$$L_s^n(E;F) \longrightarrow H^n(E;F); T \longmapsto T.\Delta$$

is a vector-space isomorphism. Further, it has as inverse the map $U: H^n(E, F) \longrightarrow L_s^n(E;F)$ defined by

$$U(p)(x_1, \ldots, x_n) = \frac{1}{2^n n!} \sum_{\substack{\epsilon_i=\pm 1\\ 1\leqslant i\leqslant n}} \epsilon_1 \cdots \epsilon_n p(\epsilon_1 x_1 + \cdots + \epsilon_n x_n),$$

where $p \in H^n(E;F)$ and $(x_1, \ldots, x_n) \in \times^n E$.

Proof. Since the map $T \longmapsto T.\Delta$ is clearly linear and is, by definition, surjective, we have only to prove its injectivity. But if $T \in L_s^n(E;F)$ and $T.\Delta = 0$, it follows immediately from the polarization lemma that $T = 0$. Hence we have injectivity.

That the map $U: H^n(E;F) \longrightarrow L_s^n(E;F)$ is the inverse to the map $T \longrightarrow T.\Delta$ is again immediate from the polarization lemma. ∎

Remark. In Proposition 1.7.11 we have deliberately refrained from saying anything about the continuity of the linear isomorphisms between $L_s^n(E;F)$ and $H^n(E;F)$. In fact, given the norm on $H^n(E;F)$, provided by Proposition 1.7.6, and the usual norm on $L_s^n(E;F)$, it can be shown that this isomorphism is continuous. This result will follow from exercises at the end of this and the following section.

Another useful application of the polarization lemma is given by

Proposition 1.7.12

Let $\mathbf{R}^p$ have basis $(e_i)_{i=1}^p$ and $P \in H^n(\mathbf{R}^p;\mathbf{R})$; then the coefficients a_m in the sum

$$P(x) = \sum_{|m|=n} \frac{n!}{m!} a_m x^m$$

are given in terms of P by

$$a_{m_1 \cdots m_p} = \frac{1}{2^n n!} \sum_{\epsilon_i = \pm 1, 1 \leqslant i \leqslant n} \epsilon_1 \cdots \epsilon_n P((\epsilon_1 + \cdots + \epsilon_{m_1})e_1 + \cdots + (\epsilon_{n-m_p+1} + \cdots + \epsilon_n)e_p).$$

Proof. Immediate from the polarization lemma and Proposition 1.7.9. ∎

Definition 1.7.13

Let E and F be normed vector spaces. If we are given elements $p_j \in H^j(E;F)$, $0 \leqslant j \leqslant n$, such that $p_n \neq 0$, then the sum $p_0 + \cdots + p_n$ is said to define a *polynomial* of degree n from E to F.

We denote the set of all polynomials of degree n from E to F, together with the zero polynomial, by $P^n(E;F)$.

We have the following isomorphisms:

$$P^n(E;F) \cong F \times H^1(E;F) \times \cdots \times H^n(E;F)$$

$$\cong F \times \mathop{\mathsf{X}}_{j=1}^{n} L_s^n(E;F).$$

Examples

1. Let $p \in P^n(\mathbf{R};\mathbf{R})$. Then $p(x) = \sum_{i=0}^n a_i x^i$, $a_i \in \mathbf{R}$, $0 \leqslant i \leqslant n$.
2. Let $P \in P^n(\mathbf{R}^s;\mathbf{R})$, $s > 1$. Then

$$P(x) = \sum_{|m| \leqslant n} \frac{|m|!}{m!} a_m x^m, \qquad \text{where } a_m \in \mathbf{R}, |m| \leqslant n.$$

Exercises

1. Let $(x_i) \in l^2$. Show that the map $(y_i) \longmapsto \sum_{i=1}^\infty x_i y_i$, $(y_i) \in l^2$, defines an element of $(l^2)^*$. Show that the norm of this linear form is equal to $\| (x_i) \|_2$.

Using Hölder's inequality, prove that an element $(x_i) \in l^r$ defines an element of $(l^{r'})^*$, where $r, r' > 1$ and $1/r + 1/r' = 1$. What is the norm of this linear form?

2. Prove Proposition 1.7.3.
3. Let E be a normed vector space over K. If $P \in H^n(E; K)$ and $Q \in H^m(E; K)$, define $PQ: E \longrightarrow K$ by $(PQ)(x) = P(x)Q(x)$. Prove $PQ \in H^{n+m}(E; K)$.
4. Find all symmetric bilinear real forms on $\mathbf{R}^2$ and hence find a basis for $L_s^2(\mathbf{R}^2; \mathbf{R})$.
5. Complete the proof of Proposition 1.7.6.
6. Find the symmetric multilinear maps which are associated with the following homogeneous polynomials on $\mathbf{R}^n$:
   (a) $p(x) = x_1^2 + \cdots + x_n^2, x = (x_1, \ldots, x_n) \in \mathbf{R}^n$.
   (b) $q(x) = x_i x_j, x = (x_1, \ldots, x_n) \in \mathbf{R}^n, i \neq j$.
7. Let $(e_i)_{i=1}^n$ be a basis for $\mathbf{R}^n$. If $T \in L_s^2(\mathbf{R}^n; \mathbf{R})$, we may define an $n \times n$ matrix $[a_{ij}]$ by setting

$$a_{ij} = T(e_i, e_j).$$

   Prove that this matrix is symmetric (that is, equal to its transpose).
8. Let E be a normed vector space. Prove that $L^2(E; K) \approx L(E, E^*)$ (use Theorem 1.6.5). Suppose now that E is a finite-dimensional real vector space with basis $(e_i)_{i=1}^n = \mathscr{E}$. If $T \in L^2(E; \mathbf{R})$, we let $\hat{T}$ denote the corresponding element of $L(E, E^*)$. Show that T is symmetric if and only if $[\hat{T}]_{\mathscr{E},\mathscr{E}^*}$ is a symmetric matrix.
9. Using the polarization lemma, prove that the 'polynomial norm' $\| \ \|'$ on $L_s^n(E; F)$ and the usual norm $\| \ \|$ on $L_s^n(E; F)$ are related by the following inequality:

$$\|T\|' \leqslant \|T\| \leqslant \left(\frac{n^n}{n!}\right)\|T\|' \qquad \text{for all } T \in L_s^n(E; F).$$

1.8 Equivalence of norms

Suppose we are given two norms $\| \ \|_1$ and $\| \ \|_2$ on a vector space E. It is natural to ask under what conditions these norms define the same topology on E. Such considerations are obviously important: For example, if E and F are normed vector spaces, the vector space $L(E, F)$ depends only on the *topologies* on E and F.

Definition 1.8.1

Two norms $\| \ \|_1$ and $\| \ \|_2$ on a vector space E are said to be *equivalent* if and only if the associated metric topologies on E are equal.

We remark that this definition defines an equivalence relation on the set of all norms on a given vector space E.

Theorem 1.8.2

Let $\| \ \|_1$ and $\| \ \|_2$ be two norms on the vector space E. Then $\| \ \|_1$ and $\| \ \|_2$ are equivalent if and only if there exist constants $M, m > 0$ such that

$$m\|x\|_1 \leqslant \|x\|_2 \leqslant M\|x\|_1 \qquad \text{for all } x \in E.$$

Proof. Let $i: (E, \| \ \|_1) \longrightarrow (E, \| \ \|_2)$ denote the identity map. $\| \ \|_1$ and $\| \ \|_2$ are equivalent if and only if the linear maps i and i^{-1} are continuous. Suppose then that $\| \ \|_1$ and $\| \ \|_2$ are equivalent. Then i is a continuous linear map and so, by

Theorem 1.5.5, $\exists M > 0$ such that

$$\|x\|_2 = \|i(x)\|_2 \leqslant M\|x\|_1 \qquad \text{for all } x \in E.$$

Again, by applying Theorem 1.5.5 to the continuous linear map i^{-1}, $\exists \tilde{m} > 0$ such that

$$\|x\|_1 = \|i^{-1}(x)\|_1 \leqslant \tilde{m}\,\|x\|_2 \qquad \text{for all } x \in E.$$

Setting $m = 1/\tilde{m}$, we obtain the desired inequality.

The sufficiency of the condition on the norms is obtained by reversing the above argument to prove that i and i^{-1} are continuous. ∎

Examples

1. The p-norms $\|\ \|_p$ on $\mathbf{R}^n$ are all equivalent. We shall prove that $\|\ \|_1$, $\|\ \|_2$ and $\|\ \|_\infty$ are equivalent and leave the general case as an exercise. Let $x = (x_1, \ldots, x_n) \in \mathbf{R}^n$. Then

$$\|x\|_\infty \leqslant \|x\|_1 \leqslant \sum_{i=1}^{n} \max_i |x_i| = n\|x\|_\infty.$$

Hence, by Theorem 1.8.2, $\|\ \|_1$ and $\|\ \|_\infty$ are equivalent. Now

$$\|x\|_2^2 = \sum_{i=1}^{n} |x_i|^2 \leqslant \left(\sum_{i=1}^{n} \max |x_i|\right)^2 = n\|x\|_\infty^2,$$

and so

$$\|x\|_\infty \leqslant \|x\|_2 \leqslant \sqrt{n}\,\|x\|_\infty,$$

proving the equivalence of $\|\ \|_\infty$ and $\|\ \|_2$. Since equivalence of norms is an equivalence relation we have proved that all three norms are equivalent.

2. Not all norms on a given vector space need be equivalent. Let us consider the sup norm on $C[0, 1]$ together with the norm $\|\ \|_1$ defined by

$$\|f\|_1 = \int_0^1 |f(u)|\,du, \qquad f \in C[0, 1].$$

We shall show that these two norms are inequivalent. First, however, note that

$$\|f\|_1 \leqslant \int_0^1 \sup_{[0,1]} |f(u)|\,du = \|f\| \qquad \text{for all } f \in C[0, 1].$$

On the other hand we claim that it is impossible to find $m > 0$ such that $m\|f\| \leqslant \|f\|_1$ for all $f \in C[0, 1]$. We define the sequence of functions $f_n \in C[0, 1]$ by the formula (Fig. 1.4)

$$f_n(t) = \begin{cases} nt, & 0 \leqslant t \leqslant 1/n \\ 2 - nt, & 1/n \leqslant t \leqslant 2/n \\ 0, & t \geqslant 2/n. \end{cases}$$

Clearly $\|f_n\| = 1$ whilst $\|f_n\|_1 = 1/n$, $n = 1, \ldots$. It follows that there can exist no $m > 0$ such that $m\|f\| \leqslant \|f\|_1$ for all $f \in C[0, 1]$.

Many properties of normed vector spaces can be reformulated in terms of equivalence classes of norms. A characteristic example is given by the following proposition.

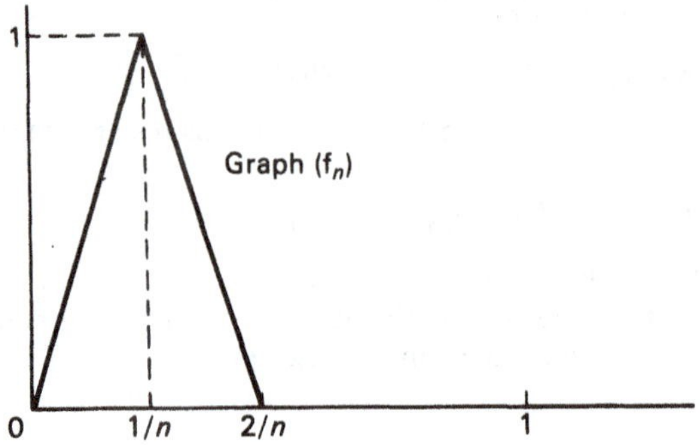

Fig. 1.4

Proposition 1.8.3

Let E and F be normed vector spaces. Then:

1. As a vector space, $L(E, F)$ depends only on the equivalence class of the given norms on E and F.
2. The norm on $L(E, F)$ given by Theorem 1.6.2 is determined up to equivalence by the equivalence classes of the norms on E and F. In particular, the topology on $L(E, F)$ depends only on the equivalence classes of the norms on E and F.

Proof. (1) is trivial. We leave the proof of (2) as an exercise. ∎

Exercises

1. Show that the norms $\| \quad \|_p$ on $\mathbf{R}^n$ are all equivalent.
2. Complete the proof of Proposition 1.8.3. What is the generalization of Proposition 1.8.3 to multilinear maps?
3. We define the norms $\| \quad \|_p, p \geqslant 1$, on $C[0, 1]$ by setting

$$\|f\|_p = \left(\int_1^0 |f(u)|^p \, du \right)^{1/p}, \qquad f \in C[0, 1].$$

By considering the sequence of functions $(f_n) \in C[0, 1]$, defined by

$$f_n(t) = \begin{cases} n^2 t^2, & 0 \leqslant t \leqslant 1/n \\ n^3(t - 2/n)^2 & 1/n \leqslant t \leqslant 2/n \\ 0, & t \geqslant 2/n, \end{cases}$$

show that the norms $\| \quad \|_1$ and $\| \quad \|_2$ are not equivalent. Generalize to show that $\| \quad \|_p$ is equivalent to $\| \quad \|_{p'}$ if and only if $p = p'$.
4. Using Exercise 9 of Section 1.7, prove that the map defined in Proposition 1.7.11 is continuous.

1.9 Complete normed spaces

By far the most important class of normed vector spaces consists of those that are *complete* with respect to the induced metric.

Definition 1.9.1

Let $(E, \| \ \|)$ be a normed vector space and $(x_n)_{n=1}^{\infty}$ be a sequence of points of E. The sequence (x_n) is said to be a *Cauchy* sequence if

$$d(x_n, x_m) = \|x_n - x_m\| \longrightarrow 0 \text{ as } n, m \longrightarrow \infty.$$

Recall that a metric space is said to be *complete* if every Cauchy sequence converges.

Definition 1.9.2

A normed vector space $(E, \| \ \|)$ is called a *Banach* space if it is complete.

An inner-product space $(E, \langle \ , \ \rangle)$ is called a *Hilbert* space if it is complete with respect to the associated norm.

Proposition 1.9.3

Let $\| \ \|_1$ and $\| \ \|_2$ be equivalent norms on the vector space E. Then $(E, \| \ \|_1)$ is Banach if and only if $(E, \| \ \|_2)$ is Banach.

Proof. From Theorem 1.8.2, $\exists M, m > 0$ such that

$$m\|x\|_2 \leqslant \|x\|_1 \leqslant M\|x\|_2 \qquad \text{for all } x \in E.$$

Suppose that $(E, \| \ \|_1)$ is Banach and $(x_n)_{n=1}^{\infty}$ is a Cauchy sequence in $(E, \| \ \|_2)$. Then

$$\|x_n - x_m\|_2 \longrightarrow 0 \qquad \text{as } n, m \longrightarrow \infty.$$

Since $\|x_n - x_m\|_2 \geqslant (1/M)\|x_n - x_m\|_1$, it follows that (x_n) is a Cauchy sequence in $(E, \| \ \|_1)$. Therefore, since $(E, \| \ \|_1)$ is assumed complete, $\exists x \in E$ such that

$$\|x - x_n\|_1 \longrightarrow 0 \qquad \text{as } n \longrightarrow \infty.$$

To complete the proof it suffices to show that $x_n \longrightarrow x$ with respect to $\| \ \|_2$. But this is obvious since $\|x - x_n\|_2 \leqslant (1/m)\|x - x_n\|_1$. ■

Our first class of examples of Banach spaces is given by

Proposition 1.9.4

$(\mathbf{R}^n, \| \ \|_p)$ is a Banach space, $1 \leqslant p \leqslant \infty$.

Proof. In the previous section we showed that $\| \ \|_1$, $\| \ \|_2$ and $\| \ \|_\infty$ were equivalent and left, as an exercise, the proof of the equivalence of all the norms $\| \ \|_p$ on $\mathbf{R}^n$. The latter result also follows from a general theorem about the equivalence of norms on finite-dimensional vector spaces given in the next section. In any case, for the purposes of this proof, we shall assume that the norms $\| \ \|_p$, $1 \leqslant p \leqslant \infty$, are all equivalent. By Proposition 1.9.3 it is, therefore, sufficient to show that $(\mathbf{R}^n, \| \ \|_\infty)$ is complete.

The metric induced on $\mathbf{R}^n$ by $\| \ \|_\infty$ is the product metric defined by

$$d((x_1, \ldots, x_n), (y_1, \ldots, y_n)) = \max_i |x_i - y_i|.$$

Let $(x^q)_{q=1}^{\infty}$ be a Cauchy sequence of points in $\mathbf{R}^n$ for this metric. Then, if we set $x^q = (x_1^q, \ldots, x_n^q)$, we have

$$\|x^p - x^q\|_\infty = \max_i |x_i^p - x_i^q| \longrightarrow 0 \qquad \text{as } p, q \longrightarrow \infty.$$

Hence, for $1 \leq j \leq n$, $(x_j^q)_{q=1}^{\infty}$ is a Cauchy sequence of points in **R**. Since **R** is complete, each of these Cauchy sequences converges. Set

$$x_j = \lim_{q \to \infty} x_j^q, \qquad 1 \leq j \leq n,$$

and let $x = (x_1, \ldots, x_n) \in \mathbf{R}^n$. We claim that $\lim_{q \to \infty} x^q = x$. Given $\epsilon > 0$, $\exists q_1, \ldots, q_n \in \mathbf{N}$ such that

$$|x_j - x_j^q| \leq \epsilon, \qquad q \geq q_j, \quad 1 \leq j \leq n.$$

Set $Q = \max_j q_j$. It follows that

$$\|x - x^q\|_\infty \leq \epsilon, \qquad q \geq Q.$$

Since $\epsilon > 0$ was arbitrary, it follows that x^q converges to x and so $(\mathbf{R}^n, \|\ \|_\infty)$ is complete. ∎

Remark. The proof of Proposition 1.9.4 is standard and may be used to show that the product of any (finite) set of complete metric spaces is complete, given the product metric.

Examples

1. The standard inner product associated to the Euclidean norm $\|\ \|_2$ gives $(\mathbf{R}^n, \|\ \|_2)$ the structure of a Hilbert space.
2. The argument given in the proof of Proposition 1.9.4 may be repeated for the complex normed spaces $(\mathbf{C}^n, \|\ \|_p)$ to show that $(\mathbf{C}^n, \|\ \|_p)$ is a complex Banach space, $1 \leq p \leq \infty$.

Before stating our next results we introduce some new notation. Let X be a metric space (more generally, a topological space) and let $(E, \|\ \|)$ be a normed vector space. We let $C_B(X; E)$ denote the set of all continuous functions from X to E which are bounded. That is, $f \in C_B(X; E)$ if and only if

1. $f: X \longrightarrow E$ is continuous.
2. $\sup_{x \in X} \|f(x)\| < \infty$.

If X is compact, every continuous function on X is bounded and we shall write $C(X; E)$ for the set of continuous functions from X to E.

If $E = \mathbf{R}$ and X is compact, we shall follow the notation of Section 1.1 and set $C(X; \mathbf{R}) = C(X)$. Similarly we set $C(X; \mathbf{C}) = C_{\mathbf{C}}(X)$.

Proposition 1.9.5

Let X be a metric or topological space and $(E, \|\ \|)$ be a normed vector space. Then $C_B(X; E)$ has the natural structure of a normed vector space.

Proof. Let $f, g \in C_B(X; E)$ and $\lambda \in K$. We define $f + g$, $\lambda f \in C_B(X; E)$ and $\|f\|$ by

$$(f + g)(x) = f(x) + g(x) \qquad \text{for all } x \in X.$$

$$(\lambda f)(x) = \lambda f(x) \qquad \text{for all } x \in X.$$

$$\|f\| = \sup_{x \in X} \|f(x)\|.$$

The proof that $(C_B(X; E), \|\ \|)$ is a normed vector space is identical to that which showed ($B(X)$, sup norm) was a normed vector space and we refer the reader to Section 1.1 for details. ■

Examples

1. Let $\mathbf{R}^n$ be given the norm $\| \ \|_2$. Then $C_B(\mathbf{R}^p; \mathbf{R}^q)$ consists of all bounded continuous functions from $\mathbf{R}^p$ to $\mathbf{R}^q$. The norm of such a function just measures the largest value, relative to $\| \ \|_2$, that the function takes.
2. Let $\mathbf{N} = \{0, 1, 2, \ldots\}$ be given the metric induced from $\mathbf{R}$:

$$d(n, m) = |n - m|, \qquad n, m \in \mathbf{N}.$$

With this metric $\mathbf{N}$ becomes a non-compact topological space with the discrete topology. In particular, every subset of $\mathbf{N}$ is open and so every function from $\mathbf{N}$ to any topological space is necessarily continuous. Let us consider the normed vector space $C_B(\mathbf{N}; \mathbf{R})$. Let $S \in C_B(\mathbf{N}; \mathbf{R})$. Then $S: \mathbf{N} \longrightarrow \mathbf{R}$ satisfies the condition

$$\sup_{n \in \mathbf{N}} |S(n)| < \infty.$$

In other words, S defines a bounded sequence of real numbers. Conversely, any sequence of real numbers which is bounded defines a unique element of $C_B(\mathbf{N}; \mathbf{R})$. Thus we have a norm-preserving isomorphism between $(C_B(\mathbf{N}; \mathbf{R}), \| \ \|)$ and $(m, \| \ \|)$ defined by mapping $S \in C_B(\mathbf{N}; \mathbf{R})$ to $(S(0), S(1), \ldots, S(n), \ldots) \in m$.
3. Generalizing the previous example, let $(E, \| \ \|)$ denote a normed vector space. Then $C_B(\mathbf{N}; E)$ represents the set of all bounded sequences of points of E. Thus, if $(x_n)_{n=1}^{\infty}$ is a bounded sequence of points of E, it defines a unique element $x \in C_B(\mathbf{N}; E)$ by the rule

$$x(n) = x_n, \qquad n \in \mathbf{N}$$

and

$$\|x\| = \max_n \|x_n\|.$$

One of the basic theorems of elementary analysis is that the space ($C[a, b]$, sup norm is complete. That is, a uniformly convergent sequence of continuous functions (and so a Cauchy sequence, relative to the sup norm) on the closed interval $[a, b]$ converges to a *continuous* function. The generalization of this theorem to vector-space valued functions on topological spaces is given by the following important theorem.

Theorem 1.9.6

Let X be a metric or topological space and $(E, \| \ \|)$ be a normed vector space. Then $(C_B(X; E), \| \ \|)$ is Banach if and only if $(E, \| \ \|)$ is Banach.

Proof. The proof of the main part of this result (the sufficiency of the condition on E) is formally identical to the proof that ($C[a, b]$, sup norm) is complete. Thus, suppose that (f_n) is a Cauchy sequence in $C_B(X; E)$. Let $x \in E$, then

$$\|f_n(x) - f_m(x)\| \leq \|f_n - f_m\| \longrightarrow 0 \qquad \text{as } n, m \longrightarrow \infty.$$

Hence $(f_n(x))$ is a Cauchy sequence of points of E. Since E is assumed complete, this Cauchy sequence $(f_n(x))$ is convergent. Let us denote the limit of the sequence $(f_n(x))$ by $f(x)$. In this way we construct a function $f: X \longrightarrow E$. We shall prove that $f \in C_B(X; E)$ and that $f_n \longrightarrow f$.

Given $\epsilon > 0$, $\exists n_0$ such that

$$\|f_n - f_m\| < \epsilon/3 \qquad \text{for } n, m \geqslant n_0.$$

Hence

$$\|f_n(x) - f_m(x)\| < \epsilon/3 \qquad \text{for } n, m \geqslant n_0 \text{ and } x \in X.$$

Letting $m \longrightarrow \infty$, we obtain

$$\|f_n(x) - f(x)\| \leqslant \epsilon/3 \qquad \text{for } n \geqslant n_0 \text{ and } x \in X.$$

Fix $x_0 \in X$. Since f_{n_0} is continuous at x_0, there exists a neighbourhood U of x_0 such that

$$\|f_{n_0}(x_0) - f_{n_0}(y)\| < \epsilon/3 \qquad \text{for all } y \in U.$$

Now

$$\begin{aligned}\|f(x_0) - f(y)\| &= \|(f(x_0) - f_{n_0}(x_0)) + (f_{n_0}(x_0) - f_{n_0}(y)) + (f_{n_0}(y) - f(y))\| \\ &< \epsilon/3 + \epsilon/3 + \epsilon/3 \qquad \text{for all } y \in U.\end{aligned}$$

Since $\epsilon > 0$ is arbitrary, it follows that f is continuous.

We also have

$$\|f(x)\| \leqslant \|f_n(x)\| + \epsilon/3 = \epsilon \text{ for all } n \geqslant n_0 \text{ and } x \in X,$$

and so $\|f\| \leqslant \|f_{n_0}\| + \epsilon/3$. Therefore f is bounded and $f \in C_B(X; E)$. Finally,

$$\|f_n - f\| \leqslant \epsilon/3, \qquad n \geqslant n_0, \text{ and so } f_n \longrightarrow f.$$

Conversely, let $C_B(X; E)$ be complete and (e_n) be a Cauchy sequence of points of E. We define a sequence of elements of $C_B(X; E)$ by setting

$$f_n(x) = e_n \qquad \text{for all } x \in X.$$

Since $\|f_n - f_m\| = \|e_n - e_m\|$, it follows that (f_n) is a Cauchy sequence in $C_B(X;E)$. Hence $\exists f \in C_B(X; E)$ such that $f_n \longrightarrow f$. We leave it as an exercise for the reader to verify that f is a constant function and that, if e is the constant value of f, $e_n \longrightarrow e$ as $n \longrightarrow \infty$. ■

As an important corollary of Theorem 1.9.6 we have

Corollary 1.9.6.1

Let X be a compact topological space. Then the normed space ($C(X)$, sup norm) is complete.

Example. From Theorem 1.9.6 it follows immediately that the normed vector space of bounded sequences m ($\cong C_B(\mathbf{N}; \mathbf{R})$) is complete.

It is natural to ask how completeness properties behave when we form spaces of continuous linear and multilinear maps. Closely associated with Theorem 1.9.6 we have

Theorem 1.9.7

Let E and F be normed vector spaces. Then $L(E, F)$ is Banach if F is Banach. More generally, if $E_1, \ldots, E_n, F$ are normed vector spaces, $L(E_1, \ldots, E_n; F)$ is Banach if F is Banach.

Proof. Let (T_n) be a Cauchy sequence in $L(E, F)$. Then, just as in the proof of Theorem 1.9.6, $(T_n(x))$ is a Cauchy sequence in F for all $x \in E$. Hence we may define $T: E \longrightarrow F$ by

$$T(x) = \lim_{n\to\infty} T_n(x) \qquad \text{for all } x \in E.$$

T is clearly linear since for all $x, y \in E$ and $k \in K$ we have

$$\begin{aligned} T(kx + y) - kT(x) - T(y) &= \lim_{n\to\infty} (T_n(kx + y) - kT_n(x) - T_n(y)) \\ &= 0, \qquad \text{since } T_n \text{ is linear.} \end{aligned}$$

Finally we must prove that T is continuous and that $T_n \longrightarrow T$ as $n \longrightarrow \infty$. Let D denote the closed unit disc $\bar{B}_1(0)$ in E. Then $(T_n \mid D)$ is a Cauchy sequence of functions in $C_B(D; F)$. By Theorem 1.9.6 $(T_n \mid D)$ is convergent to a continuous *bounded* function on D, which is clearly $T \mid D$. Hence, by Theorem 1.5.5, T is continuous. Further $T_n \mid D$ converges to $T \mid D$ in $C_B(D; F)$ if and only if T_n converges to T in $L(E; F)$, since convergence in $L(E; F)$ is *defined* as convergence in $C_B(D; F)$.

The second part of the theorem dealing with multilinear maps is proved similarly and we leave details to the exercises. ∎

Remark. The converse of Theorem 1.9.7, $L(E, F)$ Banach implies F Banach, is also true but the proof is much less trivial and requires the Hahn–Banach Theorem (cf. Section 1.18).

Example. Let E denote any, not necessarily complete, normed vector space over the field K. Then the dual space of E, $E^* = L(E, K)$, is always a Banach space. This follows immediately from Theorem 1.9.7, since K (always assumed to be **R** or **C**) is complete.

The preceding theorems might lead the reader to suspect that all normed spaces are complete. This is, however, far from the truth as the following examples show. Particular note should be taken of the second example.

Examples

1. The space s of finitely non-zero sequences is not complete in the norm induced from m. (See p. 13 for definitions and notation.) Indeed, let $(x_n)_{n=1}^{\infty}$ be the sequence of points of s defined by

$$x_n = (1, 1/2, \ldots, 1/n, 0, 0, \ldots), \qquad n = 1, 2, \ldots.$$

$(x_n)_{n=1}^{\infty}$ clearly defines a non-convergent Cauchy sequence of points of s.

2. The normed vector space $(C[0, 1], \| \ \|_1)$ is not complete. To see this we define a sequence of functions $(g_n)_{n=2}^{\infty} \subset C[0, 1]$ by the formulae (Fig. 1.5)

$$g_n(t) = \begin{cases} 1, & 0 \leqslant t \leqslant \frac{1}{2} - 1/n \\ \frac{1}{2}(n/2 + 1 - nt), & \frac{1}{2} - 1/n \leqslant t \leqslant \frac{1}{2} + 1/n \\ 0, & t \geqslant \frac{1}{2} + 1/n. \end{cases}$$

Then $\| g_n \|_1 = 1/2$, $n \geqslant 2$. The reader may easily verify, by inspection of the above diagram, that $(g_n)_{n=1}^{\infty}$ is a Cauchy sequence of functions relative to the norm $\| \ \|_1$. Suppose that the sequence converges to $g \in C[0, 1]$. Then $\| g_n - g \|_1 \longrightarrow 0$ as $n \longrightarrow \infty$. This implies that

$$\int_a^b | g_n(u) - g(u) | \, du \longrightarrow 0 \qquad \text{for all } a, b \in [0, 1].$$

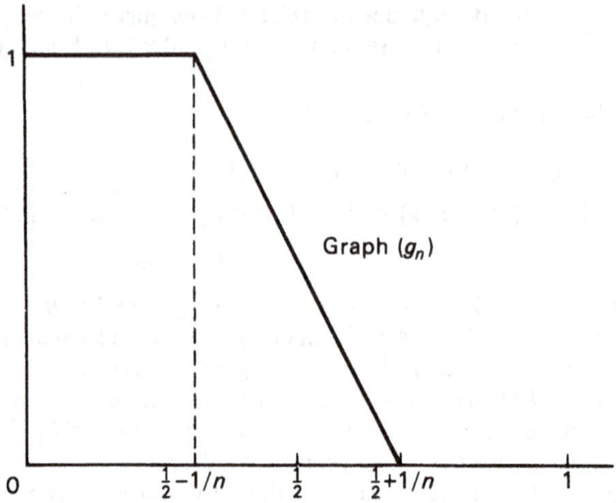

FIG. 1.5

Taking $a = 0$ and b any point $<1/2$, we see that $g(u) = 1, u < 1/2$. Taking $a > 1/2$ and $b = 1$, it follows that $g(u) = 0, u > 1/2$. Hence g cannot be continuous at $t = 1/2$. It follows that the Cauchy sequence (g_n) is not convergent in $C[0, 1]$,, proving that $C[0, 1]$ is not complete if given the norm $\| \quad \|_1$.

We end this section with an important example of an infinite-dimensional Hilbert space.

Proposition 1.9.8

$(l^2, \langle \,, \rangle)$ is a Hilbert space.

Proof. Let $(x_n)_{n=1}^{\infty}$ be a Cauchy sequence of points of l^2. We write

$$x_n = (x_{n1}, x_{n2}, \ldots, x_{nj}, \ldots) = (x_{nj})_{j=1}^{\infty}, \qquad \text{for } n = 1, 2, \ldots.$$

Thus we have

$$\|x_n - x_m\|_2^2 = \sum_{j=1}^{\infty} |x_{nj} - x_{mj}|^2.$$

Since (x_n) is Cauchy, it follows that

$$|x_{nj} - x_{mj}| \longrightarrow 0 \qquad \text{as } n, m \longrightarrow \infty, \quad \text{for } j = 1, 2, \ldots.$$

Consequently, $(x_{nj})_{n=1}^{\infty}$ is a Cauchy sequence of real numbers, $j = 1, 2, \ldots$.

Set $y_j = \lim_{n\to\infty} x_{nj}, j = 1, 2, \ldots$, and denote the sequence $(y_j)_{j=1}^{\infty}$ by y. We must prove that $y \in l^2$ and that $\lim_{n\to\infty} x_n = y$.

Given $\epsilon > 0$, $\exists m_0$ such that

$$\sum_{j=1}^{\infty} |x_{nj} - x_{mj}|^2 < \epsilon \qquad \text{for all } n, m \geqslant m_0.$$

In particular, therefore, for any positive integer N we have

$$\sum_{j=1}^{N} |x_{nj} - x_{mj}|^2 < \epsilon \qquad \text{for all } n, m \geqslant m_0.$$

Letting $m \longrightarrow \infty$, we obtain

$$\sum_{j=1}^{N} |x_{nj} - y_j|^2 \leqslant \epsilon \qquad \text{for all } n \geqslant m_0.$$

This inequality holds for all positive integers N and so, letting $N \longrightarrow \infty$, we have

$$\sum_{j=1}^{\infty} |x_{nj} - y_j|^2 \leqslant \epsilon \qquad \text{for all } n \geqslant m_0.$$

Hence $(x_{nj} - y_j)_{j=1} \in l^2$, $n \geqslant m_0$. Since l^2 is a vector space,

$$(y_j) = (x_{nj}) - (x_{nj} - y_j) \in l^2.$$

Finally, since $\epsilon > 0$ was arbitrary, it follows that $x_n \longrightarrow y$ in l^2. ∎

Exercises

1. Complete the proofs of Theorems 1.9.6 and 1.9.7.
2. Let E be a normed vector space and F be a Banach space. Prove that $L_s^n(E; F)$ and $L_a^n(E; F)$, given the norm induced from $L^n(E; F)$, are Banach spaces.
3. Prove that $(C[0, 1], \| \ \|_2)$ is not complete. What about the general case $(C[0, 1], \| \ \|_p)$, $p \geqslant 1$?
4. Prove that $(l^p, \| \ \|_p)$ is complete, $p \geqslant 1$.
5. Try to prove that if $L(E, F)$ is Banach, then F is Banach, using a method similar to that used in Theorem 1.9.6. What problem arises?
6. Let $(x_n)_{n=1}^{\infty}$ be any sequence of points of the Banach space $(E, \| \ \|)$. We define the sequence $(s_n)_{n=1}^{\infty}$ of partial sums of $(x_n)_{n=1}^{\infty}$ by the relation $s_n = \sum_{i=1}^{n} x_i$. The pair of sequences $((x_n), (s_n))$ is then called a *series*. The series is said to converge to the sum $s \in E$ if $\lim_{n \to \infty} s_n = s$. The series is said to be absolutely convergent if the sum $\sum_{i=1}^{\infty} \|x_i\|$ converges absolutely. Prove
   (a) $((x_n), (s_n))$ converges if and only if, given $\epsilon > 0$, $\exists n_0$ such that $\|s_{n+p} - s_n\| < \epsilon$ for all $n \geqslant n_0$ and $p \geqslant 0$ (Cauchy's criterion).
   (b) Absolute convergence of a series implies convergence.
   (c) If $((x_n), (s_n))$ is absolutely convergent and $\sigma: \mathbf{N}^+ \longrightarrow \mathbf{N}^+$ is a bijection then the series $(x_{\sigma(n)})$ is absolutely convergent and

$$\sum_{n=1}^{\infty} x_n = \sum_{n=1}^{\infty} x_{\sigma(n)}.$$

1.10 Equivalence of norms on finite-dimensional vector spaces

We have already indicated that the p-norms on $\mathbf{R}^n$ are all equivalent. In this section we shall show that *all* norms on a given finite-dimensional vector space are equivalent. Before proving this result, we give a technical definition and prove a lemma.

Definition 1.10.1

A subset B of a vector space E over a field K is called *balanced* if

$$tB \subset B \qquad \text{for all } t \in K \text{ satisfying } |t| \leqslant 1.$$

Example. The disc, radius $r > 0$, centre zero, is a balanced subset of a normed vector space.

Lemma 1.10.2

Let $(E, \| \ \|)$ be a normed vector space and $f: E \longrightarrow K$ be a linear map. Then f is continuous if and only if $f^{-1}(0)$ is closed.

Proof. The necessity of the condition being trivial, we prove the converse. Without loss of generality assume that $f \neq 0$. Let B denote the unit disc of E. Since B is balanced, $f(B) \subset K$ is also balanced. Indeed, if $t \in K$ and $|t| \leqslant 1$, then $tf(B) = f(tB) \subset f(B)$. If f is *discontinuous*, $f(B)$ is unbounded and so, since $f(B)$ is balanced, $f(B) = K$. Let $x \in E$. Given $\epsilon > 0$, $\exists b \in B$ such that

$$f(b) = \epsilon^{-1} f(x), \qquad \text{since } f(B) = K.$$

Hence, $f(x - \epsilon b) = 0$ or $x - \epsilon b \in f^{-1}(0)$. Since $\epsilon > 0$ was arbitrary, it follows that $x \in \overline{f^{-1}(0)}$. Since x was chosen arbitrarily, $f^{-1}(0)$ is *dense* in E. But if $f^{-1}(0)$ is *closed*, it follows that $f^{-1}(0) = E$ and so $f = 0$, contrary to assumption. Hence $f^{-1}(0)$ is not closed. ∎

Theorem 1.10.3

Let E be a finite-dimensional vector space. Then

1. Any two norms on E are equivalent.
2. If $\| \ \|$ is any norm on E, $(E, \| \ \|)$ is complete.

Proof. We shall prove the theorem for real vector spaces. The proof for complex vector spaces is practically identical.

Our proof proceeds by a double induction on the dimension of E. Thus, if $\dim(E) = n$, we let (1_n) and (2_n) be the statements

1_n. Any two norms on E are equivalent.
2_n. If $\| \ \|$ is any norm on E, $(E, \| \ \|)$ is complete.

To establish the inductive step we shall show that (1_n) implies (2_n) and (2_n) implies (1_{n+1}).

Case $n = 1$. Let $\| \ \|$ be any norm on $\mathbf{R}$. Then $\|t\| = |t| \|1\|$ for all $t \in \mathbf{R}$. Hence $\| \ \|$ and $| \ |$ are equivalent and any norm on $\mathbf{R}$ is of the form $k| \ |$, $k > 0$. This proves (1_1), and (2_1) follows immediately from this representation of norms on $\mathbf{R}$ and the fact that $\mathbf{R}$ is complete with respect to the usual metric.

(1_n) implies (2_n)

By assumption (1_n) and Proposition 1.9.3, it is sufficient to show that there exists at least one norm on E with respect to which E is complete. Let $T: E \longrightarrow \mathbf{R}^n$ be any (algebraic) linear isomorphism. We define a norm $||| \ |||$ on E by setting

$$|||x||| = \|T(x)\|_\infty$$

We claim that $(E, ||| \ |||)$ is complete: Let $(x_n)_{n=1}^\infty$ be a Cauchy sequence in E; then $(T(x_n))_{n=1}^\infty$ is obviously a Cauchy sequence in $(\mathbf{R}^n, \| \ \|_\infty)$ and so (by Proposition 1.9.4) converges to some point $y \in \mathbf{R}^n$. Let $x = T^{-1}(y)$. Then $x_n \longrightarrow x$ with respect to $||| \ |||$. Indeed,

$$|||x_n - x||| = \|T(x_n) - y\|_\infty \longrightarrow 0 \qquad \text{as } n \longrightarrow \infty$$

Hence, $(E, ||| \ |||)$ is complete, proving (2_n).

(2_n) implies (1_{n+1})

Let $\dim(E) = n + 1$ and fix once and for all a basis $\mathscr{E} = (e_i)_{i=1}^{n+1}$ for E. Let $x \in E$ have coordinates $(x_1, \ldots, x_{n+1})$ relative to E. We define a new norm $\| \quad \|_1$ on E by setting

$$\|x\|_1 = \sum_{i=1}^{n+1} |x_i|.$$

We shall show that if $\| \quad \|$ is any norm on E, then $\| \quad \|$ is equivalent to $\| \quad \|_1$.

Let f be any linear form on E. Then $f^{-1}(0)$ has dimension n and, taking the induced norm on $f^{-1}(0)$, it follows from (2_n) that $(f^{-1}(0), \| \quad \|)$ is complete. Hence $f^{-1}(0)$ is a closed subspace of $(E, \| \quad \|)$. It follows from Lemma 1.10.2 that f is continuous. Thus (2_n) implies that every linear form on E is continuous with respect to $\| \quad \|$.

Let $x = (x_1, \ldots, x_{n+1}) \in E$. Then

$$\|(x_1, \ldots, x_{n+1})\| = \|\sum_{i=1}^{n+1} x_i e_i\| \leqslant \sum_{i=1}^{n+1} |x_i| \, \|e_i\|$$

$$\leqslant K \|(x_1, \ldots, x_{n+1})\|_1, \qquad \text{where } K = \max_i \|e_i\|.$$

On the other hand,

$$x = \sum_{i=1}^{n+1} f_i(x) e_i,$$

where $f_1, \ldots, f_{n+1}$ are the linear functionals on E defined by

$$f_i(x) = x_i, \qquad 1 \leqslant i \leqslant n + 1.$$

Therefore, the identity map $I: (E, \| \quad \|) \longrightarrow (E, \| \quad \|_1)$ may be written as the sum

$$I(x) = \sum_{i=1}^{n+1} f_i(x) e_i, \qquad x \in E.$$

Since the linear functionals f_i are continuous, by what we have shown above, it follows that I is the sum of continuous linear maps and is, therefore, continuous. From Theorem 1.5.5 it follows that $\exists k > 0$ such that

$$\|x\|_1 = \|I(x)\|_1 \leqslant k \|x\| \qquad \text{for all } x \in E.$$

Consequently $\| \quad \|$ and $\| \quad \|_1$ are equivalent and the proof of the inductive step is complete. ■

Corollary 1.10.3.1

Every finite-dimensional normed vector space is complete.

Theorem 1.10.4

Let $(E, \| \quad \|)$ and $(F, \| \quad \|)$ be normed vector spaces. If E is finite dimensional, every linear map $T: E \longrightarrow F$ is continuous.

Proof. Let $(e_i)_{i=1}^n$ be a fixed basis for E. If $x \in E$, we let $(x_1, \ldots, x_n)$ denote the coordinates of x relative to this basis and define

$$\|x\|_\infty = \max_i |x_i|.$$

$\| \ \|_\infty$ is equivalent to the original norm on E by Theorem 1.10.3. Now

$$T(x) = \sum_{i=1}^n x_i T(e_i),$$

and so

$$\|T(x)\| \leqslant \sum_{i=1}^n |x_i| \|T(e_i)\| \leqslant \|x\|_\infty \left(\sum_{i=1}^n \|T(e_i)\| \right)$$

$$= K \|x\|_\infty, \qquad \text{where } K = \sum_{i=1}^n \|T(e_i)\|.$$

Hence, by Theorem 1.5.5, T is continuous. ∎

Corollary 1.10.4.1

Let E and F be finite-dimensional normed vector spaces. Then every linear map between E and F is necessarily continuous.

We end this section with the following topological characterization of finite-dimensional normed vector spaces.

Theorem 1.10.5 (F. Riesz' theorem)

Let $(E, \| \ \|)$ be a locally compact normed vector space. Then E is finite dimensional.

Proof. For some $r > 0$, $\bar{B}_r(0)$ is compact. Since scalar multiplication is continuous it follows that $\lambda \bar{B}_r(0) = \bar{B}_{\lambda r}(0)$ is compact, $\lambda \geqslant 0$. In particular, $\bar{B}_1(0)$ is compact. From the compactness of $\bar{B}_1(0)$ it follows that we can find a finite number of points $e_i \in \bar{B}_1(0)$ $1 \leqslant i \leqslant n$, such that

$$\bigcup_{i=1}^n \bar{B}_{1/2}(e_i) \supset \bar{B}_1(0).$$

We set $X = \{e_i\colon 1 \leqslant i \leqslant n\}$. Then X is a subset of $\bar{B}_1(0)$ possessing the property that $d(z, X) \leqslant 1/2$, for all $z \in \bar{B}_1(0)$.

Let F denote the finite-dimensional vector subspace of E spanned by the set X. We shall show that $F = E$. From Corollary 1.10.3.1 it follows that F is a *closed* normed linear subspace of E. Hence the F-discs $\bar{B}_r(0)$ are compact for all $r > 0$.

Since F is closed it follows that if $F \neq E$, $\exists x \in E$ such that

$$d(x, F) = \inf_{x' \in F} \|x - x'\| > 0.$$

Since closed discs are compact in F and $d(x, x')$ is a continuous function on F, for fixed x, it follows that we can attain the above infimum on F. That is, $\exists y \in F$ such that

$$d(x, F) = \|x - y\|.$$

Set $z = (x - y)/\|x - y\| \in \bar{B}_1(0)$. Observe that

$$\begin{aligned} d(z, F) &= d\left(\frac{x-y}{\|x-y\|}, F\right) \\ &= \frac{1}{\|x-y\|} d(x-y, F) \qquad \text{(scalar invariance of } d \text{ (Proposition 1.1.3))} \\ &= \frac{1}{\|x-y\|} d(x, F) \qquad \text{(translation invariance of } d\text{)} \\ &= 1. \end{aligned}$$

But, by the definition of the set X, $d(z, X) \leqslant 1/2$. Contradiction. Therefore $F = E$. ∎

Exercises

1. Let $(E, \|\ \ \|)$ be a normed vector space for which the sphere $S_r = \{x\colon \|x\| = r\}$, $r > 0$, is compact. Prove that E is finite dimensional. (Use the continuity of scalar multiplication.)
2. Prove directly that the unit sphere in l^2 is not compact and hence that l^2 is not finite dimensional.
3. Let X be a closed and bounded subset of $\mathbf{R}$. Show that $C(X)$ is finite dimensional if and only if X is finite. Generalize to closed subsets of $\mathbf{R}^n$, $n > 1$.

1.11 Completion of a normed vector space

In Section 1.9 we showed that a normed vector space need not necessarily be complete, an example being provided by $(C[0, 1], \|\ \ \|_1)$. However, it is always possible to embed a normed vector space in a Banach space so that it becomes a dense normed vector subspace. The process is similar to that in which the real numbers are constructed as a completion of the rational numbers.

Theorem 1.11.1

Let $(E, \|\ \ \|)$ be a normed vector space over K ($K = \mathbf{R}$ or $\mathbf{C}$). Then there exists a normed vector space $(\hat{E}, \|\ \ \|')$, the *completion* of E, and a continuous linear injection $i\colon E \longrightarrow \hat{E}$ such that

1. i is norm preserving: $\|i(x)\|' = \|x\|$ for all $x \in E$.
2. $i(E)$ is a dense normed vector subspace of $\hat{E}\colon \overline{i(E)} = \hat{E}$
3. $(\hat{E}, \|\ \ \|')$ is a Banach space.
4. $(\hat{E}, \|\ \ \|')$ is unique up to norm-preserving linear isomorphism.

Proof. Let Γ denote the set of all Cauchy sequences of points of E. Γ may be given the structure of a vector space if we define addition and scalar multiplication coordinatewise:

$$(x_i) + (y_i) = (x_i + y_i), \qquad (x_i), (y_i) \in \Gamma$$
$$\lambda(x_i) = (\lambda x_i), \qquad \lambda \in K \text{ and } (x_i) \in \Gamma.$$

We let Γ_0 denote the vector subspace of Γ consisting of all Cauchy sequences which are convergent to zero.

We define a function

$$\gamma : \Gamma \longrightarrow \mathbf{R}$$

by

$$\gamma((x_i)_{i=1}^{\infty}) = \lim_{n\to\infty} \|x_n\| \qquad \text{for all } (x_n) \in \Gamma.$$

γ is certainly well defined, since if (x_i) is Cauchy, then $(\|x_i\|)$ is a Cauchy sequence of real numbers which converges, since $\mathbf{R}$ is complete.

We remark the following properties of γ (which correspond closely to those of a norm).

1. $\gamma(x) \geqslant 0$ for all $x \in \Gamma$ and $\gamma(x) = 0$ if and only if $x \in \Gamma_0$.
2. $\gamma(x + y) \leqslant \gamma(x) + \gamma(y)$ for all $x, y \in \Gamma$.
3. $\gamma(\lambda x) = |\lambda| \gamma(x)$ for all $x \in \Gamma$ and $\lambda \in K$.

Property (1) is immediate from the definition of Γ_0. Properties (2) and (3) follow easily from the definition of γ. For example, if $x = (x_i), y = (y_i) \in \Gamma$, then

$$\gamma(x + y) = \lim_{n\to\infty} \|(x_n + y_n)\| \leqslant \lim_{n\to\infty} (\|x_n\| + \|y_n\|)$$
$$= \gamma(x) + \gamma(y).$$

γ is technically known as a *pseudo-norm.*

We define $\hat{E}$ to be the quotient vector space Γ/Γ_0, and we let q denote the projection map $\Gamma \longrightarrow \Gamma/\Gamma_0 = \hat{E}$.

We define a map $\| \quad \|' : \hat{E} \longrightarrow \mathbf{R}$ by setting

$$\|\hat{x}\|' = \gamma(x) \qquad \text{for any } x \in \Gamma \text{ satisfying } q(x) = \hat{x} \in \hat{E}.$$

From property (1) of γ it follows that $\| \quad \|'$ is well defined on $\hat{E}$. Since $\|\hat{x}\|' = 0$ if and only if $\hat{x} = 0$ by property (1) of γ, it follows easily from properties (2) and (3) of γ that $\| \quad \|'$ defines a norm on $\hat{E}$. Hence $(\hat{E}, \| \quad \|')$ is a normed vector space.

We have a natural map

$$i : (E, \| \quad \|) \longrightarrow (\hat{E}, \| \quad \|')$$

defined by mapping $x \in E$ to the 'constant' Cauchy sequence $(x, x, \ldots)$ and applying q. The map i is clearly a linear injection, and

$$\|i(x)\|' = \gamma(x, x, \ldots)) = \|x\|.$$

It follows that i is a continuous norm-preserving linear injection of E into $\hat{E}$. For the rest of the proof we will identify E with its image $i(E) \subset \hat{E}$.

$\bar{E}$ is clearly equal to $\hat{E}$ since, if $(x_n)_{n=1}^{\infty}$ is any Cauchy sequence in E, it converges to $q((x_n)_{n=1}^{\infty}) \in \hat{E}$. That is

$$\|x_n - q((x_i)_{i=1}^{\infty})\|' = \lim_{m\to\infty} \|x_n - x_m\| \longrightarrow 0 \qquad \text{as } n \longrightarrow \infty.$$

Next we must prove that $(\hat{E}, \| \quad \|')$ is complete. Let $(x_n)_{n=1}^{\infty}$ be any Cauchy sequence of points of $\hat{E}$. Then we can pick for each $n \geqslant 1$ an element $y_n \in E$ such that

$$\|x_n - y_n\|' < 1/2^n.$$

This is a consequence of $\bar{E} = \hat{E}$. Now $(y_n)_{n=1}^{\infty}$ is a Cauchy sequence of points of E. Indeed,

$$\|y_n - y_m\| = \|y_n - y_m\|'$$
$$= \|(y_n - x_n) + (x_n - x_m) + (x_m - y_m)\|' \longrightarrow 0 \text{ as } n, m \longrightarrow \infty.$$

Therefore, $\exists y \in \hat{E}$ such that

$$\lim_{n\to\infty} y_n = y.$$

We assert that $\lim_{n\to\infty} x_n = y$. Indeed we have

$$\|x_n - y\|' = \|(x_n - y_n) + (y_n - y)\|' \longrightarrow 0 \qquad \text{as } n \longrightarrow \infty.$$

Finally we must prove that $(\hat{E}, \|\ \ \|')$ is unique up to norm-preserving linear isomorphism. We leave this easy verification to the exercises. ∎

Remark. The construction of Theorem 1.11.1 is, in practice, not as important as obtaining a useful representation of the completion. For example, it can be shown that the completion of the space $(C[0, 1], \|\ \ \|_1)$ is isomorphic to the space of all Lebesgue-integrable functions on the interval $[0, 1]$. Similar results hold for the spaces $(C[0, 1], \|\ \ \|_p)$. At a much deeper level, it is also true that the completion of $(C[0, 1], \|\ \ \|_2)$ is isomorphic to $(l^2, \|\ \ \|_2)$. This last result shows that, for example, a square-integrable function on $[0, 1]$ can be uniquely represented (in the sense of measure theory!) by a square summable sequence of real numbers. This isomorphism is, in fact, given by Fourier series. Indeed, it was the proof of this type of result that spurred the development of Lebesgue (as opposed to Riemann) integration. A full discussion of these results unfortunately lies outside the scope of this text. Some indication of the problems that can arise in a more complete treatment is given in the exercises.

Exercises

1. Let $(E, \|\ \ \|)$ and $(F, \|\ \ \|)$ be normed vector spaces and $(\tilde{E}, \|\ \ \|^*)$, $(\tilde{F}, \|\ \ \|^*)$ be normed vector spaces which satisfy, for E and F respectively, conditions (1), (2) and (3) of Theorem 1.11.1. Prove that every linear map $A : E \longrightarrow F$ has a unique extension as a continuous linear map to $\tilde{A} : \tilde{E} \longrightarrow \tilde{F}$ and that $\|A\| = \|\tilde{A}\|$. Show by examples that
   (a) A injective does not imply $\tilde{A}$ injective.
   (b) A surjective does not imply $\tilde{A}$ surjective.
2. With the notation of Question 1, prove that if A is norm preserving (and so an injection), then so is $\tilde{A}$. Deduce the uniqueness part of Theorem 1.11.1 by taking A to be the identity map on E.
3. Let $f \in (C[0, 1], \|\ \ \|_2)$. We may define the *Fourier coefficients* a_n, b_n of f by

$$a_0 = \int_0^1 f(t)\, dt$$

$$a_n = \sqrt{2} \int_0^1 f(t) \cos 2\pi nt\, dt, \qquad n = 1, 2, \ldots$$

$$b_n = \sqrt{2} \int_0^1 f(t) \sin 2\pi nt\, dt, \qquad n = 1, 2, \ldots.$$

Since the sequences $(a_n)_{n=0}^{\infty}$ and $(b_n)_{n=1}^{\infty}$ are clearly bounded, we may define a linear map $F: C[0, 1] \longrightarrow m$ by

$$F(f) = (a_0, a_1, b_1, a_2, \ldots, a_n, b_n, \ldots)$$

(a) Find $F(\cos 2\pi nt)$ and $F(\sin 2\pi nt)$.
(b) Show $\| \cos 2\pi nt \|_2 = \| F(\cos 2\pi nt) \|_2$ and $\| \sin 2\pi nt \|_2 = \| F(\sin 2\pi nt) \|_2$.
(c) Assume that every $f \in C[0, 1]$ can be written (uniquely) in the form

$$f(t) = a_0 + \sum_{n=1}^{\infty} (a_n \cos 2\pi nt + b_n \sin 2\pi nt).$$

Verify, *formally*, that $\| F(f) \|_2 = \| f \|_2$. Deduce that

$$F: (C[0, 1], \| \quad \|_2) \longrightarrow (l^2, \| \quad \|_2)$$

is a norm-preserving linear injection with dense image. Hence show that $(l^2, \| \quad \|_2)$ is isomorphic to the completion of $(C[0, 1], \| \quad \|_2)$. Try to formalize the problems inherent in proving the above assumptions.

1.12 Hilbert spaces

Hilbert spaces form a particularly important class of normed vector spaces as we can define concepts of orthogonality, angle and rotation for such spaces. They form, geometrically, the most interesting class of vector spaces.

In the first part of this section, H will always denote a *real* Hilbert space. At the end of the section we shall show how the theory generalizes to complex Hilbert spaces.

Definition 1.12.1

Let $x, y \in H$. We say x is orthogonal to y, written $x \perp y$, if $\langle x, y \rangle = 0$.

Remarks

1. Notice that orthogonality is a symmetric relation: $x \perp y$ if and only if $y \perp x$.
2. It should be noticed from our definition that the zero vector of H is trivially orthogonal to every vector of H.

Definition 1.12.2

Let F be a vector subspace of H and $x \in H$. We say x is orthogonal to F, written $x \perp F$, if $x \perp y$ for all $y \in F$.

If G is another vector subspace of H we say G is orthogonal to F, written $G \perp F$, if $x \perp F$ for all $x \in G$.

Examples

1. Denote the standard basis of $\mathbf{R}^n$ by $(e_i)_{i=1}^n$. Taking the standard inner product on $\mathbf{R}^n$, we have $e_i \perp e_j = 0$ $i \neq j$. Moreover the coordinate axes in $\mathbf{R}^n$ defined by this basis form a set of mutually orthogonal subspaces of $\mathbf{R}^n$.
2. Consider the unit circle in $\mathbf{R}^2$. With the notation of Fig. 1.6 the coordinates of the point (x, y) are $(\cos\theta, \sin\theta)$. Denote the line defined by the point (x, y) and the origin of $\mathbf{R}^2$ by l_θ. Similarly let the coordinates of the point (x', y') be $(\cos\psi, \sin\psi)$ and the corresponding line be denoted by l_ψ. Then the line $l_\theta \perp l_\psi$ if and only if $(x, y) \perp (x', y')$. To see this, just notice that arbitrary points on the lines l_θ and l_ψ are respectively given by $(\lambda x, \lambda y)$ and $(\mu x', \mu y')$ for some $\lambda, \mu \in R$. Hence $l_\theta \perp l_\psi$ if and only if $(x, y) \perp (x', y')$. Now

$$(x, y) = (\cos\theta, \sin\theta), \qquad (x', y') = (\cos\psi, \sin\psi)$$

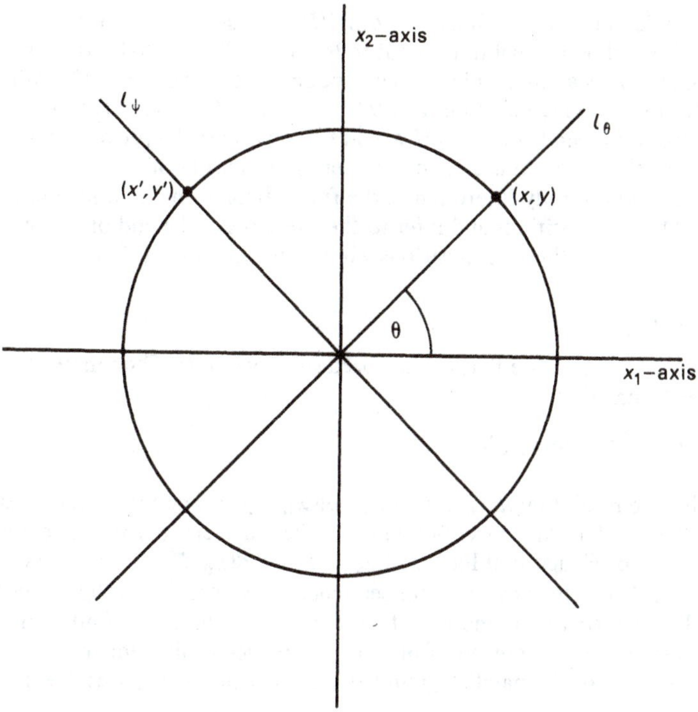

FIG. 1.6

and so

$$\langle\langle (x, y), (x', y') \rangle\rangle = \cos\theta \cos\psi + \sin\theta \sin\psi = \cos(\theta - \psi).$$

But $\cos(\theta - \psi) = 0$ if and only if $\theta - \psi = (2n + 1)\pi/2$ for some integer n. That is, $l_\psi \perp l_\theta$ if and only if the angle between l_ψ and l_θ is $\pi/2$.

3. In the (incomplete!) inner-product space $(C[0, 1], \langle\, ,\, \rangle)$, the function $f_n(t) = \cos 2n\pi t$ is orthogonal to the function $g_m(t) = \sin 2n\pi t$. Moreover, $f_n \perp f_m$ and $g_n \perp g_m, n \neq m$ (cf. Exercise 3, Section 1.11).

In Section 1.7 we remarked that, for a general normed vector space E, there is no *natural* isomorphism between E and E^*. In fact, if E is infinite dimensional, E and E^* need not be isomorphic. For Hilbert spaces, however, we do have a natural isomorphism between the space and its dual. The main aim of this section will be to prove the existence of this isomorphism. (Riesz representation theorem). Let us start by showing that if $(H, \langle\, ,\, \rangle)$ is a Hilbert space, there is a natural map $\gamma : H \longrightarrow H^*$. We define γ by the formula

$$\gamma(e)(f) = \langle f, e \rangle \qquad \text{for all } e, f \in H.$$

γ is certainly linear and $\|\gamma\| = 1$. Indeed γ is norm preserving, since

$$\begin{aligned} \|\gamma(e)\| &= \sup_{\|f\| \leqslant 1} \langle f, e \rangle \\ &= \|e\|, \qquad \text{taking } f = e/\|e\|. \end{aligned}$$

It follows immediately that γ is an injection.

If H is finite dimensional, $\dim(H) = \dim(H^*)$ and so γ is an isomorphism. However, if H is not finite dimensional it does not follow immediately that γ is an isomorphism. For example, if H is an incomplete inner product space, $\gamma: H \longrightarrow H^*$ is still defined and is a norm-preserving injection. But H^* is complete (Theorem 1.9.7) and so γ cannot be an isomorphism (otherwise H would be complete). To prove that γ is an isomorphism we shall need to use the completeness properties of H.

The reader who is only interested in the finite-dimensional case may omit the proofs of the next few propositions and refer to the exercises at the end of the section for the rather easy proofs of these propositions when H is finite dimensional.

Proposition 1.12.3

Let $x_0 \in H$ and suppose that F is a closed subspace of H. Then there exists a unique point $y_0 \in F$ such that

$$d(x_0, F) = d(x_0, y_0).$$

Proof. First we recall that $d(x_0, F) = \inf_{y \in F} \| x_0 - y \|$. The problem is to show that we can actually attain this lower bound on F. We may certainly pick a sequence of points $(y_n)_{n=1}^{\infty}$ of F such that $\lim_{n\to\infty} \| y_n - x_0 \| = d(x_0, F)$. Set $d = d(x_0, F)$. The main problem is to show that this sequence converges. This is by no means as easy as it looks at first sight and is, in fact, false for general, even finite-dimensional, normed vector spaces. In our proof of the convergence of the sequence (y_n) we shall make essential use of the parallelogram law for inner-product spaces. We have

$$\| y_n - y_m \|^2 = 2\| y_n - x_0 \|^2 + 2\| y_m - x_0 \|^2 - \| y_n + y_m - 2x_0 \|^2 \quad \text{(parallelogram law)}$$

$$= 2\| y_n - x_0 \|^2 + 2\| y_m - x_0 \|^2 - 4\| \tfrac{1}{2}(y_n + y_m) - x_0 \|^2$$

$$\leqslant 2\| y_n - x_0 \|^2 + 2\| y_m - x_0 \|^2 - 4d^2 \longrightarrow 0 \text{ as } n, m \longrightarrow \infty.$$

Therefore, (y_n) is a Cauchy sequence. Set $y_0 = \lim_{n\to\infty} y_n$. $y_0 \in F$, since F is a closed subspace of H, and $d(x_0, y_0) = \lim_{n\to\infty} \| x_0 - y_n \| = d$.

We must show that y_0 is unique. Suppose that y_0' also satisfies the conditions of the proposition. We have

$$\| y_0 - y_0' \|^2 = 2(\| y_0 - x_0 \|^2 + \| y_0' - x_0 \|^2) - 4(\| (y_0 + y_0')/2 - x_0 \|)^2, \quad \text{(parallelogram law)}$$

$$= 4d^2 - 4(\| (y_0 + y_0')/2 - x_0 \|^2) \leqslant 0.$$

Therefore, $y_0 = y_0'$. ∎

Proposition 1.12.4

Let F be a closed vector subspace of H. Then every point $x_0 \in H$ has a unique decomposition $x_0 = z_0 + y_0$, where $z_0 \perp F$ and $y_0 \in F$.

Proof. By Proposition 1.12.3 there exists a unique point $y_0 \in F$ such that $d(x_0, F) = \| x_0 - y_0 \|$. We set $z_0 = x_0 - y_0$. We claim that $z_0 \perp F$. Now $\|z_0\| = d(x_0, F)$ and $d(x_0, y) \geqslant \| z_0 \|$ for all $y \in F$. Hence for all $y \in F$ we have

$$\| z_0 - y \|^2 \geqslant d(z_0, F)^2 = \| z_0 \|^2.$$

But, expanding $\|z_0 - y\|^2$ in terms of the inner product, we have

$$\|z_0 - y\|^2 = \|z_0\|^2 + \|y\|^2 - 2\langle z_0, y\rangle \geqslant \|z_0\|^2 \qquad \text{for all } y \in F.$$

Hence

$$\|y\|^2 \geqslant 2\langle z_0, y\rangle.$$

Fix $y \in F$. Then for all $\lambda \in \mathbf{R}$ we have $\|\lambda y\|^2 \geqslant 2\langle z_0, \lambda y\rangle$. In particular, if $\lambda > 0$, we have $\lambda \|y\|^2 \geqslant 2\langle z_0, y\rangle$. Letting $\lambda \longrightarrow 0$ it follows that

$$\langle z_0, y\rangle \leqslant 0 \qquad \text{for all } y \in F.$$

It follows that, for fixed $y \in F$, we have $0 \geqslant \langle z_0, \lambda y\rangle = \lambda\langle z_0, y\rangle$ for all $\lambda \in \mathbf{R}$. This can clearly only happen if $\langle z_0, y\rangle = 0$. Since $y \in F$ was arbitrary we have, therefore, shown that $z_0 \perp F$.

We must verify that the decomposition of x_0 we have obtained is unique. Suppose then that $x_0 = y_0' + z_0'$, $y_0' \in F$ and $z_0' \perp F$. Then $y_0' + z_0' = y_0 + z_0$ and so $y_0 - y_0' = z_0' - z_0$. This implies that $z_0 - z_0' \in F$. Since $z_0 - z_0' \perp F$, it follows that $\langle z_0 - z_0', z_0 - z_0'\rangle = 0$. Hence $z_0 - z_0' = 0$ and the uniqueness is proved. ∎

Remark. Notice that it is essential that F is closed in H and H is complete for the truth of Proposition 1.12.4. For example, the reader may check that $x \perp F$ if and only if $x \perp \bar{F}$.

Definition 1.12.5

Let F be a vector subspace of H. The orthogonal complement of F, $F^\perp$, is defined by

$$F^\perp = \{x \in H : x \perp F\}.$$

Proposition 1.12.6

Let F be a vector subspace of H. Then $F^\perp$ is a closed vector subspace of H.

Proof. Let $x, y \in F^\perp$. Then

$$\langle \lambda x + y, z\rangle = \lambda\langle x, z\rangle + \langle y, z\rangle = 0$$

for all $z \in F$. Therefore, $F^\perp$ is a vector subspace of H. Let $x \in \overline{F^\perp}$. Then there exists a sequence $(x_n)_{n=1}^\infty \subset F^\perp$ such that $\lim_{n\to\infty} x_n = x$. But if $z \in F$, $\langle x, z\rangle = \lim_{n\to\infty}\langle x_n, z\rangle = 0$, by the continuity of the inner product. Thus $x \in F^\perp$ and so $F^\perp$ is closed in H. ∎

Example. Consider the x_1-axis

$$L_1 = \{(x_1, 0, \ldots, 0) : x_1 \in \mathbf{R}\}$$

in $\mathbf{R}^n$. Then

$$L_1^\perp = \{(0, x_2, \ldots, x_n) : x_2, \ldots, x_n \in \mathbf{R}\}.$$

Definition 1.12.7

Let F_1 and F_2 be closed vector subspaces of H. We define the vector sum of F_1 and F_2, $F_1 + F_2$, by

$$F_1 + F_2 = \{x + y : x \in F_1, y \in F_2\}.$$

We leave as an exercise the proof of the following proposition.

Proposition 1.12.8

Let F_1 and F_2 be be vector subspaces of H. Then $F_1 + F_2$ is a vector subspace of H a: $(F_1 + F_2)^{\perp} = F_1^{\perp} \cap F_2^{\perp}$. Moreover, if F_1 and F_2 are closed and $F_1 \perp F_2$, then $F_1 + F_2$ is a closed vector subspace of H.

Theorem 1.12.9

Let F be a closed vector subspace of H. Then

1. $H = F + F^{\perp}$.
2. Every $x \in H$ may be written uniquely in the form $y + z, y \in F, z \in F^{\perp}$.
3. $F \cap F^{\perp} = \{0\}$.

Proof. An immediate corollary of Proposition 1.12.4. Notice that from Proposition 1.12.6 $F^{\perp}$ is necessarily closed. ∎

Example. $\mathbf{R}^n$ is the vector sum of the coordinate axes. Thus, if L_j denotes the x_j-axis of $\mathbf{R}^n$, $1 \leqslant j \leqslant n$, we have

$$L_j \perp L_i, \qquad j \neq i$$

$$L_j^{\perp} = L_1 + \cdots + \hat{L}_j + \cdots + L_n, \qquad \text{where the caret denotes omission}$$

$$\mathbf{R}^n = L_1 + \cdots + L_n.$$

Hence every $x \in \mathbf{R}^n$ may be written as a sum $\sum_{j=1}^{n} x_j e_j$ of mutually orthogonal vectors. Here $(e_i)_{i=1}^{n}$ denotes the standard basis of $\mathbf{R}^n$.

Notation. Let E and F be vector subspaces of H such that $E \cap F = \{0\}$ and $E \perp F$. Then we write $E + F = E \textcircled{\perp} F$ (the orthogonal direct sum of E and F). Thus, with the assumptions of Theorem 1.12.9, we have $H = F \textcircled{\perp} F^{\perp}$.

If F is any normed vector space we will let I_F denote the identity map on F. If the underlying space is clear from the context we will sometimes drop the suffix and just write I for the identity map.

Proposition 1.12.10

Let F be a closed vector subspace of H. Then there exists a unique continuous linear map $p_F : H \longrightarrow H$ possessing the following properties.

1. Image $(p_F) = F$.
2. $p_F | F = I_F$.
3. $p_F | F^{\perp} = 0$.
4. $\| p_F \| = 1$.

p_F is called the *projection* map of H onto F.

Proof. From Theorem 1.12.9, $H = F \textcircled{\perp} F$. Thus, if $x \in H, x = y + z, y \in F,\ z \in F^{\perp}$. We define $p_F(x) = y$. We leave it to the exercises for the reader to check that p_F is linear and continuous and satisfies the conditions of the proposition. ∎

We may now prove the main theorem of this section.

Theorem 1.12.11 (Riesz representation theorem)

Let $(H, \langle\, , \rangle)$ be a Hilbert space. Then there exists a canonical norm-preserving isomorphism between H and H^*.

Proof. Let us first define a map $\gamma : H \longrightarrow H^*$. Thus, if $x \in H$, we define $\gamma(x) : H \longrightarrow \mathbf{R}$ by

$$\gamma(x)(y) = \langle y, x\rangle \qquad \text{for all } y \in H.$$

From the bilinearity of the inner product it follows that $\gamma(x) \in H^*$ for all $x \in H$ and that the map γ is linear. Further

$$\begin{aligned} \|\gamma(x)\| &= \sup_{\|y\| \leqslant 1} \langle y, x\rangle \\ &\leqslant \|x\|, \qquad \text{by Schwartz' inequality.} \end{aligned}$$

Taking $y = x/\|x\|$, we see that $\|\gamma(x)\| = \|x\|$.

Since γ is norm preserving it is clearly an injection. That is, $\gamma(x) = 0$ if and only if $\|\gamma(x)\| = \|x\| = 0$. The non-trivial part of the theorem is the surjectivity of γ. Suppose then that $f \in H^*$. If $f = 0$, then $\gamma(0) = f$. We may, therefore, suppose that $f \neq 0$. The continuity of f implies that $\text{Ker}(f)$ is a proper closed subspace of H. Let us set $L = \text{Ker}(f)$. We assert that $\dim L^\perp = 1$. Indeed, since $f \neq 0$, $\dim L^\perp \geqslant 1$. Let $z, z' \in L^\perp - \{0\}$. Then

$$z - \frac{f(z)}{f(z')} z' \in \text{Ker}(f)$$

and so

$$z = \frac{f(z)}{f(z')} z',$$

since $f^{-1}(0) = L$ intersects $L^\perp$ in the point $\{0\}$. Hence z, z' are linearly dependent and so $\dim L^\perp = 1$.

Let $z \in L^\perp$ be of norm 1. We define an element $g \in H^*$ by

$$g(x) = \langle x, f(z)z\rangle, \qquad x \in H.$$

We claim that $f = g$. Indeed, if $x \in L$, then $g(x) = 0$, since $z \in L^\perp$. On the other hand $g(z) = f(z)$ and so $g = f$ on the one-dimensional subspace $L^\perp$. Since $H = L \textcircled{1} L^\perp$, it follows that $f = g = \gamma(f(z)z)$.

Thus we have shown that γ is a continuous injective and surjective map. γ^{-1} is clearly linear and continuous: indeed $\|\gamma^{-1}\| = 1$. ∎

Example. $l^2 \approx (l^2)^*$. In particular, if $(x_i)_{i=1}^\infty$ is a real sequence, the map

$$(y_i) \longmapsto \sum_{i=1}^\infty y_i x_i, \qquad (y_i) \in l^2$$

defines a bounded linear form on l^2 if and only if $(x_i)_{i=1}^\infty \in l^2$.

Remark. Theorem 1.12.11 is not true for Banach spaces. In general, it is not true that a Banach space need even be homeomorphic to its dual. Even in the finite-dimensional case, whilst E is naturally isomorphic to E^{**}, there is no *natural* linear isomorphism between a space and its dual. Notice that even the linear map from E to E^* defined by choosing a basis for E and corresponding dual basis for E^* is not, in general, norm preserving.

Another important application of Theorem 1.12.9 is given by

Theorem 1.12.12

Let F be a vector subspace of H and let $f \in F^*$. Then there exists a unique element $\hat{f} \in H^*$ such that

1. $\hat{f}|F = f$.
2. $\hat{f}|F^\perp = 0$.
3. $\|\hat{f}\| = \|f\|$.

Proof. Without loss of generality we may assume that F is closed, since f extends uniquely to $\bar{F}$ and $F^\perp = \bar{F}^\perp$. We define $\hat{f} = f.\, p_F$. Then

$$\begin{aligned} \|\hat{f}\| &= \|f.\, p_F\| \\ &\leqslant \|f\|\,\|p_F\| \qquad \text{(Proposition 1.6.4)} \\ &= \|f\| \qquad \text{(Proposition 1.12.10).} \end{aligned}$$

Since $\hat{f}|F = f$, we must have $\|\hat{f}\| \geqslant \|f\|$ and so $\|\hat{f}\| = \|f\|$.
The uniqueness of $\hat{f}$ is trivial, since properties 1 and 2 characterize $\hat{f}$. ■

Remark. In Section 1.17 we shall show that a version of the above theorem also holds for normed vector spaces.

For the remainder of this section we wish to show how the preceding theory generalizes to complex Hilbert spaces.

Let us first note that our definitions of orthogonality generalize immediately to complex Hilbert spaces. In particular, if we denote the standard basis of $\mathbf{C}^n$ by $(e_i)_{i=1}^n$, then $e_i \perp e_j$, $i \neq j$.

The proof of Proposition 1.12.3 holds for the complex case also since the parallelogram law still holds. For the proof of Proposition 1.12.4 note that

$$\|z_0 - y\|^2 = \|z_0\|^2 + \|y\|^2 - 2 \operatorname{Re} \langle z_0, y \rangle,$$

where $\operatorname{Re}\langle z_0, y\rangle$ denotes the real part of $\langle z_0, y\rangle$. Hence $\operatorname{Re}\langle z_0, y\rangle = 0$ for all $y \in F$. Replacing y by iy, it follows also that the imaginary part of $\langle z_0, y\rangle$ vanishes for all $y \in F$ and so $\langle z_0, y\rangle = 0$ for all $y \in F$.

Theorem 1.12.9 on orthogonal complements remains valid with the same proof as fo the real case. For the Riesz representation theorem, however, we have to be rather more careful.

If $A: E \longrightarrow F$ is a real linear map between the complex vector spaces E and F we shall say that A is *conjugate complex linear* if

$$A(ie) = -iA(e) \qquad \text{for all } e \in E.$$

We may now state the complex version of the Riesz representation theorem.

Theorem 1.12.13

Let $(H, \langle\,,\rangle)$ be a complex inner-product space. Then there exists a canonical norm-preserving conjugate complex linear isomorphism between H and H^*.

Proof. We define a map $\gamma: H \longrightarrow H^*$ by

$$\gamma(x)(y) = \langle y, x\rangle \qquad \text{for all } x, y \in H.$$

Notice that $\gamma(x) \in H^*$ since

$$\gamma(x)(iy) = \langle iy, x\rangle = i\langle y, x\rangle = i\gamma(x)(y).$$

Just as in the proof of Theorem 1.12.11, γ is a real norm-preserving linear isomorphism. γ is conjugate complex linear since

$$\gamma(ix)(y) = \langle y, ix\rangle = -i\langle y, x\rangle = -i\gamma(x)(y). \quad \blacksquare$$

Exercises

1. Prove Proposition 1.12.8. Show also that $(F_1 \cap F_2)^\perp = \overline{F_1^\perp + F_2^\perp}$.
2. If F is a vector subspace of H show that $F^{\perp\perp} = \bar{F}$.
3. If p_F is the projection map corresponding to the closed vector subspace F of H, show that the projection map corresponding to $F^\perp$ is $I_H - p_F$.
4. Verify the following properties of projection maps.
   (a) $p_F^2 = p_F$; (b) $p_E p_F = 0$ if $E \cap F = \{0\}$; (c) $p_E p_F = p_E$ if $E \subset F$.
5. Complete the proof of Proposition 1.12.10.
6. Suppose that E is a complex vector space. Scalar multiplication by i defines an element $J \in L(E, E)$ by the rule: $J(e) = ie$. Show that
   (a) $J^2 = -I_E$.
   (d) If E is a real vector space and we are given an element $J \in L(E, E)$ satisfying $J^2 = -I_E$, then E may be given the structure of a complex vector space if we define $(a + ib)e = ae + bJ(e)$, $a, b \in \mathbf{R}$, $e \in E$. J is called a *complex structure* on E.
7. Show that if the complex vector space E has associated complex structure J, then the endomorphism $-J$ of E also defines a complex structure of E. We denote the vector space E, given the complex structure $-J$, by $\bar{E}$. (Caution: do not confuse with closure operation.) Prove
   (a) If E and F are complex vector spaces and $A : E \longrightarrow F$ is a real linear map, then A is conjugate complex linear if and only if A is a complex linear map between the complex vector spaces E and $\bar{F}$ (equivalently, between $\bar{E}$ and F).
   (b) If E is a complex vector space, then $\bar{E}^*$ is canonically isomorphic to $\overline{E^*}$ (*Hint:* Map $f \in E^*$ to its conjugate $\bar{f} \in \overline{E^*}$ defined by $\bar{f}(e) = \overline{f(e)}$, $e \in E$).
   (c) Show that the complex version of the Riesz representation theorem asserts the existence of a canonical *complex* linear isomorphism between H and $\bar{H}^*$.
8. Show that the uniqueness part of Proposition 1.12.3 need not hold for arbitrary norms on $\mathbf{R}^n$.
9. In this exercise we indicate the simplifications in proofs that can be made when H is a finite-dimensional inner-product space. In the text we proved the Riesz representation theorem for finite-dimensional inner-product spaces and we shall assume this in what follows.
   (a) Let L be a one-dimensional subspace of H and let $\phi \in \gamma(L) \subset H^*$, $\phi \neq 0$. Show that $L^\perp = \mathrm{Ker}(\phi)$, $L^\perp \oplus L = H$ and $L^\perp \cap L = \{0\}$. Generalize to construct $F^\perp$ for any subspace F of H.
   (b) Show, using (a), that Theorem 1.12.9 holds for a finite-dimensional inner-product space.
   (c) Deduce Proposition 1.12.3, Proposition 1.12.10 and Theorem 1.12.12 from (b).

1.13 Bases

In this section we shall start by reviewing the theory of orthogonal bases for a finite-dimensional inner-product space. We end the section by showing how the theory generalizes to infinite-dimensional (separable) Hilbert spaces. We shall not discuss the theory of (Schauder) bases for a Banach space but instead refer the interested reader to [19].

Definition 1.13.1

Let $(H, \langle\ ,\ \rangle)$ be a finite-dimensional real or complex inner-product space. A basis $\mathscr{E} = (e_i)_{i=1}^n$ for H is said to be an *orthogonal basis* if

$$e_i \perp e_j, \qquad i \neq j.$$

If, in addition, $\| e_i \| = 1$ for all $e_i \in \mathscr{E}$, $\mathscr{E}$ is said to be an *orthonormal basis* for H.

Examples

1. The standard basis $(e_i)_{i=1}^n$ for $\mathbf{R}^n$, given the usual inner product, is an orthonormal basis.
2. The standard basis for $\mathbf{C}^n$ is an orthonormal basis.

Proposition 1.13.2

Every finite-dimensional real or complex inner-product space has an orthonormal basis.

Proof. Let H be a finite-dimensional inner-product space. We prove by induction on the dimension of H. Let dimension $(H) = n$. The theorem is trivially true if $n = 1$. Suppose it is true for $n - 1$. Let L be any line through the origin of H and let $F = L^{\perp}$. Since the dimension of $F = n - 1$, it follows by the inductive hypothesis that we can choose an orthonormal basis $\mathscr{E}'$ for F. If e is any unit vector in L it follows immediately that $\{e\} \cup \mathscr{E}'$ is an orthonormal basis for H. ∎

Remark. Proposition 1.13.2 can be proved without recourse to the orthogonal-complement theorem of Section 1.12 by the following procedure: Suppose we have determined the first j elements $e_1, \ldots, e_j$ of an orthonormal basis for the inner-product space H. If $e_1, \ldots, e_j$ span the vector space F, we pick any vector e of H not lying in F, and define

$$e'_{j+1} = e - \left(\sum_{i=1}^{j} \langle e, e_i \rangle e_i \right).$$

It follows easily that e'_{j+1} is non-zero and orthogonal to F. We set $e_{j+1} = e'_{j+1}/\| e'_{j+1} \|$. This method of constructing an orthonormal basis is known as the Gram-Schmidt orthogonalization process.

Proposition 1.13.3

Let $(H, \langle\ ,\ \rangle)$ be a finite-dimensional inner-product space with orthonormal basis $\mathscr{E} = (e_i)_{i=1}^n$ and dual basis $\mathscr{E}^* = (e_i^*)_{i=1}^n$. Then, if $\gamma : H \longrightarrow H^*$ is the canonical isomorphism given by the Riesz representation theorem, γ maps the basis $\mathscr{E}$ onto the dual basis $\mathscr{E}^*$. That is

$$e_i^* = \gamma(e_i), \qquad 1 \leqslant i \leqslant n.$$

Proof. Trivial, since $\gamma(e_i)(e_j) = \langle e_j, e_i \rangle$ by definition of γ. ∎

Proposition 1.13.4

Let $(H, \langle\ ,\ \rangle)$ be a finite-dimensional inner-product space with orthonormal basis

$\mathscr{E} = (e_i)_{i=1}^n$. If $x \in H$ has coordinates $(x_1, \ldots, x_n)$ relative to this basis then

1. $x_i = \langle x, e_i \rangle, \qquad 1 \leqslant i \leqslant n.$
2. $\|x\|^2 = \sum_{i=1}^{n} |x_i|^2.$
3. If $y \in H$ has coordinates $(y_1, \ldots, y_n)$, $\langle x, y\rangle = \sum_{i=1}^n x_i \bar{y}_i$.

Proof. Since $x_i = e_i^*(x)$, (1) follows from Proposition 1.13.3. We leave the details of the proofs of (2) and (3) to the exercises. ■

Remark. Condition (3) of Proposition 1.13.4 is known as Parseval's identity.

We now wish to generalize the above results to separable Hilbert spaces. As we indicated in Section 0.1, Hammel bases are not very useful in applications to infinite-dimensional vector spaces. To get a good theory we must make use of the topological and metric structure of the space. The effect of this will be to make the bases of some of the more important Banach and Hilbert spaces *countable.* A Hammel basis of an infinite-dimensional Banach space would, however, be *uncountable.* This fact has considerable significance if one realizes that the main reason for introducing bases at all is computation: Once the basis becomes non-countable one has lost hope of doing any interesting computations with it. In other words, one always strives for some form of finiteness.

Definition 1.13.5

Let $(H, \langle\ ,\ \rangle)$ be a real or complex Hilbert space. A collection $\mathscr{E} = (e_i)_{i=1}^\infty$ of non-zero mutually orthogonal vectors of H is called an orthogonal basis for H if the set of all finite linear combinations of elements of $\mathscr{E}$ is *dense* in H.

If, in addition, every member of $\mathscr{E}$ is of unit length, $\mathscr{E}$ is said to be an *orthonormal basis* for H.

Remark. Note that Definition 1.13.5 implies that H is separable. One may also prove the converse: Any separable Hilbert space has a (countable) basis.

The reason why we used the term 'basis' in Definition 1.13.5 is clear from the following proposition.

Proposition 1.13.6

Let $(H, \langle\ ,\ \rangle)$ be a real or complex Hilbert space with orthonormal basis $\mathscr{E} = (e_i)_{i=1}^\infty$. If $x \in H$, we let $x_i = \langle x, e_i\rangle$, $i = 1, \ldots$. We have

1. $\sum_{i=1}^{\infty} |x_i|^2 = \|x\|^2.$
2. x is uniquely expressible as the infinite sum

$$x = \sum_{i=1}^{\infty} x_i e_i.$$

Proof. Let E_n denote the subspace of H spanned by $\{e_1, \ldots, e_n\}$ and let $p_n : H \longrightarrow E_n$ denote the corresponding projection map. It follows from Section 1.12 that

$$\|p_n(x)\| \leqslant \|x\|, \qquad n = 1, 2, \ldots,$$

$$\|p_n(x)\| \leqslant \|p_m(x)\|, \qquad n \leqslant m.$$

Thus $(\|p_n(x)\|)_{n=1}^{\infty}$ is an increasing sequence of positive numbers bounded above by $\|x\|$.

Now $p_n(x) = \sum_{i=1}^{n} x_i e_i$ and so, since $e_i \perp e_j$, $i \neq j$, we have the following expression for $\|p_n(x)\|$:

$$\|p_n(x)\|^2 = \sum_{i=1}^{n} |x_i|^2.$$

Hence $\sum_{i=1}^{\infty} |x_i|^2$ is convergent and

$$\sum_{i=1}^{\infty} |x_i|^2 \leqslant \|x\|^2.$$

In particular, $\sum_{i=1}^{\infty} x_i e_i$ is convergent. Finally we must show that the limit of this sum is x and that $\sum_{i=1}^{\infty} |x_i|^2 = \|x\|^2$. First notice that $p_n(x)$ is the unique vector in E_n that minimizes $\|x - y\|$, $y \in E_n$. Since the set of all finite linear combinations of elements of $\mathscr{E}$ is dense in H it follows that, given $\epsilon > 0$, we can find a finite linear combination $\sum_{i=1}^{N} y_i e_i$ such that

$$\|x - \sum_{i=1}^{N} y_i e_i\| < \epsilon.$$

But, by the above remark,

$$\|x - p_N(x)\| \leqslant \|x - \sum_{i=1}^{N} y_i e_i\|.$$

Since $\epsilon > 0$ was arbitrary, it follows that $\lim_{N\to\infty} \|x - p_N(x)\| = 0$ and hence that

$$\sum_{i=1}^{\infty} x_i e_i = x.$$

Since $\|\ \ \|$ is a continuous function on H we also have

$$\|x\|^2 = \lim_{N\to\infty} \|p_N(x)\|^2 = \lim_{N\to\infty} \sum_{i=1}^{N} |x_i|^2 = \sum_{i=1}^{\infty} |x_i|^2.$$

The uniqueness of this representation of x in terms of $\mathscr{E}$ follows from the fact that if $\sum_{i=1}^{\infty} y_i e_i$ were another representation of x we would have

$$\sum_{i=1}^{\infty} (x_i - y_i) e_i = 0,$$

and so, taking norms,

$$\sum_{i=1}^{\infty} |x_i - y_i|^2 = 0,$$

implying that $x_i = y_i$, $i = 1, 2, \ldots$. ■

Definition 1.13.7

Let $(H, \langle\ ,\ \rangle)$ be a real or complex Hilbert space with basis $\mathscr{E} = (e_i)_{i=1}^{\infty}$. We call the sequence $(x_i)_{i=1}^{\infty}$ defined by $x_i = \langle x, e_i \rangle$, $i = 1, 2, \ldots$ the coordinates of x relative to the basis $\mathscr{E}$.

Remark. From Proposition 1.13.6 it follows that the coordinates of x relative to the basis $\mathscr{E}$ are unique.

Just as for the finite-dimensional case we have the proposition:

Proposition 1.13.8

Let $(H, \langle\ ,\ \rangle)$ be a real or complex Hilbert space with orthonormal basis $(e_i)_{i=1}^{\infty}$. If x and y have respective coordinates $(x_1, x_2, \ldots)$ and $(y_1, y_2, \ldots)$, then

1. $\sum_{i=1}^{\infty} |x_i|^2 = \|x\|^2$
2. $\sum_{i=1}^{\infty} x_i \bar{y}_i = \langle x, y \rangle.$

Statements (1) and (2) are called Parseval's identities.

Proof. We have already proved (1). We leave the proof of (2) to the exercises. ∎

We have the following characterization of Hilbert spaces with (countable) basis.

Proposition 1.13.9

Let $(H, \langle\ ,\ \rangle)$ be a real or complex infinite-dimensional Hilbert space. Then

1. H is separable if and only if H has a countable orthonormal basis.
2. If H is separable it is isomorphic to l^2 by an inner-product preserving map.

Proof. Certainly if H has a (countable) basis, H is separable. We shall sketch the proof of the converse. Suppose $(x_n)_{n=1}^{\infty}$ is a dense subset of H. Remove any zero vectors occurring in this sequence. We define inductively a new sequence $(y_m)_{m=1}^{\infty}$ by requiring that $y_1 = x_1$ and that after n steps x_{n+1} is added to the sequence $y_1, \ldots, y_{k(n)}$ as $y_{k(n)+1} = y_{k(n+1)}$ if and only if x_{n+1} is linearly independent of $\{y_1, \ldots, y_{k(n)}\}$.

We leave it to the reader to check that the sequence $(y_m)_{m=1}^{\infty}$ will span a dense subset of H and that the elements of the sequence are linearly independent. We let E_m = space spanned by $\{y_1, \ldots, y_m\}$. Thus, dimension $(E_m) = m$. We construct a new sequence $(e_n)_{n=1}^{\infty}$ inductively by requiring that $e_1 = y_1/\|y_1\|$ and that e_{n+1} is a unit vector in E_{n+1} orthogonal to E_n and so to $\{e_1, \ldots, e_n\}$. It follows immediately from the definition of a basis that $(e_n)_{n=1}^{\infty}$ is a basis for H.

To prove the second part of the proposition let $\mathscr{E}$ be any basis for H. Then we define a map $\phi: H \longrightarrow l^2$ by requiring that

$$\phi(x) = (x_1, \ldots, x_n, \ldots),$$

where $(x_1, \ldots, x_n, \ldots)$ are the coordinates of x relative to the basis $\mathscr{E}$. From Proposition 1.13.6, ϕ is well defined. From Proposition 1.13.8, ϕ preserves inner products. ∎

Examples

1. If we define e_i to be the infinite sequence all of whose entries are 0 save the ith, which equals 1, then $(e_i)_{i=1}^{\infty}$ is an orthonormal basis for l^2.
2. Let us consider how Fourier series fit into the framework of orthonormal bases. Thus we consider the incomplete inner-product space $(C[0, 1], \|\ \ \|_2)$ of continuous real-valued functions on the unit interval. We define the following sequence of functions in $C[0, 1]$:

$$e_0(t) = 1$$

$$e_{2n}(t) = \sqrt{2}\cos(2\pi nt), \qquad n = 1, 2, \ldots$$

$$e_{2n+1}(t) = \sqrt{2}\sin(2\pi nt), \qquad n = 1, 2, \ldots.$$

Then we leave it to the reader to verify that $e_i \perp e_j$, $i \neq j$ and $\| e_i \| = 1$, $i = 0, 1, \ldots$. Thus $(e_i)_{i=0}^{\infty}$ forms an orthonormal set of vectors in $C[0, 1]$. Let $f \in C[0, 1]$. We may define a sequence $(x_n)_{n=0}^{\infty}$ by setting

$$x_n = \int_0^1 f(t)\, e_n(t)\, dt, \qquad n = 0, 1, \ldots.$$

Thus, if n is even,

$$x_n = \sqrt{2} \int_0^1 f(t) \cos(2\pi n t)\, dt,$$

whilst if n is odd,

$$x_n = \sqrt{2} \int_0^1 f(t) \sin(2\pi n t)\, dt,$$

and, finally, if $n = 0$,

$$x_0 = \int_0^1 f(t)\, dt.$$

We call the numbers x_n the *Fourier coefficients* of f.

Any finite combination $\Sigma_{i=0}^{n} a_i e_i$ is called a *trigonometric polynomial.* We denote the set of trigonometric polynomials by the symbol P_T. Thus P_T is a (non-closed) vector subspace of $C[0, 1]$. Let Q denote the closure of P_T in $\widehat{C[0, 1]}$. Then $(e_n)_{n=0}^{\infty}$ is an orthonormal basis for the (Hilbert) vector subspace Q of $\widehat{C[0, 1]}$. If $f \in \widehat{C[0, 1]}$ we may define the nth Fourier coefficient of f by

$$x_n = \langle f, e_n \rangle.$$

Just as in the proof of Proposition 1.13.6 it follows that

$$\| f \|^2 \leqslant \sum_{i=0}^{\infty} |x_i|^2$$

with equality if and only if $f \in Q$. As we indicated at the end of Section 1.11, to obtain a representation of $\widehat{C[0, 1]}$ as a space of functions one needs to develop the theory of Lebesgue integration. This being done, one can then represent the inner product on $\widehat{C[0, 1]}$ as an integral. However, since the inner product on $\widehat{C[0, 1]}$ does extend that on $C[0, 1]$ it follows that if $f \in C[0, 1]$

$$x_n = \langle f, e_n \rangle = \int_0^1 f(t)\, e_n(t)\, dt, \qquad n = 0, 1, \ldots$$

and the inequality $\| f \|^2 \leqslant \Sigma_{i=0}^{\infty} |x_i|^2$ still holds, with equality if and only if f lies in the closure of P_T in $C[0, 1]$. That is, with equality if and only if f can be approximated in the square-integral norm by trigonometric polynomials. Now if P_T is dense in $C[0, 1]$ it follows that $(e_i)_{i=0}^{\infty}$ is an orthonormal basis for $\widehat{C[0, 1]}$, and conversely. However, it is easy to see that $(e_i)_{i=0}^{\infty}$ is an orthonormal basis for $\widehat{C[0, 1]}$ if and only if the following condition holds:

Given $f \in C[0, 1]$, $\langle e_n, f \rangle = 0$ for all e_n implies $f = 0$. This condition is true, but the proof of the result is beyond the scope of this text. Let us *assume*, therefore, that P_T is dense in $C[0, 1]$. It follows that the map

$$F:P_T \longrightarrow l^2$$

defined by

$$F\left(\sum_{i=0}^{n} a_i e_i\right) = (a_0, a_1, \ldots, a_n, 0, 0, \ldots)$$

extends to a norm-preserving linear isomorphism between $\widehat{C[0, 1]}$ and l^2 which maps $C[0, 1]$ onto a dense subspace of l^2. The discussion of the inverse of F is by no means so straightforward. Thus, if $f \in C[0, 1]$ and $F(f) = (x_i)_{i=0}^{\infty}$ one might reasonably hope that

$$F^{-1}((x_i)_{i=0}^{\infty}) = \sum_{i=0}^{\infty} x_i e_i = f.$$

However, all that follows from the definition of F is that

$$\|f - \sum_{i=0}^{\infty} x_i e_i\| = 0,$$

and it need not follow that $\Sigma_{i=0}^{\infty} x_i e_i$ even *converges* at every point of $[0, 1]$, let alone converges to the function f! It does, however, follow that if $F^{-1}((x_n))$ is a continuous function, that is if

$$\sum_{i=0}^{\infty} x_i e_i$$

converges to a continuous function on $[0, 1]$, then $f = \Sigma_{i=0}^{\infty} x_i e_i$. This is a consequence of the observation that if g and h are continuous functions such that $\int_0^1 (g(t) - h(t))^2 \, dt = 0$, then $f = g$. For example, suppose that $(x_n)_{n=0}^{\infty} \in l^1 \subset l^2$. It follows easily by Weierstrass' M-test that $\Sigma_{i=0}^{\infty} x_i e_i$ converges uniformly on $[0, 1]$ to a continuous function. On the other hand if f is continuously differentiable on $[0, 1]$ one may easily show, integrating $f' e_n$ by parts, that the Fourier coefficients of f lie in l^1. So at least for continuously differentiable functions one can expect a unique representation by a Fourier series. However, this is really as far as one can go using Riemann integration. Indeed, the theory of uniform convergence is designed as a way of overcoming the deficiencies of Riemann integration when used in limit arguments. There is no such drawback with Lebesgue integration. For further discussion and examples of applications of Fourier series we refer to [6], [14]; for a more theoretical approach, using Lebesgue integration, we refer to [8], [21].

Exercises

1. Complete the proof of Proposition 1.13.4.
2. Complete the proof of Proposition 1.13.8.
3. Let $(1/\sqrt{3}, -1/\sqrt{3}, 1/\sqrt{3})$ and $(0, 1/\sqrt{2}, -1/\sqrt{2})$ be vectors in $\mathbf{R}^3$. Find a vector of unit length orthogonal to these two vectors and hence an orthonormal basis for $\mathbf{R}^3$.
4. Let S denote the set of solutions of the second-order differential equation $x'' = -x$. Regarding S as a subset of $C[0, 1]$ with the usual inner product, find an orthonormal basis for S.

5. It can be shown that the set of functions $\{1, t, t^2, \ldots, t^n, \ldots\}$ span a dense subspace of $C[-1, +1]$ given the square-integral norm (see below). The nth Legendre polynomial $P_n(t)$ is defined by

$$P_n(t) = \frac{1}{2^n n!}\left(\frac{\mathrm{d}}{\mathrm{d}t}\right)^n (t^2 - 1)^n.$$

Show that if we try to orthonormalize the sequence $\{1, t, \ldots, t^n, \ldots\}$ using the Gram–Schmidt orthogonalization process, then, at the nth step, we obtain P_n (up to sign) (*Hint:* Show that the degree of P_n equals n and that $\langle P_n, t^m\rangle = 0$, $m < n$).

6. (Stone-Weierstrass theorem). Suppose that X is a compact metric space and A is a subset of ($C(X)$, sup norm) which is an algebra (closed under addition, multiplication and scalar multiplication) such that A contains the constant functions an separates points of X (that is, if $a, b \in X$, $\exists f \in A$ such that $f(a) \neq f(b)$) then A is dense in $C(X)$. The proof is divided into several steps:
   (a) We show that we can approximate the function $\sqrt{t}$ on $[0, 1]$ by a uniformly convergent increasing sequence of polynomials. We define the polynomials p_n inductively by setting $p_1 = 0$ and

   $$p_{n+1}(t) = p_n(t) + \tfrac{1}{2}(t - p_n(t)^2).$$

   Show, inductively, that $p_{n+1} \geqslant p_n$ and that $p_n(t) \leqslant \sqrt{t}$, $t \in [0, 1]$. Hence show that $p_n(t)$ converges at every point of $[0, 1]$ to $\sqrt{t}$. Finally use Dini's theorem ('an increasing sequence of functions which is bounded above by a continuous function converges uniformly', for a proof see [7], p. 129) to deduce that the convergence of p_n is uniform.
   (b) $f \in A$ implies $|f| \in \bar{A}$. To prove this consider the sequence $p_n(f^2/\|f\|^2)$ of points in A. By (a) this converges to $\sqrt{(f^2/\|f\|^2)} = |f|/\|f\|$.
   (c) $f, g \in \bar{A}$ implies $\sup(f, g)$ and $\inf(f, g) \in \bar{A}$. This follows from (b) since $\sup(f, g) = \frac{1}{2}(f + g + |f - g|)$, $\inf(f, g) = \frac{1}{2}(f + g - |f - g|)$ and $\bar{A}$ is an algebra
   (d) For all $x \neq y \in X$ and all $a, b \in \mathbf{R}$, $\exists f \in \bar{A}$ such that $f(x) = a, f(y) = b$. This follows from the separation assumptions on A: we may find $f \in A$ satisfying these conditions.
   (e) For any $f \in C(X)$ and any $x \in X$ and any $\epsilon > 0$, there exists $g \in \bar{A}$ such that $g(x) = f(x)$ and $g(y) \leqslant f(y) + \epsilon$, $\forall y \in X$. To show this, notice that by (d) we can certainly find for each $z \in X$ a function $g_z \in \bar{A}$ such that $g_z(x) = f(x)$ and $g_z(z) < f(z) + \epsilon/2$. Now use continuity to get the inequality $g_z < f + \epsilon$ in a neighbourhood V_z of z. Doing this for every $z \in X$, we may use the compactness of X to find a finite number of functions g_i and corresponding neighbourhoods V_i. Now use (c) to define $g = \inf(g_i)$.
   (f) $\bar{A} = C(X)$. Apply (e) at each point $x \in X$ and again use compactness and continuity to find a finite collection of functions h_i and neighbourhoods U_i such that if $f \in \bar{A}$ we have

   $$f(y) - \epsilon \leqslant g_i(y) \leqslant f(y) + \epsilon.$$

   Finally define, using (c), $h = \sup(h_i)$. Since $\epsilon > 0$ was arbitrary the theorem follows.

7. Using the Weierstrass approximation theorem verify the assertion made in Exercise 5.

1.14 Transpose and adjoint of linear maps

Proposition 1.14.1

Let $(E, \| \ \|)$ and $(F, \| \ \|)$ be normed vector spaces over K and let $A \in L(E, F)$. Then there exists a unique linear map $A^t \in L(F^*, E^*)$ characterized by

$$A^t(f^*)(e) = f^*(A(e)) \qquad \text{for all } f^* \in F^* \text{ and } e \in E.$$

In addition,

$$\| A^t \| \leqslant \| A \|.$$

A^t is called the *transpose* of A.

Proof. The formula $A^t(f^*)(e) = f^*(A(e))$ certainly defines a map from F^* to E^* and the reader may easily verify that A^t is linear. We have

$$\begin{aligned}\sup_{\|f^*\| \leqslant 1} \| A^t(f^*) \| &= \sup_{\substack{\|f^*\| \leqslant 1 \\ \|e\| \leqslant 1}} \| A^t(f^*)(e) \| \\ &= \sup_{\substack{\|f^*\| \leqslant 1 \\ \|e\| \leqslant 1}} \| f^*(A(e)) \| \\ &\leqslant \sup_{\substack{\|f^*\| \leqslant 1 \\ \|e\| \leqslant 1}} \| f^* \| \| A(e) \| = \| A \|.\end{aligned}$$

proving both the continuity of A and the fact that $\| A^t \| \leqslant \| A \|$. ■

Proposition 1.14.2

Let E and F be finite-dimensional vector spaces with bases

$$\mathscr{E} = (e_i)_{i=1}^m \qquad \text{and } \mathscr{F} = (f_j)_{j=1}^n$$

respectively. Then, if $A \in L(E, F)$, the matrix of A^t, relative to the dual bases $\mathscr{F}^*$ and $\mathscr{E}^*$ of F^* and E^* respectively, is the transpose of the matrix of A:

$$[A^t] = [A]^t.$$

Proof. If we set $[A] = [a_{ij}]$ and $[A^t] = [\tilde{a}_{ij}]$, we have

$$\begin{aligned}\tilde{a}_{ij} &= i\text{th component of } A^t(f_j^*) \\ &= A^t(f_j^*)(e_i) = f_j^*(A(e_i)) = a_{ji}. \quad ■\end{aligned}$$

For the theory of adjoints of linear maps we shall distinguish between the real and complex cases. Suppose, therefore, that $(E, \langle \, , \rangle)$ and $(F, \langle \, , \rangle)$ are real Hilbert spaces. The Riesz representation theorem implies that there exist canonical isomorphisms $\mu : E \longrightarrow E^*$ and $\gamma : F \longrightarrow F^*$. It follows that if $A \in L(E, F)$, we may define a map $A^* \in L(F, E)$ by the formula

$$A^* = \mu^{-1} . A^t . \gamma.$$

That is, we have the following commutative diagram of maps:

$$\begin{array}{ccc} F & \xrightarrow{A^*} & E \\ \big\downarrow{\scriptstyle \gamma} & & \big\uparrow{\scriptstyle \mu^{-1}} \\ F^* & \xrightarrow{A^t} & E^* \end{array}$$

Definition 1.14.3

If E and F are real Hilbert spaces and $A \in L(E, F)$, we call the map $A^* \in L(F, E)$ the *adjoint* of A.

The characteristic property of the adjoint of a linear map is given by the following proposition.

Proposition 1.14.4

Let E and F be Hilbert spaces and let $A \in L(E, F)$. Then A^* satisfies the condition

$$\langle A(x), y\rangle = \langle x, A^*(y)\rangle \qquad \text{for all } x \in E \text{ and } y \in F.$$

Proof. By definition of μ,

$$\langle x, A^*(y)\rangle = \mu(A^*(y))(x) = (A^{\mathrm{t}}(\gamma(y)))(x) = \gamma(y)(A(x)) = \langle A(x), y\rangle. \quad \blacksquare$$

Proposition 1.14.5

Let E and F be finite-dimensional Hilbert spaces with orthonormal bases

$$\mathscr{E} = (e_i)_{i=1}^m \qquad \text{and} \qquad \mathscr{F} = (f_j)_{j=1}^n$$

respectively. Then, if $A \in L(E, F)$, the matrix of A^* relative to the bases $\mathscr{F}$ and $\mathscr{E}$ is the transpose of the matrix of A relative to the bases $\mathscr{E}$ and $\mathscr{F}$.

Proof. Follows from Proposition 1.14.2 and the fact that μ and γ have identity matrix relative to the dual orthonormal bases on E^* and F^* respectively. We leave details to the exercises. ■

A particularly important class of linear maps is given by the following definition.

Definition 1.14.6

Let $(H, \langle\ ,\ \rangle)$ be a Hilbert space and let $A \in L(H, H)$. Then A is said to be *symmetric* or *self-adjoint* if $A = A^*$. That is, A is self-adjoint if and only if

$$\langle A(x), y\rangle = \langle x, A(y)\rangle \qquad \text{for all } x, y \in H.$$

If H is finite dimensional, we have the following characterization of symmetric endomorphisms of H:

A linear endomorphism of H is symmetric if and only if it has symmetric matrix relative to any orthonormal basis of H.

For the remainder of this section we wish to study adjoints of complex linear maps. Suppose, therefore, that $(E, \langle\ ,\ \rangle)$ and $(F, \langle\ ,\ \rangle)$ are complex Hilbert spaces. The complex version of the Riesz representation theorem implies that we have conjugate complex linear isomorphisms:

$$\mu: E \longrightarrow E^*, \qquad \gamma: F \longrightarrow F^*.$$

It follows that if $A \in L(E, F)$, we may define a map $\bar{A}^* \in L(F, E)$ by the formula

$$\bar{A}^* = \mu^{-1} . A^{\mathrm{t}} . \gamma.$$

That is, we have the following commutative diagram of maps:

$$\begin{array}{ccc} F & \xrightarrow{\bar{A}^*} & E \\ \big\downarrow{\scriptstyle \gamma} & & \big\uparrow{\scriptstyle \mu^{-1}} \\ F^* & \longrightarrow & E^* \end{array}$$

Notice that, since both γ and μ are conjugate complex linear, $\bar{A}^*$ is complex linear.

Definition 1.14.7

Let E and F be complex Hilbert spaces and let $A \in L(E, F)$. We shall call the map $\bar{A}^* \in L(F, E)$ the *Hermitian adjoint* of A.

Just as for the real case, the characteristic property of the Hermitian adjoint is given by

Proposition 1.14.8

Let E and F be complex Hilbert spaces and let $A \in L(E, F)$. Then $\bar{A}^*$ satisfies the condition

$$\langle A(x), y \rangle = \langle x, \bar{A}^*(y) \rangle \qquad \text{for all } x \in E \text{ and } y \in F.$$

However, in matrix terms, we now have

Proposition 1.14.9

Let E and F be finite-dimensional complex Hilbert spaces with orthonormal bases

$$\mathscr{E} = (e_i)_{i=1}^m \qquad \text{and} \qquad \mathscr{F} = (f_j)_{j=1}^n$$

respectively. Then, if $A \in L(E, F)$, the matrix of $\bar{A}^*$, relative to $\mathscr{F}$ and $\mathscr{E}$, is given as the conjugate of the transpose of the matrix of A. That is, if $[A] = [a_{ij}]$, the (i, j)th element of $[\bar{A}^*]$ is equal to $\bar{a}_{ji}$, $1 \leqslant i \leqslant m$, $1 \leqslant j \leqslant n$.

Proof. Left to the exercises. ∎

Definition 1.14.10

Let $(H, \langle \ , \ \rangle)$ be a complex Hilbert space and let $A \in L(H, H)$. Then A is said to be *Hermitian adjoint* or just *Hermitian* if $A = \bar{A}^*$. That is, A is Hermitian if and only if

$$\langle A(x), y \rangle = \langle x, A(y) \rangle \qquad \text{for all } x, y \in H.$$

If H is of finite dimension n and $A \in L(H, H)$, then it follows from Proposition 1.14.9 that A will be Hermitian if and only if the matrix of A relative to an orthonormal basis of H is Hermitian. That is, if $[A] = [a_{ij}]$, we must have $a_{ij} = \bar{a}_{ji}$, $1 \leqslant i, j \leqslant n$. In particular, the diagonal elements must be *real.*

Exercises

1. Complete the proofs of Propositions 1.14.5 and 1.14.9.
2. Show that, if E and F are Hilbert spaces and $A \in L(E, F)$, A is injective if and only if the image of A^* is *dense* in E.
3. Let E and F be Hilbert spaces and let $A \in L(E, F)$. With the obvious notation, prove that $A^{**} = A$.
4. Let H be a Hilbert space and let $A \in L(H, H)$. Show that $A + A^*$ is always self-adjoint.
5. Show that $\| A \| = \| A^* \|$ for all $A \in L(H, H)$.
6. Let E and F be Hilbert spaces and let $A \in L(E, F)$. Show that both $A \,.\, A^*$ and $A^* \,.\, A$ are self-adjoint.
7. Let H be a complex Hilbert space and $A \in L(H, H)$ be Hermitian. Show that $\langle A(x), x \rangle$ is real for all $x \in H$.

8. Let H be a complex Hilbert space and $A \in L(H, H)$. Prove the polarization identity

$$a\bar{b}\langle A(x), y\rangle + \bar{a}b\langle A(y), x\rangle$$
$$= \langle A(ax + by), ax + by\rangle - |a|^2\langle A(x), x\rangle - |b|^2\langle A(y), y\rangle$$

where $a, b \in \mathbf{C}$ and $x, y \in H$.

Hence show that $A = 0$ if and only if $\langle A(x), x\rangle = 0$ for all $x \in H$. Is this result true for real linear maps on real Hilbert spaces?

9. Let H be a complex Hilbert space and $A \in L(H, H)$. Then A is Hermitian if and only if $\langle A(x), x\rangle$ is real for all $x \in H$ (*Hint*: Use the result of Exercise 8).

10. Let H be a finite-dimensional inner-product space. Show that the function trace $(A \,.\, B^*)$, $A, B \in L(H, H)$, defines an inner product on $L(H, H)$. (For the definition of trace see Section 1.15.)

11. Let A be a self-adjoint operator on the Hilbert space H. Show that H has the orthogonal decomposition

$$H = \text{Ker}\,(A) \oplus \overline{\text{Im}\,(A)}$$

12. Let $K: [0, 1] \times [0, 1] \longrightarrow \mathbf{R}$ be a continuous function. K induces a linear endomorphism $\tilde{K}$ of $C[0, 1]$ by the rule

$$\tilde{K}(f)(t) = \int_0^1 K(s, t) f(s)\, dt, \qquad f \in C[0, 1].$$

Show that

(a) Relative to the norm $\| \;\; \|_2$ on $C[0, 1]$, K is continuous.
(b) $\langle \tilde{K}(f), g\rangle = \langle f, \tilde{K}^*(g)\rangle$, where $K^*(s, t) = K(t, s)$ for all $s, t \in [0, 1]$.
(c) Show that if

$$K(t, s) = K(s, t), \qquad s, t \in [0, 1],$$

then $\tilde{K}$ extends to a self-adjoint map on the completion $\widehat{C[0, 1]}$.

1.15 Eigenvalues

In this section we shall make a brief review of the theory of eigenvalues of a linear map with a view to applications.

Definition 1.15.1

Let E be a normed vector space over the field K and suppose that $A \in L(E, E)$. Then an element $\lambda \in K$ is called an *eigenvalue* of A if

$$\dim\,(\text{Ker}\,(A - \lambda I)) \geq 1.$$

Any non-zero element of Ker $(A - \lambda I)$ is called an *eigenvector* of A (corresponding to the eigenvalue λ).

If $\lambda \in K$, we shall set $E_\lambda = \text{Ker}\,(A - \lambda I)$. We shall also let $E(A)$ denote the set of eigenvalues of A.

Remarks

1. If $x \in \text{Ker}\,(A - \lambda I)$, then $A(x) = \lambda x$. Conversely, if $\lambda \in K$ and there exists

$x \neq 0$ such that $A(x) = \lambda x$, then λ is an eigenvalue of A and x is an eigenvector of A corresponding to the eigenvalue λ.

2. If $K = \mathbf{R}$, then $E(A)$ may be empty (for example, if A is the rotation of $\mathbf{R}^2$ through $90°$).

We leave the proof of the following proposition to the exercises.

Proposition 1.15.2

Let E be a normed vector space and let $A \in L(E, E)$. Then

1. E_λ is a closed vector subspace of E for all $\lambda \in K$.
2. $E_\lambda \cap E_\mu = \{0\}$, $\lambda \neq \mu$.
3. If E is finite dimensional, then the number of eigenvalues of A is not greater than the dimension of E.

Remark. Notice that if $\dim_K E = n$ and there are n distinct eigenvalues of $A \in L(E, E)$ we must have

$$E = \bigoplus_{\lambda \in E(A)} E_\lambda .$$

However, this relation is not true for every endomorphism of E.

In the remainder of this section we shall be chiefly interested in finite-dimensional vector spaces over the real or complex numbers. The material we present does, however, have a generalization to Hilbert spaces but it is by no means a straightforward one. Notice, for example, that if E is finite dimensional and 0 is not an eigenvalue of a map $A \in L(E, E)$, then A is a linear isomorphism. This is no longer true, however, if E is not finite dimensional.

We recall that if E is a finite-dimensional vector space over the field K we have a natural map

$$\det_K : L(E, E) \longrightarrow K$$

defined by taking the determinant of linear maps. Thus, if $A \in L(E, E)$, $\det_K(A)$ may be defined by taking any basis for E (over K) and evaluating the determinant of the matrix of A relative to this basis. Since the matrix of A changes according to the law $[S]^{-1}[A][S]$, where S is the coordinate change map induced from a change of basis, this definition of $\det_K(A)$ is independent of the choice of basis on E. In the sequel we shall usually drop the subscript K from $\det_K$ if the underlying field is clear from the context. We recall the following characteristic properties of det.

1. $\det(A \,.\, B) = (\det(A))(\det(B))$ for all $A, B \in L(E, E)$.
2. $\det(A) \neq 0$ if and only if A is a linear isomorphism.
3. $\det(I_E) = 1$.
4. $\det(kA) = k^n \det(A)$, where $k \in K$ and n = dimension of E.

Definition 1.15.3

Let E be a vector space of dimension n over K. Then, if $A \in L(E, E)$, the characteristic polynomial of A is the nth-order polynomial with coefficients in K defined by

$$\det(A - \lambda I).$$

We shall usually denote the characteristic polynomial of A by the symbol p_A. Thus, $p_A(\lambda) = \det(A - \lambda I)$.

We denote the set of complex roots of p_A by ch (A). Any element of ch (A) is called a *characteristic root* of A.

Remarks

1. Expanding $\det(A - \lambda I)$, relative to any basis on E, we find that

$$\det(A - \lambda I) = (-1)^n(\lambda^n - \text{trace}(A)\lambda^{n-1} + \cdots + (-1)^n \det(A)),$$

where trace (A) is the sum of the diagonal elements of the matrix of A. Since the characteristic equation of A is invariantly defined it follows that all the coefficients of p_A are invariants of A (that is, defined independent of choice of basi on E). We remark also the identities:

$$\text{trace}(A) = \sum_{\lambda \in \text{ch}(A)} \lambda, \qquad \det(A) = \prod_{\lambda \in \text{ch}(A)} \lambda,$$

which follow from the usual relations between the coefficients of a polynomial aı its roots.

2. In all the material in this text the underlying field K is always the real or complex field. Thus it makes sense to demand that the characteristic polynomial has complex roots. If K had been an arbitrary field we would have required that the roots of the characteristic equation lay in the algebraic closure of K (that is, the smalle: field containing K and the roots of all polynomial equations with coefficients in K).

Proposition 1.15.4

Let E be a real finite-dimensional vector space and let $A \in L(E, E)$. Then ch $(A) \cap \mathbf{R} = E(A)$.

If E is a complex finite-dimensional vector space then ch $(A) = E(A)$, for all $A \in L(E, E)$.

Proof. Trivial, since $\text{Ker}(A - \lambda I) \neq \{0\}$ if and only if $A - \lambda I$ is not a vector-space isomorphism. ∎

Remark. It follows from Proposition 1.15.4, together with the fundamental theorem of algebra, that if E is a finite-dimensional complex vector space and $A \in L(E, E)$ then $E(A) \neq \phi$.

The following technical lemma will prove to be useful.

Lemma 1.15.5

Let E be a finite-dimensional vector space and $A \in L(E, E)$. Suppose that we can write E as a product of subspaces $E_1 \times E_2$ such that E_1 is left invariant by $A : A(E_1) \subset E$ Then, if we define $A_1 = A \mid E_1$ and $A_2 = p_2 . (A \mid E_2)$, we have

$$p_A = p_{A_1} p_{A_2}.$$

Here p_2 denotes the projection of $E_1 \times E_2$ onto E_2.

Proof. Choose bases $(e_1, \ldots, e_p)$, $(f_1, \ldots, f_q)$ for E_1 and E_2 respectively. Then $(e_1, \ldots, e_p, f_1, \ldots, f_q)$ defines a basis for E, and the matrix of A, relative to this basis, takes the form

$$\left[\begin{array}{c|c} [A_1] & [X] \\ \hline 0 & [A_2] \end{array}\right].$$

It follows that

$$\det(A - \lambda I) = \det([A_1] - \lambda[I_{E_1}])\det([A_2] - \lambda[I_{E_2}]). \quad ■$$

Suppose now that E is an n-dimensional real or complex vector space and that $A \in L(E, E)$. Let us suppose that there are m distinct roots of the characteristic equation of A. Thus we set

$$\text{ch}(A) = \{\lambda_1, \ldots, \lambda_m\}.$$

Let us denote the multiplicity of the root λ_j by q_j, $1 \leqslant j \leqslant m$. Thus,

$$\sum_{j=1}^{m} q_j = n,$$

The characteristic polynomial of A therefore takes the form

$$p_A(\lambda) = (-1)^n \prod_{j=1}^{m} (\lambda - \lambda_j)^{q_j}.$$

If E is a real vector space and λ_j is not a real root of $p_A(\lambda)$, there will be no corresponding eigenspace of A.

If E is a real or complex vector space, we claim that if $\lambda_j \in \text{ch}(A) \cap E(A)$, then we have the inequality

$$1 \leqslant \dim(E_{\lambda_j}) \leqslant q_j.$$

It follows from Proposition 1.15.4 that $\dim(E_{\lambda_j}) \geqslant 1$ if

$$\lambda_j \in \text{ch}(A) \cap E(A).$$

Let F be any complement of E_{λ_j} in E. Then E_{λ_j} is left invariant by A, and setting $A_1 = A \mid E_{\lambda_j}$ and $A_2 = p_2 . (A \mid F)$, we may apply Lemma 1.15.5 to obtain

$$p_A(\lambda) = p_{A_1}(\lambda) p_{A_2}(\lambda).$$

But

$$p_{A_1}(\lambda) = \pm(\lambda - \lambda_j)^{\dim(E_{\lambda_j})},$$

And so the multiplicity of λ_j in $p_A(\lambda)$ is at least $\dim(E_{\lambda_j})$. Summing up the above argument in a proposition, we have

Proposition 1.15.6

Let E be a finite-dimensional vector space and let $A \in L(E, E)$. Then

$$1 \leqslant \dim(E_{\lambda_j}) \leqslant q_j \qquad \text{for all } \lambda_j \in \text{ch}(A) \cap E(A).$$

In the sequel we shall be most interested in the case when

$$\dim(E_{\lambda_j}) = q_j, \qquad 1 \leqslant j \leqslant m.$$

It then follows that

$$\bigoplus_{j=1}^{m} E_{\lambda_j} = E$$

and that the matrix of A can be given the particularly simple form

$$\begin{bmatrix} \lambda_1 & 0 & \dots & 0 & 0 \\ 0 & \lambda_1 & \dots & 0 & 0 \\ \vdots & \vdots & & \vdots & \vdots \\ 0 & 0 & \dots & \lambda_m & 0 \\ 0 & 0 & \dots & 0 & \lambda_m \end{bmatrix}$$

relative to a basis of E of eigenvectors. Each eigenvalue is repeated in the above matrix a number of times equal to its multiplicity.

In the next two sections we shall study two classes of maps for which it is possible to obtain this special representation.

Exercises

1. Let $A \in L(\mathbf{R}^2, \mathbf{R}^2)$ have matrix

$$\begin{bmatrix} 0 & -1 \\ 1 & 0 \end{bmatrix}.$$

relative to the standard basis of $\mathbf{R}^2$. Find ch (A) and hence show that $E(A) = \emptyset$. Describe A geometrically.

2. Let $A \in L(\mathbf{R}^2, \mathbf{R}^2)$ have matrix

$$\begin{bmatrix} 1 & 1 \\ 0 & 1 \end{bmatrix}$$

relative to the standard basis of $\mathbf{R}^2$. Show that A has a single characteristic root of multiplicity 2 but that the corresponding eigenspace is only of dimension 1.

3. Show that 'trace' defines a map trace : $L(E, E) \longrightarrow K$ which satisfies the followin; properties:

(a) Trace $(A + B)$ = trace (A) + trace (B) for all $A, B \in L(E, E)$.
(b) Trace $(A) = 0$ if and only if the sum of the characteristic roots of A is zero.
(c) Trace (I_E) = dimension of E.

1.16 Self-adjoint maps and quadratic forms

We start this section with the definition of a particularly important class of self-adjoint maps on a Hilbert space.

Definition 1.16.1

Let $(H, \langle\ ,\ \rangle)$ be a real Hilbert space and suppose $A \in L(H, H)$ is symmetric. A is said to be *positive definite* if

$$\langle A(x), x\rangle > 0 \qquad \text{for all } x \neq 0.$$

Similarly, if $(H, \langle\ ,\ \rangle)$ is a complex Hilbert space and $A \in L(H, H)$ is Hermitian, we say that A is positive definite if

$$\langle A(x), x\rangle > 0 \qquad \text{for all } x \neq 0.$$

Notation. If H is a Hilbert space (real or complex) we shall let $A(H)$ denote the subset of $L(H, H)$ consisting of self-adjoint maps. We also let $A^+(H)$ denote the subset of $A(H)$ consisting of positive-definite maps.

Remark. If H is a real Hilbert space, then $A(H)$ is a closed vector subspace of $L(H, H)$. If H is complex, however, $A(H)$ is only a real subspace of $L(H, H)$. This is a consequence of the fact that if $T = \overline{T}^*$ then $\overline{(iT)}^* = -iT$ (iT is *skew-adjoint*).

In this section we shall be exclusively interested in *finite*-dimensional inner-product spaces. Recall that all finite-dimensional inner-product spaces are necessarily Hilbert spaces (Section 1.10). In fact the theory that we shall present has a generalization to infinite-dimensional Hilbert spaces. We indicate some of the problems that can arise in undertaking such a generalization in the exercises.

As we shall soon see, every symmetric linear map naturally defines a Hermitian map on a complex inner-product space in such a way that the study of the real case may be reduced to the study of the complex case. For the moment, therefore, we shall restrict attention to Hermitian maps.

Proposition 1.16.2

Let $(H, \langle\ ,\ \rangle)$ be a complex (finite-dimensional) inner-product space and let $A \in A(H)$. Then $\langle A(x), x\rangle$ is real for all $x \in H$.

Proof. Since $A = \overline{A}^*$, it follows that for all $x \in H$ we have

$$\langle A(x), x\rangle = \langle x, A(x)\rangle = \overline{\langle A(x), x\rangle}$$

and so $\langle A(x), x\rangle$ is real. ∎

Remark. The converse to Proposition 1.16.2 is also true: see the exercises at the end of Section 1.15.

Proposition 1.16.3

Let H be a complex inner-product space and let $A \in A(H)$. Then we have the following properties of A.

1. $E(A) \subset \mathbf{R}$. In particular, all characteristic roots of A are real.
2. $|\lambda| \leqslant \|A\|$ for all $\lambda \in E(A)$.
3. $H_\lambda \perp H_\mu$, $\lambda \neq \mu$.
4. If A is positive definite, all eigenvalues of A are strictly positive. In particular, A is a linear isomorphism.

Proof. Let $\lambda \in E(A)$ and $x \in H_\lambda - \{0\}$. Then $\langle A(x), x\rangle = \lambda\langle x, x\rangle$. Since, by Proposition 1.16.2, $\langle A(x), x\rangle$ is real, it follows that λ is real. This proves (1). (2) follows by Schwartz' inequality:

$$|\lambda|\langle x, x\rangle = |\langle A(x), x\rangle| \leqslant \|A(x)\|\,\|x\| \leqslant \|A\|\langle x, x\rangle \qquad \text{for all } x \in H_\lambda.$$

For (3), let $x \in H_\lambda$ and $y \in H_\mu$; then

$$\lambda\langle x, y\rangle = \langle A(x), y\rangle = \langle x, A(y)\rangle = \mu\langle x, y\rangle.$$

Since $\lambda \neq \mu$ it follows that $x \perp y$.

(4) follows using the identity $\langle A(x), x\rangle = \lambda\langle x, x\rangle$, $x \in H_\lambda$. ∎

Remark. It is natural to ask how far Proposition 1.16.3 generalizes to infinite-dimensional Hilbert spaces. We can no longer define the characteristic equation of A but, with this proviso, (1) certainly holds with the same proof. Likewise, (2), (3) and the first part of (4) all hold with the same proof. For the correct version of the second part of (4) we refer to the exercises.

Our main theorem in this section is

Theorem 1.16.4 (Spectral theorem)

Let H be a complex inner-product space and let $A \in A(H)$. Denoting the distinct eigenvalues of A by $\lambda_1, \ldots, \lambda_m$, we have

1. The multiplicity of $\lambda_j = \dim(H_{\lambda_j})$, $1 \leqslant j \leqslant m$.
2. $H = \oplus_{j=1}^m H_{\lambda_j}$.
3. $A = \sum_{j=1}^m \lambda_j p_j$, where p_j denotes the projection map associated with the subspace H_{λ_j}, $1 \leqslant j \leqslant m$.

Proof. Let us define the closed subspace $\tilde{H}$ of H by

$$\tilde{H} = \bigoplus_{j=1}^{m} H_{\lambda_j}.$$

Suppose that $\tilde{H} \neq H$. We let $F = \tilde{H}^\perp$. Then $\tilde{H}$ is clearly left invariant by A. So also is F, since if $x \in F$, $y \in \tilde{H}$ we have $\langle x, A(y)\rangle = \langle A(x), y\rangle = 0$, by the A-invariance of $\tilde{H}$.

Since F is A-invariant, any eigenvector of $A \mid F$ is also an eigenvector of A. From Proposition 1.15.4 it follows that if $\dim(F) > 0$, $A \mid F$ has at least one eigenvalue and so non-zero eigenvectors. But by assumption $\oplus H_{\lambda_j} \cap F = \{0\}$. Therefore $F = \{0\}$ and $\tilde{H} = H$.

This proves parts (1) and (2) of Theorem 1.16.4. To complete the proof we notice that any $x \in H$ has a unique orthogonal decomposition as

$$x = p_1(x) + \cdots + p_m(x).$$

This follows from (2) and part (3) of Proposition 1.16.3. But

$$A(p_j(x)) = \lambda_j p_j(x), \qquad 1 \leqslant j \leqslant m,$$

since H_{λ_j} is A-invariant. Hence

$$A(x) = \sum_{j=1}^{m} \lambda_j p_j(x). \quad ∎$$

Corollary 1.16.4.1

Let H be a complex inner-product space and let $A \in A(H)$. Then we may find an orthonormal basis for H such that the matrix of A, relative to this basis, assumes the diagonal form

$$\begin{bmatrix} \lambda_1 & 0 & \dots & 0 & 0 \\ 0 & \lambda_1 & \dots & 0 & 0 \\ \vdots & \vdots & & \vdots & \vdots \\ 0 & 0 & \dots & \lambda_m & 0 \\ 0 & 0 & \dots & 0 & \lambda_m \end{bmatrix}$$

where λ_j, $1 \leqslant j \leqslant m$, denote the characteristic roots of A, each root repeated a number of times equal to its multiplicity.

Proof. Immediate from Theorem 1.16.4. ∎

Before stating the version of Theorem 1.16.4 for real vector spaces we wish to introduce the idea of complexification. Suppose that E is any real vector space. We shall let ${}_cE$ denote the product vector space $E \times E$ with complex multiplication defined by

$$(a + ib)(x, y) = (ax - by, ay + bx), \qquad a, b \in \mathbf{R}.$$

We leave it as an exercise for the reader to check that ${}_cE$ is a complex vector space with this definition of complex multiplication (think of (x, y) as $x + iy$). We call ${}_cE$ the *complexification* of E.

We shall often regard E as identified with the real subspace $\{(x, 0), x \in E\}$ of ${}_cE$. If we let iE denote the subspace $\{(0, x): x \in E\}$ of ${}_cE$ we see that

$${}_cE = E \oplus (iE).$$

Notice that scalar multiplication by i maps the subspace E isomorphically onto iE.

If $A \in L(E, E)$, A induces a *complex* linear map ${}_cA \in L({}_cE, {}_cE)$ by the rule

$${}_cA(x, y) = (A(x), A(y)), \qquad (x, y) \in {}_cE.$$

Notice that the subspaces E and iE are both left invariant by ${}_cA$.

Suppose that E has a real inner product $\langle\, ,\rangle$. Then $\langle\, ,\rangle$ extends to a complex inner product ${}_c\langle\, ,\rangle$ on ${}_cE$ defined by

$${}_c\langle(x, y), (x', y')\rangle = (\langle x, x'\rangle + \langle y, y'\rangle) + i(\langle x, y'\rangle - \langle x', y\rangle),$$

where $(x, y), (x', y') \in {}_cE$.

We leave it to the exercises for the reader to verify that ${}_c\langle\, ,\rangle$ is a complex inner product on ${}_cE$. Finally, we remark that the real dimension of E equals the complex dimension of ${}_cE$.

We have the following lemma.

Lemma 1.16.5

Let $(H, \langle\, ,\rangle)$ be a real inner-product space and let $A^+ \in L(H, H)$. Then

1. $\mathrm{ch}(A) = \mathrm{ch}({}_cA)$.
2. $A \in A(H)$ if and only if ${}_cA \in A({}_cH)$.
3. $A \in A^+(H)$ if and only if ${}_cA \in A({}_cH)$.

Here ${}_cH$ is given the complex inner product ${}_c\langle\, ,\rangle$.

Proof. We shall prove (1) and leave the proofs of (2) and (3) to the exercises. To prove (1) it is sufficient to show that the characteristic equation of A is equal to the characteristic equation of ${}_{\mathbf{c}}A$. That is

$$\det_{\mathbf{R}} (A - \lambda I) = \det_{\mathbf{C}} ({}_{\mathbf{c}}A - \lambda I).$$

But notice that if $(e_i)_{i=1}^n$ is a real basis for H, then the set $((e_i, 0))_{i=1}^n$ is a complex basis for ${}_{\mathbf{c}}H$. It follows that the matrix of A relative to the real basis $(e_i)_{i=1}^n$ is equal to the matrix of ${}_{\mathbf{c}}A$ relative to the complex basis $((e_i, 0))_{i=1}^n$, and so the characteristic equations are equal. ∎

It follows from Lemma 1.16.5 and Proposition 1.16.3 that if $A \in A(H)$, then all the characteristic values of A are real and therefore that $\text{ch}(A) = E(A)$. We clearly then have $E(A) = E({}_{\mathbf{c}}A)$. Suppose that $\lambda \in E({}_{\mathbf{c}}A)$. We let $\tilde{H}_\lambda$ denote the corresponding eigenspace of ${}_{\mathbf{c}}A$. Then $\tilde{H}_\lambda$ is a complex vector subspace of ${}_{\mathbf{c}}H$, and we obviously have

$$\tilde{H}_\lambda = {}_{\mathbf{c}}H_\lambda = H_\lambda \times H_\lambda \subset {}_{\mathbf{c}}H.$$

It follows that

$$\dim_{\mathbf{R}} H_\lambda = \dim_{\mathbf{C}} \tilde{H}_\lambda.$$

By Theorem 1.16.4, the dimension of $\tilde{H}_\lambda$ is equal to the multiplicity of the root λ in the characteristic equation of ${}_{\mathbf{c}}A$. Since the real dimension of H equals the complex dimension of ${}_{\mathbf{c}}H$ it follows easily from Theorem 1.16.4 that the multiplicity of any characteristic root λ of A is equal to the dimension of H_λ.

The above observations, together with Proposition 1.16.3 and Theorem 1.16.4, imply

Theorem 1.16.6 (Spectral theorem for self-adjoint maps)

Let H be a real inner-product space and let $A \in A(H)$. Denoting the distinct eigenvalues of A by $\lambda_1, \ldots, \lambda_m$, we have

1. $\text{ch}(A) = E(A)$ and A is positive definite if and only if all eigenvalues of A are strictly positive.
2. $H_{\lambda_i} \perp H_{\lambda_j}$, $i \neq j$.
3. The multiplicity of $\lambda_j = \dim(H_{\lambda_j})$, $1 \leqslant j \leqslant m$.
4. $H = \textcircled{\perp}_{j=1}^m H_{\lambda_j}$.
5. $A = \Sigma_{j=1}^m \lambda_j p_j$, where p_j denotes the projection map associated with the subspace H_{λ_j}, $1 \leqslant j \leqslant m$.

Corollary 1.16.6.1

Let H be a real inner-product space and let $A \in A(H)$. Then we may find an orthonormal basis for H such that the matrix of A, relative to this basis, assumes the diagonal form

$$\begin{bmatrix} \lambda_1 & 0 & \ldots & 0 & 0 \\ 0 & \lambda_1 & \ldots & 0 & 0 \\ \vdots & \vdots & & \vdots & \vdots \\ 0 & 0 & \ldots & \lambda_m & 0 \\ 0 & 0 & & 0 & \lambda_m \end{bmatrix},$$

where λ_j, $1 \leqslant j \leqslant m$, denote the characteristic roots of A, each root repeated a number of times equal to its multiplicity.

Remark. The geometric content of the above theorem may be described in the following way. Suppose $\mathbf{R}^n$ is given the standard inner product and we pick any set of n mutually orthogonal lines $L_1, \ldots, L_n$ through the origin of $\mathbf{R}^n$ and any set of n real numbers $\lambda_1, \ldots, \lambda_n$. Then there exists a unique symmetric endomorphism of $\mathbf{R}^n$ which, restricted to L_j, multiplies vectors by the constant λ_j, $1 \leqslant j \leqslant m$. Thus, if $n = 2$, $\lambda_1 = 2$, $\lambda_2 = -1/2$ and L_1 and L_2 are as shown in Fig. 1.7.

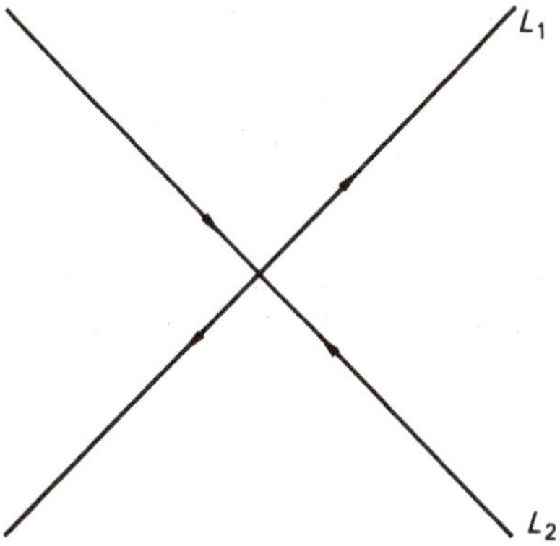

FIG. 1.7

We assert the existence of a unique symmetric linear map A on $\mathbf{R}^2$ which expands L_1 by the factor 2 and contracts L_2 by the factor 1/2 and reflects about the L_1 direction. Thus, $\mathbf{R}^2$ is stretched in direction L_1 and contracted in direction L_2 and reflected about the L_1-axis. This type of geometric description is characteristic of inner-product spaces.

We do not have a similar uniqueness statement for normed vector spaces. But notice, though, that if $L_1, \ldots, L_n$ are linearly independent lines in $\mathbf{R}^n$, we can always choose an inner product on $\mathbf{R}^n$ with respect to which these lines are mutually orthogonal.

Examples

1. Consider the linear map $A \in L(\mathbf{R}^3, \mathbf{R}^3)$ whose matrix relative to the standard basis of $\mathbf{R}^3$ is given as

$$\begin{bmatrix} 1 & 0 & 3 \\ 0 & -1 & 1 \\ 3 & 1 & 2 \end{bmatrix}.$$

It follows from Proposition 1.14.5 that A is a symmetric map. Let us find the eigenvalues of A: We must solve the equation

$$\det \begin{bmatrix} 1-\lambda & 0 & 3 \\ 0 & -1-\lambda & 1 \\ 3 & 1 & 2-\lambda \end{bmatrix} = 0.$$

Expanding and simplifying we find that

$$p_A(\lambda) = -(\lambda^3 - 2\lambda^2 - 11\lambda - 6) = -(\lambda + 2)(\lambda^2 - 4\lambda - 3).$$

Hence the characteristic roots of A are given by

$$\lambda_1 = -2, \quad \lambda_2 = 2 - \sqrt{7}, \quad \lambda_3 = 2 + \sqrt{7}.$$

Let us denote the corresponding eigenspaces by L_1, L_2, L_3. To find L_j we must solve the linear equation

$$(A - \lambda_j I)(x) = 0.$$

We leave it to the reader to verify that the lines L_1, L_2, L_3 are defined by

$$L_1 = \{(-t, -t, t) \colon t \in \mathbf{R}\},$$
$$L_2 = \{(\tfrac{1}{2}(\sqrt{7} - 1)t, \tfrac{1}{2}(3 - \sqrt{7})t, t) \colon t \in \mathbf{R}\},$$
$$L_3 = (-\tfrac{1}{2}(\sqrt{7} + 1)t, \tfrac{1}{2}(3 + \sqrt{7})t, t) \colon t \in \mathbf{R}\},$$

and that they are mutually orthogonal. It follows that, relative to the orthonormal basis defined by L_1, L_2 and L_3, the matrix of A has the form

$$\begin{bmatrix} -2 & 0 & 0 \\ 0 & 2 - \sqrt{7} & 0 \\ 0 & 0 & 2 + \sqrt{7} \end{bmatrix}.$$

2. We shall show in this example how the device of complexification can give additional information about a linear map. We claim that if E is a finite-dimensional vector space and $A \in L(E, E)$, then there exists either a one-dimensional subspace of E invariant by A or a two-dimensional invariant subspace. We already know that if A has any real characteristic roots, then it has real eigenvalues, and so one-dimensional invariant subspaces. Let us suppose that all the characteristic roots of A are complex. Let $\lambda \in \mathrm{ch}(A)$. Then λ is a characteristic root, and hence an eigenvalue, of ${}_cA$. If $X = (x, y)$ is a (non-zero) eigenvector of ${}_cA$ associated with the eigenvalue λ it follows from the definition of ${}_cA$ that if we set $\lambda = a + ib$ we have

$$A(x) = ax - by, \qquad A(y) = bx + ay.$$

Hence the subspace of E spanned by x and y is A-invariant.

For the remainder of this section we wish to discuss part of the theory of quadratic forms on finite-dimensional inner-product spaces. This material is of importance in the theory of critical points of differentiable functions (see, for example, Section 2.12). We shall start with the theory of quadratic forms on real vector spaces, leaving to the end of the section a sketch of the theory for complex spaces. We remind the reader that all vector spaces are assumed to be of finite dimension. In particular, all multilinear maps are continuous.

Definition 1.16.7

Let E be a real normed vector space. An element of $H^2(E; \mathbf{R})$ is called a *quadratic form* on E. We denote the space of quadratic forms on E by $Q(E)$.

If the quadratic form q satisfies the condition

$$q(x) = 0 \qquad \text{if and only if } x = 0,$$

q is said to be a *non-degenerate* quadratic form.

If, in addition,

$$q(x) > 0, \qquad x \neq 0,$$

we say that q is a *positive-definite* quadratic form.

Proposition 1.16.8

Let $(H, \langle\ ,\ \rangle)$ be a real inner-product space. Then there exists a canonical isomorphism

$$\phi : Q(H) \approx A(H).$$

Moreover, we have

1. $q \in Q(H)$ is non-degenerate if and only if $\phi(q)$ is an isomorphism.
2. $q \in Q(H)$ is positive definite if and only if $\phi(q) \in A^{+}(H)$.

Proof. We have

$$L^2(H; \mathbf{R}) \approx L(H, H^*) \qquad \text{(Proposition 1.6.5)}$$

$$\approx L(H, H) \qquad \text{(Riesz representation theorem).}$$

Denote this isomorphism between $L^2(H; \mathbf{R})$ and $L(H, H)$ by ϕ. Then, if $A \in L^2(H; \mathbf{R})$ and $x, y \in H$, ϕ is characterized by

$$\langle \phi(A)(x), y \rangle = A(x, y).$$

Clearly A is symmetric if and only if

$$\langle \phi(A)(x), y \rangle = \langle \phi(A)(y), x \rangle$$

for all x and y in H. That is, if and only if $\phi(A) = \phi(A)^*$. Hence ϕ restricts to an isomorphism between $L_s^2(H; \mathbf{R})$ and $A(H)$. But, from Proposition 1.7.11, $L_s^2(H; \mathbf{R}) \approx H^2(H; \mathbf{R}) = Q(H)$.

Suppose that $q \in Q(H)$ is equal to $\phi^{-1}(A)$, where $A \in A(H)$. Then q is non-degenerate if and only if

$$\langle A(x), x \rangle \neq 0 \qquad \text{for all } x \neq 0.$$

Hence $A(x) \neq 0$ for all $x \neq 0$ and so A is injective. Since H is assumed to be finite dimensional, it follows that A is an isomorphism.

Part 2 of the proposition is just the definition of a positive-definite linear map. ∎

Remark. The reader should notice that the isomorphism of Proposition 1.16.8 depends on the inner product on H. In general, if E is a normed vector space of finite dimension, we can always find an inner product on E (not, in general, inducing the given norm on E) and so represent the quadratic forms on E as symmetric linear maps. The representation will, however, depend on the choice of inner product.

Proposition 1.16.8 may be written in terms of coordinates.

Proposition 1.16.9

Let H be a real inner-product space with a fixed orthonormal basis. Then every quadratic form on H may be written uniquely in the form

$$q(x) = [x]^t[A]\,[x], \qquad x \in H,$$

where $[A]$ is a symmetric matrix.

If we write $q(x)$ in coordinates as

$$q(x) = \sum_{i \leqslant j} q_{ij}x_i x_j, \qquad x = (x_1, \ldots, x_n) \in H,$$

then the components of $[A]$ are given uniquely in terms of the coefficients q_{ij} by the rules

$$\begin{aligned} a_{ij} &= \tfrac{1}{2}q_{ij}, \qquad & i \neq j \\ &= q_{ij}, \qquad & i = j. \end{aligned}$$

q is non-degenerate if and only if $\det([A]) \neq 0$, and positive definite if and only if all characteristic roots of $[A]$ are strictly positive.

Proof. Left to the exercises. ■

Examples. Suppose that $\mathbf{R}^3$ is given the standard inner product. Find the quadratic form on $\mathbf{R}^3$ associated with the symmetric matrix

$$\begin{bmatrix} 2 & -1 & 3 \\ -1 & 0 & 1 \\ 3 & 1 & -1 \end{bmatrix}.$$

From Proposition 1.16.9 it follows that the form is given by

$$\begin{aligned} q(x_1, x_2, x_3) &= [x_1, x_2, x_3]\begin{bmatrix} 2 & -1 & 3 \\ -1 & 0 & 1 \\ 3 & 1 & -1 \end{bmatrix}\begin{bmatrix} x_1 \\ x_2 \\ x_3 \end{bmatrix} \\ &= 2x_1^2 - x_3^2 - 2x_1x_2 + 6x_1x_3 + 2x_2x_3. \end{aligned}$$

A simple verification shows that q is non-degenerate but not positive definite.

We have the following fundamental theorem.

Theorem 1.16.10

Let q be a quadratic form on the real inner product space $(H, \langle\,,\rangle)$. Then we may find an orthonormal basis of H and real numbers $\mu_1, \ldots, \mu_n$, where $n =$ dimension (H), such that relative to this basis q has the form

$$q(x) = \sum_{i=1}^{n} \mu_i x_i^2, \qquad x = (x_1, \ldots, x_n) \in H.$$

Further, q is non-degenerate if and only if all the constants μ_i are non-zero, and q is positive definite if and only if the constants μ_i are all strictly positive.

Proof. From Proposition 1.16.8 there exists a unique element $A \in A(H)$ such that

$$q(x) = \langle A(x), x\rangle \qquad \text{for all } x \in H.$$

From Corollary 1.16.6.1 we may choose an orthonormal basis for H such that the matrix of A, relative to this basis, is of the form

$$\begin{bmatrix} \mu_1 & 0 & \dots & 0 & 0 \\ 0 & \mu_2 & \dots & 0 & 0 \\ \vdots & \vdots & & \vdots & \vdots \\ 0 & 0 & \dots & \mu_{n-1} & 0 \\ 0 & 0 & \dots & 0 & \mu_n \end{bmatrix}$$

where each μ_i is a characteristic root of A. Applying Proposition 1.16.9 the result follows. ■

Since the notions of quadratic form, non-degenerate and positive definite are defined independently of any inner-product structure, it is natural to ask how far the constants μ_i obtained in Theorem 1.16.10 depend on q.

Definition 1.16.11

Let $(H, \langle\ ,\ \rangle)$ be a real inner-product space. Suppose that $q \in Q(H)$, let $\mu_1, \dots, \mu_n$ be the constants defined by Theorem 1.16.10. Then we define the positive integers k, m, p by

$$k = \text{number of } \mu_i\text{'s which are zero,}$$

$$m = \text{number of } \mu_i\text{'s which are strictly negative,}$$

$$p = \text{number of } \mu_i\text{'s which are strictly positive.}$$

Theorem 1.16.12

Let $(H, \langle\ ,\ \rangle)$ be a real inner-product space and let $q \in Q(H)$. Then the integers k, m, p defined above are independent of the inner product $\langle\ ,\ \rangle$ on H and depend only on q. Furthermore, we can choose an inner product on H such that the constants μ_i given by Theorem 1.16.10 are equal to 0, -1 or $+1$.

Proof. Suppose that $\langle\ ,\ \rangle'$ is another inner product on H and that k', m', p' are the corresponding integers given by Definition 1.16.11.

Now from Theorem 1.16.10 we have

$$\begin{aligned} q(x) &= \langle A(x), x\rangle, && \text{where } A \in A(H), \\ &= \langle A'(x), x\rangle' && \text{where } A' \in A(H)', \end{aligned}$$

where we have written $A(H)'$ to indicate the dependence on $\langle\ ,\ \rangle'$.

If $A \in A(H)$, it follows from Theorem 1.16.6 that H decomposes as the orthogonal direct sum

$$K \oplus M \oplus P,$$

where $K = \text{Ker}(A)$, M = orthogonal direct sum of those eigenspaces of A corresponding to negative eigenvalues and P = orthogonal direct sum of those eigenspaces of A corresponding to positive eigenvalues.

From Theorem 1.16.10 we have

$$k = \dim(K) \qquad m = \dim(M) \qquad p = \dim(P).$$

Similarly, $H = K' \oplus' M' \oplus' P'$, where K', M' and P' are defined relative to A' and $\langle\, ,\rangle'$, and $k' = \dim(K')$, $m' = \dim(M')$ and $p' = \dim(P')$. We remark that the integers k, m, p, k', m', p' are related by

$$k + m + p = k' + m' + p' = \text{dimension}(H) = n, \qquad \text{say.}$$

Suppose, for example, that $p > p'$. Consider the vector subspaces P and $K' \oplus' M'$ of H. Since $p + k' + m' > n$, it follows that

$$\dim(P \cap (K' \oplus M')) \geqslant 1.$$

But if $x \in P$ and $x \neq 0$, we have $q(x) > 0$, whilst, if $x \in K' \cap M'$, $q(x) \leqslant 0$. Hence, $P \cap (K' \oplus' M')$ must be zero. Contradiction. Therefore p must equal p'. Similarly, $m = m'$ and, since

$$k = n - (m + p) = n - (m' + p') = k',$$

it follows that $k = k'$.

The remaining part of the theorem follows easily from Theorem 1.16.10 and we leave it to the exercises. ∎

Definition 1.16.13

Let E be a real normed vector space and let $q \in Q(E)$. Then the rank of q, rank (q), is the number $n + m$, where p and m are defined according to Definition 1.16.11 and we have chosen any real inner product on E.

If q is non-degenerate, we define the *index* of q to be the integer $p - m$.

Remark. Clearly if $q \in Q(E)$, then q is non-degenerate if and only if rank (q) = dimension (E). For the case when rank $(q) <$ dimension (E), the rank of q is easily seen to be the rank of the matrix $[A]$ given by Proposition 1.16.9 (that is, rank (q) = dimension of image of A).

For the remainder of this section we wish to briefly sketch the theory of Hermitian quadratic forms. The archetypal example of a Hermitian quadratic form is given by a complex inner product. One of the problems in studying complex inner products is the fact that, whilst complex linear in the first variable, they are conjugate complex linear in the second variable. This means that, for example, we cannot directly apply the canonical isomorphisms of Section 1.6. To overcome this difficulty and present a more natural exposition of Hermitian quadratic forms, we shall start by defining the idea of the conjugate-complex structure of a complex vector space (cf. the exercises at the end of Section 1.15).

Suppose then that E is a complex vector space. Then scalar multiplication by i defines an endomorphism $J : E \longrightarrow E$ by the rule

$$J(x) = ix \qquad \text{for all } x \in E.$$

Clearly $J^2 = -I$. On the other hand, suppose that E is a real vector space and we are given an element $J \in L(E, E)$ satisfying the condition $J^2 = -I$. Then we may define the structure of a complex vector space on E by setting

$$(a + ib)x = ax + bJ(x) \qquad \text{for all } x \in E \text{ and } a, b \in \mathbf{R}.$$

We call J a *complex structure* on E. Clearly any complex vector space determines a unique complex structure by the above rules, and conversely.

Definition 1.16.14

Let H be a complex vector space with associated complex structure J. We define the conjugate of H to be the complex vector space $\bar{H}$ defined by giving the vector space H the complex structure $-J$. Thus, if $x \in H$ and $a + ib \in \mathbf{C}$, we define

$$(a + ib)x = ax - bJ(x).$$

$-J$ is called the *conjugate complex structure* on H and is denoted by $\bar{J}$.

Example. Recall that $\mathbf{C}^m$ is the space of m-tuples of complex numbers with scalar multiplication defined by

$$(a + ib)(z_1, \ldots, z_m) = ((a + ib)z_1, \ldots, (a + ib)z_m), (z_1, \ldots, z_m) \in \mathbf{C}^m.$$

$\overline{\mathbf{C}^m}$ is then the space of m-tuples of complex numbers with scalar multiplication defined by

$$(a + ib)(z_1, \ldots, z_m) = ((a - ib)z_1, \ldots, (a - ib)z_m), (z_1, \ldots, z_m) \in \overline{\mathbf{C}^m}.$$

We have the following trivial lemmas.

Lemma 1.16.15

Let E and F be complex vector spaces with respective complex structures J and J'. Then a real linear map $A: E \longrightarrow F$ is complex linear if and only if

$$A . J = J' . A.$$

Lemma 1.16.16

Let E and F be complex vector spaces and let $A: E \longrightarrow F$ be a real linear map. Then the following conditions on A are equivalent.

1. A is conjugate complex linear as a map between E and F.
2. $A: \bar{E} \longrightarrow F$ is complex linear.
3. $A: E \longrightarrow \bar{F}$ is complex linear.
4. $A: \bar{E} \longrightarrow \bar{F}$ is conjugate complex linear.

It follows from Lemma 1.16.16 that a bilinear map $A: E \times E \longrightarrow \mathbf{C}$, which is complex linear in the first variable and conjugate complex linear in the second variable, may be regarded as an element of $L(E, \bar{E}; \mathbf{C})$ and conversely. By $L(E, \bar{E}; \mathbf{C})$ is meant, of course, the space of *complex* bilinear maps from $E \times \bar{E}$ to $\mathbf{C}$.

Definition 1.16.17

A Hermitian bilinear form A on the complex normed vector space E is an element of $L(E, \bar{E}; \mathbf{C})$ which satisfies the additional condition

$$A(x, y) = \overline{A(y, x)} \qquad \text{for all } x, y \in E.$$

We denote the complex vector space of all Hermitian bilinear forms on E by the symbol

$$L_h^2(E; \mathbf{C}).$$

Remark. Notice that if $A \in L_h^2(E; \mathbf{C})$, then $A(x, x) \in \mathbf{R}$ for all $x \in E$.

Definition 1.16.18

Let E be a complex normed vector space and let $q: E \longrightarrow \mathbf{R}$. If q can be expressed in the form

$$q(x) = A(x, x) \qquad \text{for all } x \in E,$$

for some Hermitian bilinear form A on E, we say that q is a *Hermitian quadratic form* on E. We denote the set of such forms by $Q_h(E)$.

As in the case of real quadratic forms we may easily show that

$$Q_h(E) \approx L_h^2(E; \mathbf{C}).$$

Suppose that $(H, \langle\ ,\ \rangle)$ is a complex inner-product space. By Theorem 1.6.5 we have

$$L(H, \bar{H}; \mathbf{C}) \approx L(H, \bar{H}^*).$$

The complex version of the Riesz representation theorem gives us a complex linear isomorphism $\gamma: H \longrightarrow \bar{H}^*$ defined by

$$\gamma(e)(f) = \langle e, f \rangle, \qquad e, f \in H.$$

(Notice that $\gamma(e): \bar{H} \longrightarrow \mathbf{C}$ is complex linear according to this definition.) Combining these two natural isomorphisms gives us a natural isomorphism

$$\tilde{\phi}: L(H, \bar{H}; \mathbf{C}) \longrightarrow L(H, H)$$

characterized by

$$\langle \tilde{\phi}(A)(x), y \rangle = A(x, y) \qquad \text{for all } A \in L(H, \bar{H}; \mathbf{C}).$$

It follows, just as in Proposition 1.16.8, that we have

Proposition 1.16.19

Let $(H, \langle\ ,\ \rangle)$ be a complex inner-product space. Then there exists a canonical isomorphism

$$\phi: Q_h(H) \approx A(H)$$

such that if $q \in Q_h(H)$ and $\phi(q) = A$ we have

$$q(x) = \langle A(x), x \rangle \qquad \text{for all } x \in H.$$

Moreover,

1. $q \in Q_h(H)$ is non-degenerate if and only if $\phi(q)$ is an isomorphism.
2. $q \in Q_h(H)$ is positive definite if and only if $\phi(q) \in A^+(H)$.

In matrix terms, if q is a Hermitian quadratic form on the complex inner-product space $(H, \langle\ ,\ \rangle)$ and we take a fixed orthonormal basis on H, then there exists a unique Hermitian matrix $[A]$ such that

$$q(x) = [x]^t [A] [x] \qquad \text{for all } x \in H.$$

Just as for the real case, Theorem 1.16.4 implies

Theorem 1.16.20

Let q be a Hermitian quadratic form on the complex inner-product space $(H, \langle\ ,\ \rangle)$. Then we may find an orthonormal basis of H and *real* numbers $\mu_1, \ldots, \mu_n$, where $n =$ dimension (H), such that relative to this basis q has the form

$$q(z) = \sum_{i=1}^{n} \mu_i z_i \bar{z}_i, \qquad z = (z_1, \ldots, z_n) \in H.$$

Further, q is non-degenerate if and only if all the constants μ_i are non-zero, and q is positive definite if and only if all the constants μ_i are strictly positive.

Finally, the number of μ_i which are zero, strictly positive, strictly negative, are invariants of q and we may find an inner product and orthonormal basis for H such that, relative to this basis, q has the form

$$q(z) = \sum_{i=1}^{p} |z_i|^2 - \left(\sum_{j=p+1}^{p+m} |z_j|^2 \right).$$

Exercises

1. Let $A \in L(l^2, l^2)$ be positive definite according to Definition 1.16.1. Show that A may have dense image in l^2 and not be an isomorphism. Show that if A satisfies the additional condition that there exists a $c > 0$ such that $\langle A(x), x\rangle \geqslant c\langle x, x\rangle$ for all $x \in l^2$, then A is an isomorphism (with continuous inverse). (*Hint*: Prove $\operatorname{Im}(A)$ is closed.)
2. Let $(H, \langle\ ,\ \rangle)$ be a complex Hilbert space. We say that a map $A \in L(H, H)$ is skew Hermitian if $\bar{A}^* = -A$. Show that every endomorphism of H may be written as the sum of a Hermitian and a skew-Hermitian map. Show also that if A is skew Hermitian, all eigenvalues of A are purely imaginary. Show also that $A \in L(H, H)$ is skew Hermitian if and only if iA is Hermitian.
3. Using the results of Exercise 2, show that if H is finite dimensional, one can prove a version of Theorem 1.16.4 for skew-Hermitian maps.
4. Using the results of Exercise 1, obtain the generalization of Proposition 1.16.3 to complex Hilbert spaces.
5. Prove Lemma 1.16.5 and verify the other assumptions made in the text about complexification.
6. Write out the details of the proof of Theorem 1.16.6.
7. Consider the linear maps from $\mathbf{R}^3$ to $\mathbf{R}^3$ whose matrices, with respect to the standard basis of $\mathbf{R}^3$, are given by

$$\text{(a)} \begin{bmatrix} 0 & 0 & 0 \\ 0 & 2 & 1 \\ 0 & 1 & 2 \end{bmatrix} \qquad \text{(b)} \begin{bmatrix} 2 & 1 & 1 \\ 1 & 4 & 1 \\ 1 & 1 & 2 \end{bmatrix} \qquad \text{(c)} \begin{bmatrix} 0 & 1 & 3 \\ 1 & 0 & 2 \\ 3 & 2 & 0 \end{bmatrix}.$$

   In each case find the eigenvalues and associated eigenspace of the linear maps.
8. If the matrices of Exercise 7 define quadratic forms according to the pattern of Proposition 1.16.9, find in each case the constants μ_1, μ_2, μ_3 given by Theorem 1.16.10 and the corresponding expression for the quadratic form. State also the rank and index of the form.
9. Prove Proposition 1.16.9.
10. Complete the proof of Theorem 1.16.12.
11. (Simultaneous reduction of quadratic forms.) Show that Theorem 1.16.10 may be interpreted in the following way: Given two quadratic forms on a finite-dimensional vector space E, one of which is positive definite, we may find a basis for E with respect to which the non-degenerate form may be expressed as

$$\sum_{i=1}^{n} x_i^2,$$

   whilst the other form may be expressed as the sum

$$\sum_{i=1}^{n} \mu_i x_i^2,$$

   for some real constants μ_i, $1 \leqslant i \leqslant n = \text{dimension}\,(E)$.

12. Show that if J is a complex structure on the vector space E, then $\bar{\bar{J}} = J$ and $\bar{\bar{E}} = E$. Verify also that

$$L(\bar{E}, \bar{F}) \approx \overline{L(E, F)} \qquad \text{(isomorphism of complex vector spaces).}$$

13. Taking the standard complex inner product on $\mathbf{C}^2$, find the orthonormal basis of $\mathbf{C}^2$ for which the following Hermitian quadratic forms take the form given by Theorem 1.16.20:

(a) $[z_1 \quad z_2]\begin{bmatrix} 2 & i \\ -i & -3 \end{bmatrix}\begin{bmatrix} z_1 \\ z_2 \end{bmatrix}$ (b) $[z_1 \quad z_2]\begin{bmatrix} 1 & i \\ -i & 1 \end{bmatrix}\begin{bmatrix} z_1 \\ z_2 \end{bmatrix}$

(c) $[z_1 \quad z_2]\begin{bmatrix} 2 & 1+i \\ 1-i & 3 \end{bmatrix}\begin{bmatrix} z_1 \\ z_2 \end{bmatrix}$

14. Let H be a finite-dimensional real inner-product space. Show, using Theorem 1.16.6, that if $A \in L(H, H)$ is positive definite, then there exists a *unique* positive definite map $B \in L(H, H)$ satisfying $B^2 = A$ (B is called the 'square root' of A). What happens if A is just symmetric? Or if A is Hermitian and H is a complex inner product space?

1.17 Isometries

In this section we shall make a study of linear maps on inner-product spaces which preserve the inner product. As in Section 1.16 on self-adjoint maps, there will be a strong geometric flavour to much of what we do. Unless stated to the contrary, we assume that all inner-product spaces are *finite* dimensional.

Definition 1.17.1

Let $(H, \langle\ ,\ \rangle)$ be a finite-dimensional real inner-product space. An element $A \in L(H,H)$ is called an *orthogonal* linear map if

$$\|A(x)\| = \|x\| \qquad \text{for all } x \in H.$$

Proposition 1.17.2

Let $(H, \langle\ ,\ \rangle)$ be a real inner-product space and let $A \in L(H, H)$. The following conditions on A are equivalent.

1. A is orthogonal.
2. $\langle A(x), A(y)\rangle = \langle x, y\rangle$ for all $x, y \in H$.
3. $A^*.A = I_H$.

Further, if any one of the above conditions is satisfied, we have $\det(A) = \pm 1$, $A.A^* = I_H$ and A is an isomorphism with $A^{-1} = A^*$.

Proof. That (2) and (3) are equivalent is obvious from the definition of A^*. That (1) implies (2) is a consequence of the identity

$$\langle A(x), A(y)\rangle = \tfrac{1}{2}\{\|A(x+y)\|^2 - \|A(x)\|^2 - \|A(y)\|^2\}.$$

Since $\det(A) = \det(A^*)$ it follows that $\det(A)^2 = 1$ and so $\det(A) = \pm 1$. The remaining properties are straightforward consequences of the theory of linear maps between finite-dimensional vector spaces. ∎

Remark. If, in Proposition 1.17.2, H had been infinite-dimensional (1), (2) and (3) would still have been equivalent, but, in general, A would not have been an isomorphism. For example, if we define $U: l^2 \longrightarrow l^2$ by $U(x_1, \ldots,) = (0, x_1, \ldots)$, it is easy to check that U is orthogonal but U is certainly not an isomorphism and $U . U^* \neq I$.

We shall say that an orthogonal linear map is an *isometry* if it is a linear isomorphism. From Proposition 1.17.2 it follows that every orthogonal linear map on a finite-dimensional inner-product space is an isometry. It follows from the closed graph theorem that an orthogonal map on a Hilbert space is an isometry if it is surjective. (For the closed-graph theorem see [19]. Also see Section 3.1 for the correct definition of linear isomorphism.)

Our aim in this section is to obtain a classification of the isometries of finite-dimensional inner-product spaces.

Examples

1. Let $A \in L(\mathbf{R}^2, \mathbf{R}^2)$ be orthogonal. We let S^1 denote the unit circle in $\mathbf{R}^2$, centre the origin. Let e_1, e_2 denote the standard basis elements of $\mathbf{R}^2$. Then, with the notation of Fig. 1.8, we have

$$A(e_1) = (\cos\theta, \sin\theta)$$
$$A(e_2) = (\cos\psi, \sin\psi)$$
$$A(e_1) \perp A(e_2).$$

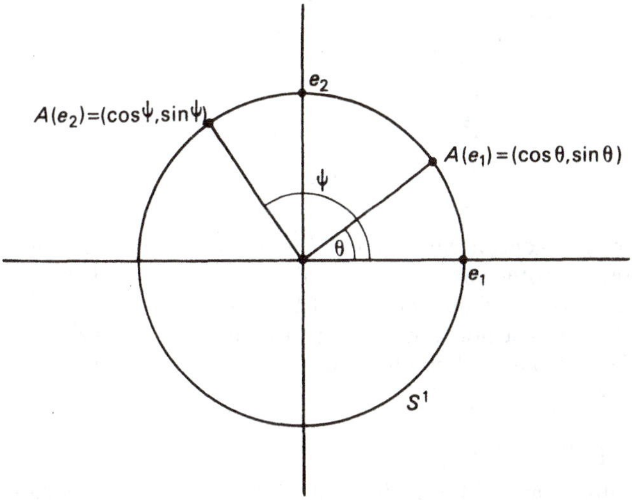

FIG. 1.8

Thus, if the matrix of A, relative to the basis $\{e_1, e_2\}$, is written $[a_{ij}]$, we must have

$$a_{11} = \cos\theta \qquad a_{21} = \sin\theta \qquad a_{12} = \cos\psi \qquad a_{22} = \sin\psi.$$

Since $A(e_1) \perp A(e_2)$, we also have the relation

$$\cos\theta\cos\psi + \sin\theta\sin\psi = 0.$$

Since

$$\cos\theta\cos\psi + \sin\theta\sin\psi = \cos(\theta - \psi),$$

we must have $\psi = \theta + \pi/2$ or $\psi = \theta - \pi/2$. In the first case the matrix of A is of the form

$$\begin{bmatrix} \cos\theta & -\sin\theta \\ \sin\theta & \cos\theta \end{bmatrix}$$

and corresponds to a rotation about the origin through the angle θ. In the second case the matrix of A is of the form

$$\begin{bmatrix} \cos\theta & \sin\theta \\ \sin\theta & -\cos\theta \end{bmatrix}.$$

In this case it may easily be verified by direct computation that A corresponds to reflection in the line $\theta/2$ = constant:

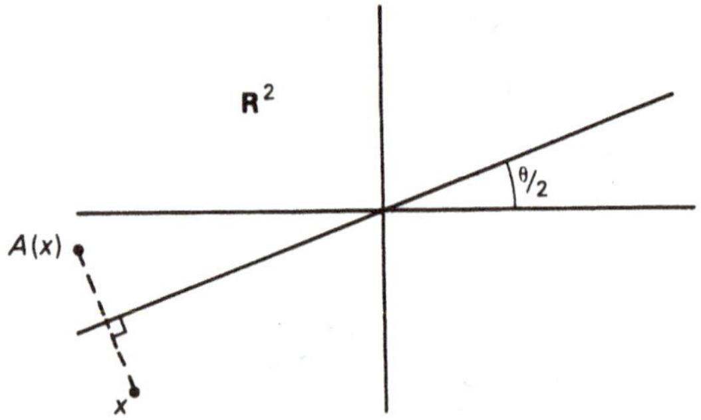

FIG. 1.9

Thus, an orthogonal linear map on $\mathbf{R}^2$ is either a rotation or a reflection. The two cases are distinguished by noting that A is a rotation if and only if $\det(A) = +1$ and a reflection if and only if $\det(A) = -1$.

It should also be noted that if A is a rotation through angle θ, then the matrix of A relative to *any* orthonormal basis of $\mathbf{R}^2$ is of the form

$$\begin{bmatrix} \cos\theta & -\sin\theta \\ \sin\theta & \cos\theta \end{bmatrix}.$$

On the other hand, if A is a reflection, we may choose an orthonormal basis for $\mathbf{R}^2$ such that the matrix of A relative to this basis has the especially simple form

$$\begin{bmatrix} 1 & 0 \\ 0 & -1 \end{bmatrix}.$$

In the sequel we shall always let R_θ denote the unique anticlockwise rotation of the plane through θ, $\theta \in [0, 2\pi)$.

We also note that the set $\{R_\theta : \theta \in [0, 2\pi)\}$ forms a multiplicative group under the operation of composition. Indeed, $(R_\theta)^{-1} = R_{-\theta}$ and

$$R_\theta \,.\, R_\phi = R_{\theta+\phi} \qquad \text{for all } \theta, \phi \in [0, 2\pi).$$

2. Let P denote a fixed plane through the origin of $\mathbf{R}^3$. Suppose that $S:\mathbf{R}^3 \longrightarrow \mathbf{R}^3$ is an orthogonal map that leaves P invariant: $S(P) = P$. We claim that S leaves the line $L = P^{\perp}$ invariant. This follows since $\langle S(x), S(y)\rangle = \langle x, y\rangle$ for all $x, y \in \mathbf{R}^3$ and, in particular, when $x \in P$ and $y \in L$. Let us start by considering the action of

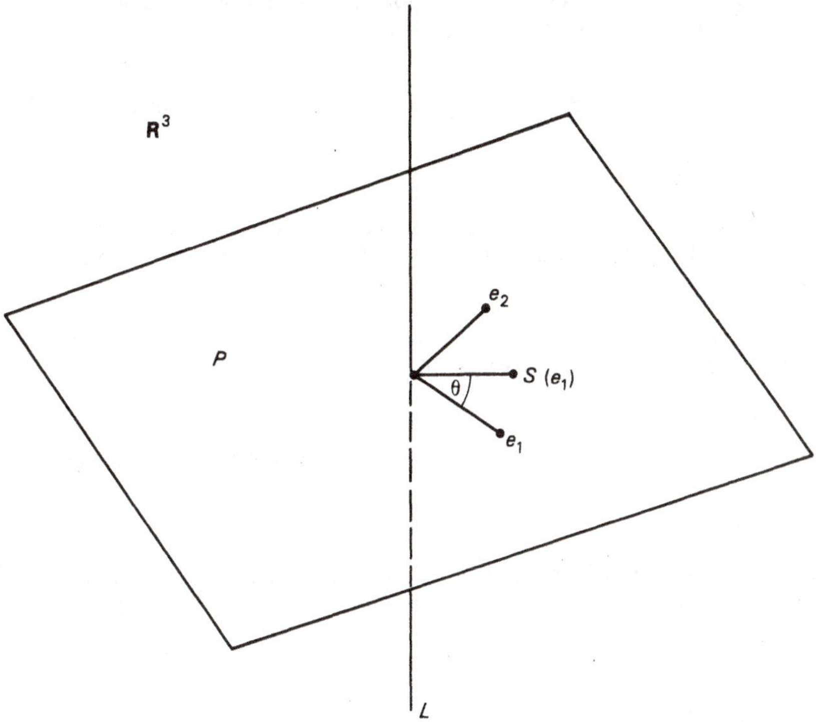

FIG. 1.10

$S|P{:}P \longrightarrow P$. Suppose that $\det(S|P) = +1$. Since $\|S(x)\| = \|x\|$ for all $x \in P$, $S|P$ acts like a rotation of P into itself. Let $\{e_1, e_2\}$ be an (ordered) orthonormal basis for P and let θ be the angle between e_1 and $S(e_1)$ measured anticlockwise in the sense given by the basis $\{e_1, e_2\}$ ($\{e_2, e_1\}$ gives the opposite sense). The matrix of $S|P$ relative to the basis $\{e_1, e_2\}$ is

$$\begin{bmatrix} \cos\theta & -\sin\theta \\ \sin\theta & \cos\theta \end{bmatrix}.$$

Let us now consider the action of $S|L$. Since $\det(S|L) = \pm 1$ it follows that $S|L = \pm I_L$. If $S|L = I_L$, then S acts on $\mathbf{R}^3$ as an anticlockwise rotation of $\mathbf{R}^3$ about the axis L through the angle θ. (The sense of the rotation is determined by our choice of basis $\{e_1, e_2\}$.) If $S|L = -I_L$, then S acts like a rotation of $\mathbf{R}^3$ through θ about the axis L followed by a reflection in the plane P.

Let $e \in L$ be of unit length. Then $\{e, e_1, e_2\}$ is an orthonormal basis of $\mathbf{R}^3$ and the matrix of S is given by

$$\begin{bmatrix} \pm 1 & 0 & 0 \\ 0 & \cos\theta & -\sin\theta \\ 0 & \sin\theta & \cos\theta \end{bmatrix}.$$

Notice that the determinant of this matrix is positive or negative according as to whether we have a pure rotation or rotation with reflection. We may make a similar analysis of the case when $\det(S \mid P) = -1$, and we shall leave this to the reader. Notice how the choice of an orthonormal basis in P determines the sense of rotation. A different choice of orthonormal basis gives the same sense of rotation if and only if one basis can be rotated (without reflection) into the other.

Our aim is now to extend the above results on orthogonal maps of $\mathbf{R}^2$ and $\mathbf{R}^3$ to all isometries of $\mathbf{R}^n$, $n \geqslant 1$.

Proposition 1.17.3

Let $(H, \langle\, ,\rangle)$ be a finite-dimensional real inner-product space and let $A \in L(H, H)$ be an isometry. Then every characteristic root of A has absolute value 1 and complex roots occur in conjugate pairs.

Proof. A induces a map ${}_{c}A : {}_{c}H \longrightarrow {}_{c}H$ which preserves ${}_{c}\langle\, ,\rangle$. If $\lambda \in \mathrm{ch}\,({}_{c}A)$ and x is a corresponding non-zero eigenvector, we have

$$ {}_{c}\langle {}_{c}A(x), {}_{c}A(x)\rangle = \lambda\bar{\lambda}\,{}_{c}\langle x, x\rangle $$

and so $|\lambda|^2 = 1$. Since $\mathrm{ch}\,(A) = \mathrm{ch}\,({}_{c}A)$ the result follows. We leave full details to the exercises. ∎

Theorem 1.17.4

Let A be an isometry of the finite-dimensional real inner-product space $(H, \langle\, ,\rangle)$. Then there exists a unique orthogonal decomposition $F_1 \oplus F_2 \oplus F_3$ of H which is left invariant by A such that

1. $A \mid F_1 = I_{F_1}$.
2. $A \mid F_2 = -I_{F_2}$.
3. $F_3 = \oplus_{j=1}^{k} P_j$, where P_j is a two-dimensional subspace of A, left invariant by A, such that $A \mid P_j$ is a non-trivial rotation, $1 \leqslant j \leqslant k$.

Finally, the dimensions of F_1, F_2 and F_3 are given in terms of the characteristic roots of A by

$\dim(F_1)$ = multiplicity of the characteristic root 1 of A.

$\dim(F_2)$ = multiplicity of the characteristic root -1 of A.

$\dim(F_3)$ = the number of non-real characteristic roots of A, counting multiplicities.

Proof. We define

$$ F_1 = \{x \in H : A(x) = +x\} $$

$$ F_2 = \{x \in H : A(x) = -x\} $$

$$ F_3 = (F_1 \oplus F_2)^{\perp}. $$

F_1 and F_2 are clearly A-invariant subspaces of H. Since A maps orthogonal vectors to orthogonal vectors, F_3 is also A-invariant.

Let us set $A_3 = A \mid F_3$. If $\dim(F_3) = 0$ we are done. If $\dim(F_3) = 1$ the characteristic root of A_3 would be real (Proposition 1.17.3) and hence ± 1. This would contradict the fact that $F_3 \perp (F_1 \oplus F_2)$. We may assume, therefore, that $\dim(F_3) \geqslant 2$. From the example preceding Definition 1.16.7, A_3 has at least a one- or two-dimensional invariant subspace. As for the case when $\dim(F_3) = 1$, A_3 cannot have any one-dimensional invariant subspaces. Suppose then that A_3 leaves the two-dimensional subspace P_1 invariant. But, since A_3 is an isometry, $P_1^{\perp}$ is also invariant.

An easy inductive argument gives the asserted orthogonal decomposition of F_3. ∎

Corollary 1.17.4.1

Let A be an isometry of the finite-dimensional real inner-product space H. Then we may find an orthonormal basis for H such that the matrix of A takes the form

$$\begin{bmatrix} [I_p] & & & & \\ & [-I_q] & & 0 & \\ & & [R_{\theta_1}] & & \\ & 0 & & \ddots & \\ & & & & [R_{\theta_k}] \end{bmatrix}.$$

Here p = multiplicity of the characteristic root 1 of A; q = multiplicity of the characteristic root -1 of A. If we denote the complex characteristic roots of A by $\lambda_1, \bar{\lambda}_1, \ldots, \lambda_k, \bar{\lambda}_k$, then the angles θ_j are determined by the relations

$$e^{i\theta_j} = \lambda_j, \qquad 1 \leqslant j \leqslant k.$$

Proof. Follows from Theorem 1.17.4 together with the matrix form for isometries of $\mathbf{R}^2$ given by a previous example. ∎

Definition 1.17.5

If, with the notation of Theorem 1.17.4, $\dim(F_1) = \dim(H) - 2$ and $\dim(F_3) = 2$ we shall say that A is a *simple rotation.* Geometrically, a simple rotation of H consists of a rotation of a fixed plane of H into itself which keeps the orthogonal complement of the plane fixed.

If we have $\dim(F_1) = \dim(H) - 1$ and $\dim(F_2) = 1$ then we shall say that A is a *simple reflection.* Geometrically, A reflects vectors in the hyperplane F_1.

Remarks

1. It follows from Theorem 1.17.4 that every isometry of H is the composite of simple reflections and simple rotations.
2. It follows from Theorem 1.17.4 that any isometry of $\mathbf{R}^3$ is either a simple rotation or the composite of a reflection with a simple rotation.

We wish to study simple rotations further. Let us, for the moment, remove our assumptions on the finite dimensionality of H.

Definition 1.17.6

Let $(H, \langle\ ,\ \rangle)$ be a real Hilbert space and let x and y be any two non-zero vectors of H. We define the angle between x and y to be the unique solution in $[0, \pi]$ of the equation

$$\cos\theta = \langle x, y\rangle / \|x\|\,\|y\|.$$

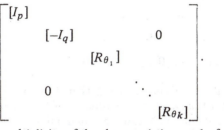

Remark. Notice that the angle between x and y is equal to the angle between y and x.

Examples

1. Definition 1.17.6 generalizes the definition of angle in the plane. To see this, suppose that x_1 and x_2 are unit vectors in $\mathbf{R}^2$ with respective coordinates $(\cos\phi_1, \sin\phi_1)$ and $(\cos\phi_2, \sin\phi_2)$, where $\phi_1, \phi_2 \in [0, 2\pi)$. Then the angle between x_1 and x_2 is given by

$$\begin{aligned} &|\phi_1 - \phi_2| && \text{if } |\phi_1 - \phi_2| \leqslant \pi \\ 2\pi - &|\phi_1 - \phi_2| && \text{if } |\phi_1 - \phi_2| \geqslant \pi \end{aligned}$$

Now according to Definition 1.17.6 this angle is given by

$$\cos\theta = \cos\phi_1 \cos\phi_2 + \sin\phi_1 \sin\phi_2 = \cos(\phi_1 - \phi_2).$$

The unique solution of this equation in $[0, \pi]$ is given precisely by the above relations.

2. Let x and y be unit vectors in the real Hilbert space H. Then if the angle between x and y is θ and $\theta \neq \pi$, there exists a unique simple rotation of H through angle θ which maps x into y. If $\theta = \pi$ and $\dim(H) > 2$, we may find infinitely many such simple rotations which take x into y.

Finally we remark that the theory of isometries of real vector spaces has its counterpart for complex vector spaces. We leave to the exercises the details of the theory.

Exercises

1. Complete the proof of Proposition 1.17.3.
2. Let P and Q be linearly independent points on the unit sphere S^2 of $\mathbf{R}^3$. Let A be a linear endomorphism of $\mathbf{R}^3$ with positive determinant satisfying $A(S^2) \subset S^2$. Show that $\{A(P), A(Q)\}$ determines A uniquely.
3. Show that isometries preserve angles between vectors.
4. Let A be a linear isometry of $\mathbf{R}^3$ with matrix

$$\begin{bmatrix} \sqrt{3}/2 & 1/2\sqrt{2} & -1/2\sqrt{2} \\ 0 & 1/\sqrt{2} & 1/\sqrt{2} \\ \frac{1}{2} & -\sqrt{3}/2\sqrt{2} & \sqrt{3}/2\sqrt{2} \end{bmatrix}.$$

Find an orthonormal basis of $\mathbf{R}^3$ such that the matrix of A has the form given by Corollary 1.17.4.1.

5. Let $\mathscr{E} = (e_i)_{i=1}^n$ be an orthonormal basis of the inner-product space H. Show that if A is an isometry of H, then both the rows and columns of the matrix of A, relative to $\mathscr{E}$, define orthonormal bases of $\mathbf{R}^n$. Conversely, find necessary and sufficient conditions on an $n \times n$ matrix in order that it may define an isometry of $\mathbf{R}^n$.
6. Let $O(n)$ denote the set of all isometries of $\mathbf{R}^n$ and $SO(n)$ the subset of $O(n)$ consisting of isometries with positive determinant. Show that $O(n)$ has the structure of a group if we define composition to be composition of linear maps. Show also that $SO(n)$ is a subgroup of $O(n)$ and that $O(n)/SO(n) \cong \mathbf{Z}_2$ (cyclic group of order 2). ($O(n)$ and $SO(n)$ are respectively called the *orthogonal* and *special orthogonal* groups of order n.)

7. We have a natural inclusion of $\mathbf{R}^n$ into $\mathbf{R}^{n+1}$ defined by mapping $(x_1, \ldots, x_n)$ to $(x_1, \ldots, x_n, 0)$. Show that this inclusion induces inclusions of groups $O(n) \longrightarrow O(n+1)$ and $SO(n) \longrightarrow SO(n+1)$. Let x be any non-zero vector in $\mathbf{R}^{n+1}$ and define

$$SO(n+1)_x = \{A \in SO(n+1) : A(x) = x\}.$$

Show that $SO(n+1)_x$ is isomorphic, as a group, to $SO(n)$. (*Hint:* Start by proving the result when $x = (0, 0, 0, \ldots, 0, 1)$.) Hence show that we may put the quotient $SO(n+1)/SO(n)$ into 1 : 1 correspondence with the unit sphere S^n of $\mathbf{R}^{n+1}$.

8. Using Corollary 1.17.4.1, show that $O(n)$ contains an Abelian subgroup isomorphic to the product of $SO(2)$ k times, $\times^k SO(2)$, where $k = n/2$, if n is even and $k = (n-1)/2$ if n is odd.

   Could $O(n)$ contain an Abelian subgroup isomorphic to $\times^{k+1} SO(2)$? (k is called the *rank* of the group $O(n)$.)

9. Let $(H, \langle\ ,\ \rangle)$ be a finite-dimensional complex inner-product space. We say that an endomorphism A of H is *unitary* if $\|A(x)\| = \|x\|, \forall x \in H$. Generalize the theorems and definitions of Section 1.17 to unitary operators. In particular, show that every unitary operator on H has a *diagonal* matrix relative to a suitable orthonormal basis of H.

10. Let us denote the space of square-integrable complex-valued functions on $[0, 1]$ by $C_{\mathbf{C}}[0, 1]$. Thus, if $f, g \in C_{\mathbf{C}}[0, 1]$ we have

$$\langle f, g\rangle = \int_0^1 f(t)\overline{g(t)}\,\mathrm{d}t.$$

Show that the map $U: C_{\mathbf{C}}[0, 1] \longrightarrow C_{\mathbf{C}}[0, 1]$ defined by

$$U(f)(t) = \int_0^1 f(s)e^{its}\,\mathrm{d}s, \qquad f \in C_{\mathbf{C}}[0, 1],$$

is unitary (i.e. preserves the inner product).

(U is an example of a 'Fourier transform' and it may be shown to extend to an isomorphism of $\widetilde{C[0, 1]}$.)

1.18 The Hahn–Banach theorem

Definition 1.18.1

Let F be a closed subspace of the normed vector space E. A topological complement of F in E is a *closed* vector subspace G of E such that

1. $F \cap G = \{0\}$.
2. $F \oplus G = E$.

Remark. In general, a closed subspace of a normed or Banach space need not have a topological complement. Indeed it has been shown that if every closed subspace of a Banach space $(E, \|\ \ \|)$ has a topological complement, then there exists an inner product on E whose associated norm is equivalent to $\|\ \ \|$.

The reader may recall that in Theorem 1.12.12 we showed that a continuous linear form defined on a closed subspace of a Hilbert space had a (unique) extension as a *continuous* linear form to the whole space. The Hahn–Banach theorem tells us that we have such extensions also for normed vector spaces. The proof of Theorem 1.12.12 depended on the fact that a closed subspace of a Hilbert space always has a topological complement. For the case of normed or Banach spaces we no longer can make use of the existence of a topological complement: as pointed out in the above remark, we have to assume that there are subspaces with no topological complement. Even when the subspace has a topological complement we do not have any hope of establishing any type of *unique* extension. We can, however, control the norm of the extension and in fact require that it be as small as possible. Suppose that H and G are two Hilbert spaces and that $A: F \longrightarrow G$ is a continuous linear map defined on the closed subspace F of H. Then A has extension to H. Indeed the map $A \,.\, p_F$ defines just such an extension. We have no such theorem for normed vector spaces. Indeed suppose that F is a closed subspace of the normed vector space E. If $T: E \longrightarrow F$ was an extension of the identity map $I: F \longrightarrow F$, then the kernel of T, Ker (T), would be a topological complement of F in E. This cannot happen in general, since F need not have a topological complement.

Theorem 1.18.2 (Hahn–Banach theorem)

Let E be a real normed vector space and let F be a vector subspace of E. Then, given any continuous linear form $f \in F^*$, there exists a form $\hat{f} \in E^*$ such that

1. $\hat{f} \mid F = f$.
2. $\|\hat{f}\| = \|f\|$.

Proof. We defer the proof to the appendix at the end of this section. However, we note in passing that we may always assume that F is a closed subspace of E since f extends uniquely to $\bar{F}$. Similarly it is sufficient to prove the result for E a Banach space since $E^* = \hat{E}^*$ (Section 1.11). Finally, we remark that (2) of the theorem is not even trivial for *finite*-dimensional vector spaces. ∎

In the rest of this section we wish to give a number of important corollaries of the Hahn–Banach theorem.

Corollary 1.18.2.1

Let E be a normed vector space. Then $E^* = \{0\}$ if and only if $E = \{0\}$. In addition, if $x, y \in E$ satisfy

$$f(x) = f(y) \qquad \text{for all } f \in E^*,$$

then $x = y$. In other words, there are 'plenty' of continuous linear forms on a normed vector space.

Proof. It is clearly sufficient to show that if $x \in E \setminus \{0\}$, then $\exists f \in E^*$ such that $f(x) \neq 0$. Let F be the one-dimensional subspace of E spanned by x and define $f': F \longrightarrow \mathbf{R}$ by the condition

$$f'(tx) = t, \qquad t \in \mathbf{R}.$$

By the Hahn–Banach theorem, f' extends to $f \in E^*$. In particular, $f(x) = f'(x) = 1 \neq 0$. ∎

Remark. It should be carefully noted that if E is infinite dimensional we have no means of showing that E has *any* continuous linear forms (apart from the trivial form) without using the Hahn–Banach theorem. In point of fact the first condition of the Hahn–

Banach theorem only uses the fact that the origin has a base of *convex* neighbourhoods and this part of the theorem can in fact be generalized to a much larger class of topological vector spaces.

Corollary 1.18.2.2

Every finite-dimensional subspace of a normed vector space E has a topological complement.

Proof. Let F be a finite-dimensional subspace of E. We remark that it follows from Corollary 1.10.3.1 that F is a closed subspace. Suppose first that $\dim(F) = 1$ and that F has basis $\{e\}$. Let $f \in F^*$ be any non-zero form. By the Hahn–Banach theorem, f extends to $\hat{f} \in E^*$. Set $G = \operatorname{Ker}(\hat{f})$. Since $\hat{f}$ is continuous, G is closed. Further, any $x \in E$ may be written as a sum

$$x = \hat{f}(x)e/\|\hat{f}(e)\| + (x - \hat{f}(x)e/\|\hat{f}(e)\|).$$

The first term in this sum lies in F whilst the second lies in G. Hence G is a topological complement to F.

More generally, suppose that F has basis $(e_i)_{i=1}^n$. Let $(e_i^*)_{i=1}^n$ be the corresponding dual basis. Each of the forms e_i^* extends to a form $f_i \in E^*$. We define

$$G = \bigcap_{i=1}^{n} \operatorname{Ker}(f_i).$$

We leave it to the reader to verify that G is a complement to F. ■

Corollary 1.18.2.3

Let E and F be normed vector spaces, and suppose that F is finite dimensional. Then, if G is a closed vector subspace of E, any continuous linear map $T: G \longrightarrow F$ has an extension to E as a continuous linear map.

Proof. Let $(b_i)_{i=1}^n$ be a basis for $\operatorname{Im}(T)$. We may pick n linearly independent vectors $f_1, \ldots, f_n$ in G such that

$$T(f_i) = b_i, \qquad i = 1, \ldots, n.$$

Then the set $(f_i)_{i=1}^n$ forms a basis for a subspace F' of G. We let $(f_i^*)_{i=1}^n$ denote the dual basis of F'^*. We claim that $\operatorname{Ker}(T)$ is a topological complement to F' in G. Indeed, if $x \in G$, we may write x as the sum $x' + x''$, where

$$x' = (T|F')^{-1}(T(x)) \in F',$$

$$x'' = x - (T|F')^{-1}(T(x)) \in \operatorname{Ker}(T).$$

Each of the forms f_i^* extends naturally to a form on G if we set

$$\tilde{f}_i^*(x) = f_i^*(x'), \qquad i = 1, \ldots, n.$$

It follows that we may write T in the form

$$T(x) = \sum_{i=1}^{n} b_i \tilde{f}_i^*(x), \qquad x \in G.$$

The corollary now follows if we apply the Hahn–Banach theorem to each of the forms $\tilde{f}_i^*$ to define a linear map $\hat{T}: E \longrightarrow E$ by means of the above formula. ■

Corollary 1.18.2.4

Let F be a closed vector subspace of the complex normed vector space E. Then, given any continuous complex linear form $f \in F^*$, there exists a complex form $\tilde{f} \in E^*$ such that

1. $\tilde{f} | F = f$.
2. $\|\tilde{f}\| = \|f\|$.

Proof. Since $\mathbf{C}$ is isomorphic as a real vector space to $\mathbf{R}^2$, it follows from Corollary 1.18.2.3 that we can find a real linear form $\hat{f}: E \longrightarrow \mathbf{C}$ such that

1. $\hat{f} | F = f$.
2. $\|\hat{f}\| = \|f\|$.

We now define a map $\tilde{f}: E \longrightarrow \mathbf{C}$ by the condition

$$\tilde{f}(x) = \frac{\hat{f}(x) - i\hat{f}(ix)}{2}, \qquad x \in E.$$

Then the reader may check that $\tilde{f}$ is complex linear and that $\tilde{f} | F = f$. Since $\tilde{f} | F = f$, we must have $\|\tilde{f}\| \geqslant \|f\|$. On the other hand

$$\|\tilde{f}\| \leqslant \tfrac{1}{2}\Big(\sup_{\|x\| \leqslant 1} \|\hat{f}(x)\| + \sup_{\|x\| \leqslant 1} \|i\hat{f}(ix)\|\Big) = \|\hat{f}\| = \|f\|$$

and so we have equality. ■

Corollary 1.18.2.5

Let E and F be normed vector spaces. Then F is Banach if and only if $L(E, F)$ is Banach.

Proof. We showed in Theorem 1.9.7 that F Banach implies $L(E, F)$ Banach. We now prove the converse. Suppose, therefore, that $(e_n)_{n=1}^{\infty}$ is a Cauchy sequence in F and that $L(E, F)$ is complete. Let e be a unit vector in E and let K_e be the one-dimensional subspace of E spanned by e. Applying Corollary 1.18.2.2 we may assume that K_e has a topo logical complement, say L. For each $n = 1, 2, \ldots$ we define the continuous linear map $S_n: E \longrightarrow F$ by

$$S_n(te + y) = te_n, \qquad t \in K, \quad y \in L.$$

We leave it to the reader to verify that $(S_n)_{n=1}^{\infty}$ is a Cauchy sequence in $L(E, F)$ and that, if we denote the limit of $(S_n)_{n=1}^{\infty}$ by S, $\lim_{n\to\infty} e_n = S(e)$. ■

Corollary 1.18.2.6

Let E be a normed vector space with dual E^*. Then the bilinear pairing

$$E \times E^* \longrightarrow K; (e, e^*) \longmapsto e^*(e),$$

has norm equal to 1.

Proof. Left to the reader. ■

Exercise

1. Let E^{**} denote the dual of the dual, E^*, of the normed vector space E. Show that the natural map $\phi: E \longrightarrow E^{**}$ defined by

$$\phi(x)(f^*) = f^*(x) \qquad \text{for all } x \in E \text{ and } f^* \in E^*$$

is injective.

Show that ϕ is an isomorphism if (a) E is finite dimensional, (b) E is $(l^p, \| \ \|_p)$ $1 < p < \infty$.

Appendix: Proof of the Hahn–Banach theorem

For the proof of the Hahn–Banach Theorem we need *Zorn's lemma.*

Definition

Let E be a partially ordered set and let $x \in E$. We say x is *maximal* in E if and only if $y \in E$ and $y \geqslant x$ implies $y = x$.

We shall say that E is *inductive* if every subset C of E on which there is a total ordering has an upper bound in E. That is, $\exists\, z \in E$ such that $y \leqslant z$ for all $y \in C$.

Zorn's lemma states that

If E is an inductive partially ordered set, then E contains a maximal element.

Zorn's lemma is equivalent to either of the axiom of choice or the well-ordering theorem. We shall assume Zorn's lemma here without further comment and refer the interested reader to [11], especially Chapter 0.

Theorem (Hahn–Banach theorem)

Let E be a real normed vector space and let M be a vector subspace of E. Suppose that $f \in M^*$. Then $\exists\, F \in E^*$ such that

1. $\|F\| = \|f\|$.
2. $F|M = f$.

Proof. We divide the proof of the theorem into two steps.

Step 1

Assume that $M \neq E$ and pick $x \notin M$. Let N denote the vector subspace of E spanned by x and M. That is,

$$N = \{y + tx : y \in M, t \in \mathbf{R}\}.$$

Let $y, z \in M$. Then

$$f(z) - f(y) = f(z - y) \leqslant \|f\|\,\|z - y\| \leqslant \|f\|\,\|x + z\| + \|f\|\,\|-x - y\|.$$

Hence for all $y, z \in M$ we have the inequality

$$-f(y) - \|f\|\,\|-x - y\| \leqslant -f(z) + \|f\|\,\|x + z\|.$$

It follows that

$$\sup_{y \in M} \{-f(y) - \|f\|\,\|-x - y\|\} \leqslant \inf_{z \in M} \{-f(z) + \|f\|\,\|x + z\|\}.$$

Choose $r \in \mathbf{R}$ such that

$$\sup_{y \in M} \{-f(y) - \|f\| \|-x-y\|\} \leqslant r \leqslant \inf_{z \in M} \{-f(z) + \|f\| \|x+z\|\}.$$

Any point $u \in N$ may be written uniquely in the form $y + tx$, for some $t \in \mathbf{R}$ and $y \in M$. We define $F \in N^*$ by

$$F(u) = f(y) + tr.$$

Clearly $F|M = f$. We claim that $|F(u)| \leqslant \|f\| \|u\|$ for all $u \in N$. Now $u = y + tx$. If $t = 0$, $u \in M$ and there is nothing to prove. Suppose first that $t > 0$:

$$|F(u)| = |f(y) + tr| = t|f(t^{-1}y) + r|, \qquad \text{since } t > 0.$$

By construction $r \leqslant -f(t^{-1}y) + \|f\| \|t^{-1}y + x\|$, and substituting, we obtain

$$|F(u)| \leqslant t(\|f\| \|t^{-1}y + x\|) = \|f\| \|u\|, \qquad \text{since } t > 0.$$

On the other hand if $t < 0$, we have

$$|F(u)| = |-t(-f(t^{-1}y) - r)|.$$

By construction, $r \geqslant -f(t^{-1}y) - \|f\| \|-t^{-1}y - x\|$, and substituting, we obtain the result in this case also.

Hence we have constructed an extension $F \in N^*$ of $f \in M^*$ and $\|F\| = \|f\|$.

Step 2

We say that a continuous linear form g defined on a subspace $M(g)$ of E is an *admissible extension of f* if

$$M \subset M(g);\ g|M = f;\ \|g\| = \|f\|.$$

Let Φ denote the set of admissible extensions of f. We define a partial order $\leqslant$ on Φ as follows:

Say $g \leqslant h$ if and only if $M(g) \subset M(h)$ and $h|M(g) = g$.

We claim that every totally ordered subset of Φ has an upper bound in Φ. Let C be a totally ordered subset of Φ. Let $N = \bigcup_{g \in C} M(g)$. If $x \in N, \exists g \in C$ such that $x \in M(g)$. Define $G(x) = g(x)$. This definition is unambiguous since if $x \in M(g')$, $g(x) = g'(x)$. Then G is clearly an admissible extension of f and $g \leqslant G$ for all $g \in C$. Hence G is an upper bound for Φ.

Therefore, by Zorn's lemma, there exists a maximal element $F \in \Phi$. We claim that $M(F) = E$. If not, pick $x \notin M(F)$ and apply step 1 of the proof. This gives a proper extension of F which is an admissible extension of f and hence in Φ. Contradiction. Therefore $M(F) = E$ and F satisfies the conditions of the theorem. ∎

Miscellaneous exercises

1. Let E and F be Banach spaces. Suppose that $(e_i^*)_{i=1}^{\infty}$ and $(f_i)_{i=1}^{\infty}$ are sequences in E^* and F respectively and that the following relation is satisfied:

$$\sum_{i=1}^{\infty} \|e_i^*\| \|f_i\| < \infty.$$

Show that the expression

$$T(x) = \sum_{i=1}^{\infty} e_i^*(x) f_i$$

defines a unique continuous linear map from E to F of norm less than or equal to

$$\sum_{i=1}^{\infty} \| e_i^* \| \| f_i \|.$$

2. Let $L_N(E, F)$ denote the set of all linear maps from E to F that can be expressed (not necessarily uniquely) in the form given in Exercise 1. Show
   (a) $L_N(E, F)$ is a vector subspace of $L(E, F)$.
   (b) We can define a norm $\| \quad \|_N$ on $L_N(E, F)$ by defining $\| T \|_N$ to be the infimum of the sums $\Sigma_{i=1}^{\infty} \| e_i^* \| \| f_i \|$ taken over all representations of T of the form given in Exercise 1.
   (c) With the norm $\| \quad \|_N$, $L_N(E, F)$ is a Banach space (E and F are assumed to be Banach).

   $L_N(E, F)$ is called the space of *nuclear* linear maps from E to F.
3. Continuing with the assumptions of Exercises 1 and 2, show that if E is finite dimensional, $L_N(E, F) = L(E, F)$. Show also that if $B_r(0)$ is the disc centre 0 radius r in E, then $\overline{T(B_r(0))}$ is a compact subset of F for all $T \in L_N(E, F)$. Deduce that $L(l^2, l^2) \neq L_N(l^2, l^2)$. (Use Riesz' theorem.)
4. Let E and F be Banach spaces and let $L_c(E, F)$ denote the subset of $L(E, F)$ consisting of maps satisfying the following property:

   $\overline{T(B_1(0))}$ is a compact subset of F.

   Show that

   (a) If $T \in L_c(E, F)$, then $T(B)$ is relatively compact for all bounded subsets of E. (Use the continuity of scalar multiplication.)
   (b) $L_c(E, F)$ is a closed vector subspace of $L(E, F)$.
   (c) $L_F(E, F) \subseteq L_N(E, F) \subseteq L_c(E, F)$, where $L_F(E, F)$ denotes the subspaces of $L(E, F)$ consisting of linear maps of finite rank (that is, finite-dimensional image). An element of $L_c(E, F)$ is called a *compact* linear map.
5. Show that if $p > 1$, $l^{p*} \cong l^{p'}$, where $1/p + 1/p' = 1$ (see the exercises at the end of Section 1.7). Deduce that if $p > 1$, l^p is *reflexive*

   $$l^{p**} \cong l^p.$$

CHAPTER 2

Differentiation and calculus on vector spaces

2.1 Introduction

Let us review carefully what is meant by saying that the function $f:\mathbf{R} \longrightarrow \mathbf{R}$ is differentiable at x_0 with derivative $f'(x_0)$. In purely analytic terms we assert the existence of the limit

$$\lim_{x \to x_0} \frac{f(x) - f(x_0)}{x - x_0}$$

and this limit, if it exists, is defined to be the derivative of f at $(x_0, f'(x_0))$. Whilst this definition of derivative is perfectly satisfactory from an *analytic* point of view, it offers little insight into why one should wish to study the derivative of a function. A *geometric* description of the derivative of a function is, however, rather more illuminating (though technically less precise): A function $f:\mathbf{R} \longrightarrow \mathbf{R}$ is differentiable at x_0 if and only if the graph of f has a unique tangent at x_0.

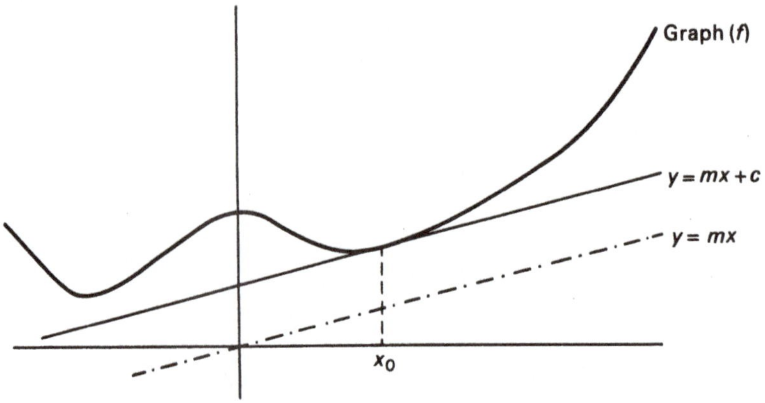

FIG. 2.1

What is meant mathematically by tangency can be answered by analysis; what is meant intuitively by tangency is, however, perfectly clear. In the geometric context, what then should be the derivative of f at x_0? Notice that the tangent line to the graph of f at x_0 defines the graph of an affine linear map $y = mx + c$, and that this map provides an affine linear approximation to f at x_0. A natural candidate for the derivative of f at x_0 might be this affine linear approximation to f. In practice it turns out to be more useful and practical to concentrate on the *linear* part of the affine linear approximation to f and we define the derivative of f at x_0 to be the linear map $y = mx$ (we can easily recover the affine linear approximation from the derivative by noting that $c = f(x_0) - mx_0$).

How do we reconcile these two approaches to differentiation? The analytic point of view defines the derivative as a point in **R**, whilst the geometric approach gives the derivative as an element of $L(\mathbf{R}, \mathbf{R})$. The two approaches are compatible, however, since, for the vector space **R**, we have a canonical isomorphism $L(\mathbf{R}, \mathbf{R}) \approx \mathbf{R}$ defined by mapping $A \in L(\mathbf{R}, \mathbf{R})$ to $A(1) \in \mathbf{R}$. In other words, if $f'(x_0) \in \mathbf{R}$ is the (analytically defined) derivative of f at x_0 it corresponds to the linear map $t \longmapsto f'(x_0)t$ (the geometrically defined derivative of f at x_0).

Now the idea of defining the derivative of a function as a *linear approximation* is fundamental to the proper understanding of the idea of a derivative. To define, mathematically, what is meant by 'linear approximation' or 'tangency' requires analysis, but the geometric concepts are always easy to describe. Let us investigate, for example, how one might attempt to define the derivative of a function $f: \mathbf{R}^2 \longrightarrow \mathbf{R}$ at a point $x_0 \in \mathbf{R}^2$. Suppose the graph of f is of the form displayed in Fig. 2.2

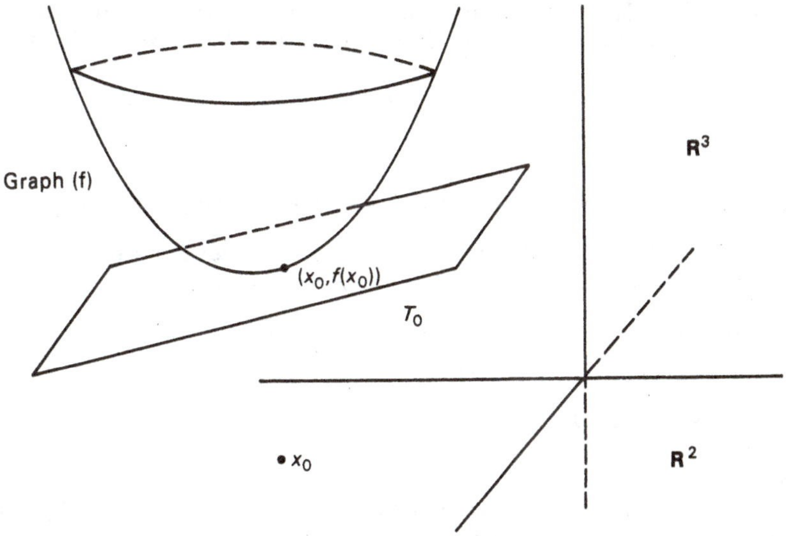

FIG. 2.2

The point $(x_0, f(x_0)) \in \text{graph}(f) \subset \mathbf{R}^3$. Provided the surface in $\mathbf{R}^3$ defined by graph (f) is 'smooth' enough near $(x_0, f(x_0))$ it is reasonable to suppose that there is a unique tangent plane T_0 to the surface at $(x_0, f(x_0))$. Now T_0 is the *graph* of an affine linear map from $\mathbf{R}^2$ to **R**. This affine linear map is clearly the natural candidate for an affine linear approximation to f at x_0. Suppose this affine linear map is given by

$x \longmapsto A(x) + c$, where $A \in L(\mathbf{R}^2, \mathbf{R})$ and $c \in \mathbf{R}$. Just as in the case of maps from $\mathbf{R}$ to $\mathbf{R}$, one then defines the derivative of f at x_0 to be the linear part of this affine linear map, namely A. In other words if f is a map from $\mathbf{R}^2$ to $\mathbf{R}$, we should expect the derivative of f at $x_0 \in \mathbf{R}^2$ to be a point in $L(\mathbf{R}^2, \mathbf{R})$. Referring to the above diagram, the graph of the derivative of f at x_0 is the unique two-dimensional *linear* subspace of $\mathbf{R}^3$ parallel to T_0. That is, the graph of the derivative is the parallel translation of T_0 by $-(x_0, f(x_0))$.

More generally, if $f: \mathbf{R}^m \longrightarrow \mathbf{R}^n$, the graph of f is a subset of $\mathbf{R}^{m+n}$ and the tangent 'plane' to the graph of f at $x_0 \in \mathbf{R}^m$ would consist of an m-dimensional affine linear subspace of $\mathbf{R}^{m+n}$. This subspace would be the graph of the affine linear approximation to f at x_0. The linear part of this affine linear approximation would lie in $L(\mathbf{R}^m, \mathbf{R}^n)$ and is the natural candidate for the derivative of f at x_0.

A natural question to ask is why we should restrict ourselves to *linear* approximations to a function, why not polynomial approximations? Taylor's theorem does in fact give us polynomial approximations to functions from $\mathbf{R}$ to $\mathbf{R}$ and we shall show later in this chapter that we may prove a Taylor's theorem for functions of several variables. In general, however, we shall restrict attention to *linear* approximations. There are several reasons why we do this: Firstly, the linear functions form the smallest class of functions that are closed under the operations of composition and addition and, at the same time, contain enough information to be able to determine properties of functions that they linearly approximate. Secondly, the linear functions are relatively easy to study: the polynomials are not! Finally, the set of linear functions between two vector spaces has a very rich structure; the set of affine linear or polynomial maps does not. A linear map may be described numerically by means of a matrix. We shall see later in this chapter that the components of the matrix of the derivative of a function correspond to the familiar 'partial derivatives' of the function.

2.2 *o* notation

In this section we shall introduce a piece of technical notation that will be useful in the sequel.

Suppose that $(E, \| \ \|)$ and $(F, \| \ \|)$ are normed vector spaces and that U is an open subset of E containing the origin. If $g: U \longrightarrow F$, we shall write

$$g(h) = o(h)$$

if

$$\lim_{h \to 0} g(h)/\|h\| = 0.$$

Remarks

1. Suppose that $g(h) = o(h)$. If we define $r(h) = g(h)/\|h\|$, $h \neq 0$, and $r(0) = 0$, then r is continuous at $h = 0$. Conversely, if r is continuous at $h = 0$, $g(h) = o(h)$.
2. o depends only on the equivalence class of the norms on E and F.

Proposition 2.2.1

Let E and F be normed vector spaces and let $T: E \longrightarrow F$ be a linear map. Then $T(x) = o(x)$ if and only if $T = 0$.

Proof. Clearly if $T = 0$, $T(x) = o(x)$. Suppose then that $T(x) = o(x)$ and that $T \neq 0$.

If $T \neq 0$, $\exists y \in E$ such that $\|T(y)\| = a \neq 0$. Let $b \in \mathbf{R}$, $b \neq 0$, and consider

$$\|T(by)\|/\|by\| = a/\|y\| \neq 0.$$

Now as $b \longrightarrow 0$, $\|by\| \longrightarrow 0$. But $\|T(by)\|/\|by\| = a/\|y\| =$ constant, not equal to zero. This contradicts the assumption that $T(x) = o(x)$. Hence $T = 0$. ∎

Exercises

1. Prove the assertions made in the remarks preceding Proposition 2.2.1.
2. Which of the following functions are $o(x)$?

   (a) $f: \mathbf{R} \longrightarrow \mathbf{R}; x \longmapsto x, x \longmapsto x^2, x \longmapsto x \log(1+x)$.
   (b) $f: \mathbf{R}^2 \longrightarrow \mathbf{R}; (x,y) \longmapsto x+y, (x,y) \longmapsto y \sin x$.
   (c) $f: \mathbf{C} \longrightarrow \mathbf{C}^2; z \longmapsto (z^2, z)$.

3. Let $f: \mathbf{R}^2 \longrightarrow \mathbf{R}$. Suppose, for fixed y, that $f(x,y) = o(x)$. Does this imply that $f(x,y) = o((x,y))$? What if f is independent of y?
4. Prove that if $f(h) = o(h)$ and $g(h) = o(h)$ and $t \in \mathbf{R}$, then $(f + tg)(h) = o(h)$.

2.3 Differentiation of functions on normed vector spaces

First let us recall that a function $f: \mathbf{R} \longrightarrow \mathbf{R}$ is differentiable at $x_0 \in \mathbf{R}$, with derivative $f'(x_0)$, if

$$\lim_{h \to 0} \frac{f(x_0 + h) - f(x_0)}{h} = f'(x_0).$$

Multiplying through by h we may equivalently write this condition as

$$f(x_0 + h) - f(x_0) - hf'(x_0) = r(h),$$

where $r(h) = o(h)$. In other words, f is differentiable at x_0, with derivative $f'(x_0)$, if and only if

$$f(x_0 + h) = f(x_0) + hf'(x_0) + o(h).$$

As remarked in Section 2.1, we may regard $f'(x_0)$ as an element of $L(\mathbf{R}, \mathbf{R})$ by making use of the canonical isomorphism between $\mathbf{R}$ and $L(\mathbf{R}, \mathbf{R})$. Let us denote the linear map in $L(\mathbf{R}, \mathbf{R})$ corresponding to $f'(x_0)$ by Df_{x_0}. Thus $Df_{x_0}(1) = f'(x_0)$. We may then rewrite the above equation to obtain

$$f(x_0 + h) = f(x_0) + Df_{x_0}(h) + o(h).$$

Notice that, since Df_{x_0} is linear,

$$Df_{x_0}(h) = hDf_{x_0}(1) = hf'(x_0).$$

If we set $x_0 + h = x$ and eliminate h from this equation, we find that

$$f(x) = (f(x_0) - Df_{x_0}(x_0)) + Df_{x_0}(x) + o(x - x_0).$$

In other words, the affine linear map $g: \mathbf{R} \longrightarrow \mathbf{R}$ defined by

$$g(x) = (f(x_0) - Df_{x_0}(x_0)) + Df_{x_0}(x), \qquad x \in \mathbf{R},$$

is the affine linear approximation to f at x_0 in the sense that

$$(f - g)(x) = o(x - x_0).$$

Summarizing the above we have the following proposition.

Proposition 2.3.1

A function $f:\mathbf{R} \longrightarrow \mathbf{R}$ is differentiable at x_0 if and only if there exists an affine linear map $g:\mathbf{R} \longrightarrow \mathbf{R}$ such that

$$(f - g)(x) = o(x - x_0).$$

The function g is necessarily unique and, if we write $g = A + c$, where $A \in L(\mathbf{R}, \mathbf{R})$ and c is a constant function, the derivative of f at x_0 is defined to be A, the linear part of g.

We now wish to generalize this proposition to maps between normed vector spaces. First, however, we wish to remark that there are many possible notations used for denoting derivatives. For example: $f'(x)$, $\mathrm{d}f/\mathrm{d}x$, $\mathrm{d}y/\mathrm{d}x$, $Df(x)$, Df_x, etc. In the sequel we shall usually use the notation Df_x to denote the derivative of the function f at x (when regarded as a linear map).

In this chapter we will continue to assume that normed vector spaces are over a fixed field K, where K is either $\mathbf{R}$ or $\mathbf{C}$.

Definition 2.3.2

Let E and F be normed vector spaces and let A be an open subset of E. Suppose that $f:A \longrightarrow F$ is a continuous map and that $x_0 \in A$. We say that f is differentiable at x_0 if there exists a linear map $Df_{x_0}:E \longrightarrow F$ such that

$$f(x_0 + h) - f(x_0) - Df_{x_0}(h) = o(h), \qquad h \in A - x_0.$$

Df_{x_0} is called the *derivative* of f at x_0.

Proposition 2.3.3

With the assumptions and notations of Definition 2.3.2 it follows that Df_{x_0} is both continuous and unique. That is, Df_{x_0} is the unique element of $L(E, F)$ satisfying the equation

$$f(x_0 + h) = f(x_0) + Df_{x_0}(h) + o(h), \qquad h \in A - x_0.$$

Proof. Notice that the derivative of f at x_0 satisfies the equation

$$Df_{x_0}(h) = f(x_0 + h) - f(x_0) + o(h), \qquad h \in A - x_0.$$

Since f is assumed continuous, it follows that $f(x_0 + h) - f(x_0)$ is a continuous function of h. Further, the term $o(h)$ is necessarily continuous at $h = 0$ (in fact $o(h)/\|h\|$ is continuous at $h = 0$, see Section 2.2). Therefore $Df_{x_0}(h)$ is continuous at $h = 0$. By Lemma 1.5.4 it follows that Df_{x_0} is continuous.

For the uniqueness of Df_{x_0} suppose that $T \in L(E, F)$ satisfies the defining equation for Df_{x_0}. Subtracting, we obtain

$$T(h) - Df_{x_0}(h) = o(h) - o(h) = o(h).$$

Therefore, $(T - Df_{x_0})(h) = o(h)$. From Proposition 2.2.1 it follows immediately that $T = Df_{x_0}$. ■

Remarks

1. Differentiability is invariant modulo equivalent norms on E and F. This follows since $o(h)$ is so invariant.
2. Suppose that $f:A \subset E \longrightarrow F$ is differentiable at $x_0 \in A$, with derivative

$Df_{x_0} \in L(E, F)$. The affine linear approximation to f at x_0 is given by the map $g: E \longrightarrow F$ defined by

$$g(x) = (f(x_0) - Df_{x_0}(x_0)) + Df_{x_0}(x), \qquad x \in E.$$

The reader may easily verify that g satisfies the approximation condition: $(g - f)(x) = o(x - x_0)$. This remark, together with Proposition 2.3.3, shows how our definition of derivative generalizes that given in Proposition 2.3.1. Finally we remark that graph(g) is an affine linear subspace of $E \times F$ which is (by definition) the tangent hyperplane to the surface graph(f).

Examples

1. Let U be an open subset of $\mathbf{R}$ and let $f: U \longrightarrow \mathbf{R}$ be differentiable at x_0 in the sense of Definition 2.3.2. Then $\exists Df_{x_0} \in L(\mathbf{R}, \mathbf{R})$ such that

$$f(x_0 + h) - f(x_0) - Df_{x_0}(h) = o(h), h \in U - x_0.$$

Now $h = h1$ and so $Df_{x_0}(h) = hDf_{x_0}(1)$ by the linearity of Df_{x_0}. The above equation may be rewritten as

$$f(x_0 + h) - f(x_0) - hf'(x_0) = o(h),$$

where $f'(x_0) = Df_{x_0}(1) \in \mathbf{R}$. In other words, f is differentiable according to Definition 2.3.2 if and only if it is differentiable in the classical sense described at the beginning of this section.

2. Let $A \in L(E, F)$, where E and F are normed vector spaces. Since A is linear we should expect that the derivative of A (that is, the linear approximation to A) at x is just A. To see that this is so notice that the linearity of A implies the relation

$$A(x + h) - A(x) - A(h) = 0 \qquad \text{for all } h \in E.$$

Since the constant function 0 is obviously $o(h)$, it follows immediately from the definition that the derivative of A at x is A. That is, A is differentiable at all points of E and $DA_x = A$ for all $x \in E$.

3. Let $c \in F$ and let $f: E \longrightarrow F$ denote the constant map taking the value c. Then, for all $x, h \in E$, we have the identity

$$f(x + h) - f(x) - o(h) = 0$$

It follows immediately that f is differentiable at every point of E with derivative the zero linear map.

4. Let E_1, E_2, and F be normed vector spaces. If $T \in L(E_1, E_2; F)$ we may regard T as a continuous map from the normed vector space $E_1 \times E_2$ to F. Fix $x = (x_1, x_2) \in E_1 \times E_2$. Then if $h = (h_1, h_2) \in E_1 \times E_2$ we have the identity

$$T(x + h) = T(x_1 + h_1, x_2 + h_2) = T(x_1, x_2) + T(h_1, x_2) + T(x_1, h_2) + T(h_1, h_2)$$

We define the *linear* map $S \in L(E_1 \times E_2, F)$ by the relation

$$S(h_1, h_2) = T(x_1, h_2) + T(h_1, x_2), \qquad (h_1, h_2) \in E_1 \times E_2.$$

We claim that $DT_x = S$. To prove this we must show that

$$T(h_1, h_2) = o((h_1, h_2)),$$

since

$$T(x+h) - T(x) - S(h) = T(h).$$

Since T is continuous it follows from Theorem 1.5.5 that $\exists K \geqslant 0$ such that $\|T(h_1, h_2)\| \leqslant K \|h_1\| \|h_2\|$. But $\|(h_1, h_2)\| = \max(\|h_1\|, \|h_2\|)$ and so

$$\|T(h_1, h_2)\| \leqslant K \|(h_1, h_2)\|^2.$$

It follows that $T(h_1, h_2) = o((h_1, h_2))$ and so

$$T(x+h) = T(x) + S(h) + o(h), \qquad h \in E_1 \times E_2$$

proving that $DT_x = S$. Hence T is differentiable at every point of E.

5. Let $E_1, \ldots, E_n$ and F be normed vector spaces. If $T \in L(E_1, \ldots, E_n; F)$ we may regard T as a map from the normed vector space $E_1 \times \cdots \times E_n$ to F. Let

$$x = (x_1, \ldots, x_n) \in E_1 \times \cdots \times E_n.$$

Just as in Exercise 4 above it follows that T is differentiable at x with derivative $DT_x \in L(E_1 \times \cdots \times E_n, F)$ defined by

$$DT_x(h_1, \ldots, h_n) = \sum_{i=1}^{n} T(x_1, \ldots, x_{i-1}, h_i, x_{i+1}, \ldots, x_n).$$

6. Let E be a normed vector space and let $f: E \longrightarrow K$ be differentiable at x. Then $Df_x \in L(E, K) = E^*$. On the other hand if $f: K \longrightarrow E$ is differentiable at x, $Df_x \in L(K, E)$. Now $L(K, E) \approx E$. The canonical isomorphism is defined by mapping $A \in L(K, E)$ to $A(1) \in E$. Using this canonical isomorphism we may regard the derivative as lying in E rather than $L(K, E)$. This, of course, is just what we do when $K = E = \mathbf{R}$ in the classical case.
7. Suppose $f: \mathbf{C} \longrightarrow \mathbf{C}$. If f is differentiable at $z \in \mathbf{C}$, $Df_z \in L(\mathbf{C}, \mathbf{C})$ (the space of continuous *complex* linear maps from $\mathbf{C}$ to $\mathbf{C}$). As in Exercise 6, $L(\mathbf{C}, \mathbf{C}) \approx \mathbf{C}$. Thus, just as for real-valued functions on $\mathbf{R}$, the (complex) derivative of a complex-valued function on $\mathbf{C}$ may be regarded as a point in the complex plane. Of course, if we regard $\mathbf{C}$ as a real vector space (isomorphic to $\mathbf{R}^2$), f may be differentiable as a function between real vector spaces without having a complex derivative. Indeed the space $L(\mathbf{R}^2, \mathbf{R}^2)$ contains $L(\mathbf{C}, \mathbf{C})$ as a (real) two-dimensional subspace. In particular, a map $u: \mathbf{C} \longrightarrow \mathbf{C}$ which is linear over the real numbers is (complex) differentiable if and only if it is complex linear.

Proposition 2.3.4

Let E and F be normed vector spaces and let A be an open subset of E. If f and g are continuous maps from A to F which are differentiable at some point $x \in A$, then the map $f + g$ is differentiable at x and

$$D(f+g)_x = Df_x + Dg_x.$$

Proof. Left to the exercises. ■

Definition 2.3.5

Let E and F be normed vector spaces and let A be an open subset of E. Suppose that the map $f:A \longrightarrow F$ is differentiable at every point of A; then f is said to be differentiable on A. Moreover, if the map

$$Df:A \longrightarrow L(E,F); x \longmapsto Df_x$$

is continuous with respect to the norm topologies on E and $L(E,F)$, we say that f is continuously differentiable, or C^1, on A.

Examples

1. Let $T \in L(E,F)$. Then T is differentiable at every point of E and $DT:E \longrightarrow L(E,F)$ is the constant map $x \longmapsto T$ which is obviously continuous. Therefore T is C^1.
2. Let E_1, E_2 and F be normed vector spaces and let $T \in L(E_1, E_2; F)$. Then T is differentiable at every point of $E_1 \times E_2$ and $DT:E_1 \times E_2 \longrightarrow L(E_1 \times E_2, F)$ is the map defined by

$$DT_{(x_1,x_2)}(e_1,e_2) = T(x_1,e_2) + T(e_1,x_2).$$

   DT is thus a linear map from $E_1 \times E_2$ to $L(E_1 \times E_2, F)$. It follows easily that DT is continuous and hence that T is C^1.
3. Let $(E, \| \ \|)$ be a normed vector space. Then $\| \ \|: E \longrightarrow \mathbf{R}$. We claim that $\| \ \|$ is never differentiable at $x = 0$. For suppose that $\| \ \|$ were differentiable at $x = 0$. Then we could find a linear map $A \in L(E, \mathbf{R})$ such that

$$\|x\| = A(x) + o(x), \qquad x \in E,$$

   and so we have

$$\begin{aligned} 2\|x\| = \|x\| + \|-x\| &= A(x) + A(-x) + o(x) \\ &= o(x) \qquad \text{(linearity of } A). \end{aligned}$$

   Dividing by $\|x\|, x \neq 0$, we obtain $o(x)/\|x\| = 2$. Letting $x \longrightarrow 0$ we obtain the desired contradiction.

Exercises

1. Prove Proposition 2.3.4.
2. Let E and F be normed vector spaces and let $T \in L_s^n(E;F)$. Show that if P is the homogeneous polynomial corresponding to T, P is C^1 with derivative given by: $DP_x(y) = nT(x, x, \ldots, x, y)$ for all $x, y \in E$.

2.4 The composite-mapping formula

Theorem 2.4.1 (The composite-mapping formula)

Let E, F, G be normed vector spaces. Suppose that A is an open subset of E and $f:A \longrightarrow F$ is differentiable at $x_0 \in A$, and B is an open subset of F and $g:B \longrightarrow G$ is differentiable at $f(x_0) \in B$.

Let $p:A \cap f^{-1}(B) \longrightarrow G$ denote the composite of f and g. Thus, $p(x) = g(f(x))$, $x \in A \cap f^{-1}(B)$. Then p is differentiable at x_0 and

$$Dp_{x_0} = Dg_{f(x_0)} \cdot Df_{x_0}.$$

Proof. Let $f(x_0) = y_0$. We define the remainder terms r_1 and r_2 by the equations

$$f(x_0 + s) = f(x_0) + Df_{x_0}(s) + r_1(s) \tag{A}$$

$$g(y_0 + t) = g(y_0) + Dg_{y_0}(t) + r_2(t). \tag{B}$$

Let us consider $p(x_0 + s)$. We have

$$p(x_0 + s) = g(y_0 + Df_{x_0}(s) + r_1(s)), \qquad \text{from eq. (A)}$$

$$= g(y_0) + Dg_{y_0} . Df_{x_0}(s) + r(s),$$

where

$$r(s) = Dg_{y_0}(r_1(s)) + r_2\,(Df_{x_0}(s) + r_1(s)).$$

This follows from eq. (B) with $t = Df_{x_0}(s) + r_1(s)$. We shall show that $r(s) = o(s)$.

Given ϵ, $0 < \epsilon < 1$, $\exists\ \delta > 0$ such that for $\|s\|$, $\|t\| \leqslant \delta$ we have

$$\|r_1(s)\| \leqslant \epsilon \|s\| \qquad \text{and} \qquad \|r_2(t)\| \leqslant \epsilon \|t\|.$$

Also, since Df_{x_0} and Dg_{y_0} are continuous, $\exists\ a, b \geqslant 0$ such that

$$\|Df_{x_0}(s)\| \leqslant a \|s\| \qquad \text{and} \qquad \|Dg_{y_0}(t)\| \leqslant b \|t\|.$$

Since $\epsilon < 1$, it follows that for $\|s\| \leqslant \delta$ we have the inequality

$$\|Df_{x_0}(s) + r_1(s)\| \leqslant (a + 1)\|s\|.$$

Hence, if $\|s\| \leqslant \delta/(a + 1)$, we have

$$\|Df_{x_0}(s) + r_1(s)\| \leqslant \delta,$$

and so

$$\|r_2(Df_{x_0}(s) + r_1(s))\| \leqslant (a + 1)\,\epsilon\,\|s\| \qquad \|s\| \leqslant \delta/(a + 1).$$

This estimates the second term in the expression for $r(s)$. For the first term we have the estimate

$$\|Dg_{y_0}(r_1(s))\| \leqslant b\,\epsilon\,\|s\|, \qquad \|s\| \leqslant \delta/(a + 1).$$

Combining these two inequalities we have

$$\|r(s)\| \leqslant \epsilon\,(a + b + 1)\|s\|, \qquad \|s\| \leqslant \delta/(a + 1).$$

Since $\epsilon > 0$ was arbitrary, it follows that $r(s) = o(s)$. Hence p is differentiable at x_0 with derivative $Dg_{y_0} . Df_{x_0}$. ■

Remark. The composite-mapping formula essentially says that the linear approximation of the composite of two differentiable functions is the composite of their linear approximations. More precisely, with the notation of Theorem 2.4.1, the linear approximations to f and g at x_0 and $y_0 = f(x_0)$ are respectively given by

$$\hat{f}(x) = (f(x_0) - Df_{x_0}(x_0)) + Df_{x_0}(x),$$

and

$$\hat{g}(y) = (g(y_0) - Dg_{y_0}(y_0)) + Dg_{y_0}(y).$$

The reader may verify that the composite of the affine maps $\hat{f}$ and $\hat{g}$ is given by

$$\hat{g} . \hat{f}(x) = (gf(x_0) - Dg_{y_0} . Df_{x_0}(x_0)) + Dg_{y_0} . Df_{x_0}(x),$$

which is precisely the affine linear approximation to gf at x_0, given by Theorem 2.4.1.

We also remark that regarding derivatives as linear approximations leads to a very natural proof of Theorem 2.4.1. In the case of real functions defined on **R**, the proof of the composite-mapping formula, using the limit definition of derivative, poses considerable problems – Newton's proof was incorrect – and correct proofs of this theorem were lacking for many years.

Corollary 2.4.1.1

Let E, F, G be normed vector spaces. Suppose that

$$A \text{ is an open subset of } E \text{ and } f: A \longrightarrow F \text{ is } C^1 \text{ on } A,$$

and

$$B \text{ is an open subset of } F \text{ and } g: B \longrightarrow G \text{ is } C^1 \text{ on } B.$$

Then the composite map $p = gf: A \cap f^{-1}(B) \longrightarrow G$ is C^1 on $A \cap f^{-1}(B)$.

Proof. It follows from Theorem 2.4.1 that p is differentiable on $A \cap f^{-1}(B)$. It is therefore sufficient to show that the map

$$D(gf): A \cap f^{-1}(B) \longrightarrow L(E, G); x \longmapsto Dg_{f(x)} . Df_x$$

is continuous. To this end we observe that the map $D(gf)$ can be written as the composite of the functions displayed in the diagram

$$\begin{array}{ccc} A \cap f^{-1}(B) & \xrightarrow{\phi} & A \cap f^{-1}(B) \times B \\ \Big\uparrow D(gf) & & \Big\downarrow Df \times Dg \\ L(E, G) & \xleftarrow{\text{comp}} & L(E, F) \times L(F, G) \end{array}$$

Here

$$\phi: A \cap f^{-1}(B) \longrightarrow A \cap f^{-1}(B) \times B$$

is the map defined by $\phi(x) = (x, f(x))$ and

$$\text{comp}: L(E, F) \times L(F, G) \longrightarrow L(E, G)$$

is the composition map defined by $\text{comp}(S, T) = T . S$.

Since f is continuous, so is ϕ. Also, $Df \times Dg$ is continuous, since both f and g are assumed to be C^1. Finally, comp is continuous by Proposition 1.6.4. Therefore, $D(gf)$ is the composite of continuous maps and is continuous. ■

Examples

1. Let E and F be normed vector spaces and let $T \in L_s^n(E; F)$. Let P denote the polynomial associated with T. We claim that P is a C^1 function on E with derivative at $x \in E$ given by

$$DP_x(y) = nT(x^{n-1}y).$$

By definition $P(x) = T . \Delta(x), x \in E$. Now the diagonal map

$$\Delta: E \longrightarrow \mathsf{X}^n E$$

is clearly continuous linear and is therefore C^1 on E with constant derivative Δ.

The map $T\colon \mathsf{X}^n E \longrightarrow F$ is easily shown to be C^1 (see Section 2.3) and the derivative of T at

$$x = (x_1, \ldots, x_n) \in \mathsf{X}^n E$$

is given by

$$DT_x(y) = \sum_{i=1}^{n} T(x_1, \ldots, x_{i-1}, y_i, x_{i+1}, \ldots, x_n),$$

where $y = (y_1, \ldots, y_n) \in \mathsf{X}^n E$.

Applying Corollary 2.4.1.1 it follows that P is C^1 with derivative at $x \in E$ given by

$$\begin{aligned} DP_x(y) &= DT_{\Delta(x)} \cdot \Delta(y) \\ &= DT_{(x, x, \ldots, x)}(y, \ldots, y) \\ &= nT(x^{n-1}y), \qquad \text{using the symmetry of } T. \end{aligned}$$

2. Let $(H, \langle\ ,\ \rangle)$ be a real inner-product space. We claim that the norm $\|\ \|$ associated with $\langle\ ,\ \rangle$ is C^1 on $H \setminus \{0\}$. We have $\|x\| = \langle x, x\rangle^{1/2}$ for all $x \in H$. Set $\mathbf{R}^+ = \{t \in \mathbf{R} : t \geqslant 0\}$. We define $\gamma : H \longrightarrow \mathbf{R}$ and $\mathrm{Sq} : \mathbf{R}^+ \longrightarrow \mathbf{R}$ by $\gamma(x) = \langle x, x\rangle$, $x \in H$, and $\mathrm{Sq}\,(t) = \sqrt{t}$, $t \in \mathbf{R}^+$, respectively. Then $\|x\| = \mathrm{Sq}\,\gamma(x)$ for all $x \in H$. By Example 1 above, γ is C^1 on H with derivative given by

$$D\gamma_x(y) = 2\langle x, y\rangle.$$

Sq is well known to be C^1 on $\mathbf{R}^+ \setminus \{0\}$ with derivative (regarded as a linear map!) given by

$$D\,\mathrm{Sq}_x(t) = t/(2\sqrt{x}).$$

From Corollary 2.4.1.1 it follows that $\|\ \|$ is C^1 on

$$H \setminus \gamma^{-1}(0) = H \setminus \{0\},$$

with derivative given by

$$\begin{aligned} D(\|\ \|)_x(y) &= D\,\mathrm{Sq}_{\gamma(x)} \cdot D\gamma_x(y) \\ &= D\,\mathrm{Sq}_{\langle x, x\rangle}(2\langle x, y\rangle) \\ &= \frac{\langle x, y\rangle}{\|x\|} \qquad \text{for all } x \in H \setminus \{0\}. \end{aligned}$$

Exercises

1. Let $f : E \longrightarrow F$ be a homeomorphism from the normed vector space E to the normed vector space F. Suppose that f and f^{-1} are differentiable at x and $f(x)$ respectively. Prove that Df_x is a linear homeomorphism. Show, by finding an example when $E = \mathbf{R}$, that f differentiable at x does not, in general, imply f^{-1} differentiable at $f(x)$, even if f is a homeomorphism.
2. Let $(H, \langle\ ,\ \rangle)$ be a real inner-product space. Show that the function $g(x) = \|x\|^{3/2}$ is C^1 on $H \setminus \{0\}$. What is its derivative?

2.5 Differentiable maps into products of normed vector spaces

In this section we wish to consider functions $f:A \longrightarrow F_1 \times \cdots \times F_n$, where A is an open subset of a normed vector space E and $F_1, \ldots, F_n$ are normed vector spaces. Given such a function f, we may write it in component form as

$$f = (f_1, \ldots, f_n),$$

where $f_i : A \longrightarrow F_i$, $1 \leqslant i \leqslant n$, is the ith component of f. We may write the components of f explicitly in terms of f by using projection maps:

$$f_i = p_i f, \qquad 1 \leqslant i \leqslant n.$$

Here p_i denotes the projection onto the factor F_i, $1 \leqslant i \leqslant n$.

Proposition 2.5.1

Let $E, F_1, \ldots, F_n$ be normed vector spaces and let A be an open subset of E. Suppose $f:A \longrightarrow F_1 \times \cdots \times F_n$. Then f is differentiable at $x \in A$ if and only if each component function f_i is differentiable at x. Moreover the derivatives of f and its components are related by

$$Df_x = \sum_{j=1}^{n} I_j . Df_{j,x}$$

$$Df_{j,x} = p_j . Df_x, \qquad 1 \leqslant j \leqslant n.$$

Here $Df_{j,x}$ denotes the derivative of f_j at x, $1 \leqslant j \leqslant n$.

Proof. Suppose that f is differentiable at $x \in A$. Now the projection map $p_i : F_1 \times \cdots \times F_n \longrightarrow F_i$ is continuous linear and, therefore, differentiable with constant derivative p_i, $1 \leqslant i \leqslant n$. It follows from the composite-mapping formula that $f_i = p_i f$ is differentiable at x and that the derivative of f_i at x is given by

$$Df_{i,x} = p_i . Df_x, \qquad 1 \leqslant i \leqslant n.$$

Hence f differentiable at x implies each component function is differentiable at x.

Suppose, on the other hand, that f_i is differentiable at x, $1 \leqslant i \leqslant n$. Then we have, for $1 \leqslant i \leqslant n$,

$$f_i(x+h) = f_i(x) + Df_{i,x}(h) + r_i(h), \qquad h \in A - x,$$

where $r_i(h) = o(h)$, $1 \leqslant i \leqslant n$. It follows that

$$f(x+h) = f(x) + T(h) + r(h), \qquad h \in A - x,$$

where

$$T(h) = (Df_{1,x}(h), \ldots, Df_{n,x}(h)) = \left(\sum_{j=1}^{n} I_j . Df_{j,x} \right)(h)$$

$$r(h) = (r_1(h), \ldots, r_n(h)).$$

Since $\|r(h)\| = \max \|r_i(h)\|$, it follows that $r(h) = o(h)$. Hence f is differentiable at x with derivative given by

$$Df_x = \sum_{j=1}^{n} I_j . Df_{j,x}. \quad \blacksquare$$

Corollary 2.5.1.1

Ler $E, F_1, \ldots, F_n$ be normed vector spaces and let A be an open subset of E. Suppose

$$f = (f_1, \ldots, f_n) : A \longrightarrow F_1 \times \cdots \times F_n.$$

Then f is C^1 on A if and only if each component f_i of f is C^1 on A.

Proof. Left to the exercises. ■

Notation. Suppose that U is an open subset of $\mathbf{R}$ and that $f: U \longrightarrow F_1 \times \cdots \times F_n$ is differentiable at some point $x \in U$. Now, as explained in Example 6 of Section 2.3, $L(\mathbf{R}, F)$ is canonically isomorphic to F for any normed vector space F. The canonical isomorphism being defined by mapping $A \in L(\mathbf{R}, F)$ to $A(1) \in F$. By analogy with the special case $F = \mathbf{R}$, we shall set

$$f'(x) = Df_x(1) \in F_1 \times \cdots \times F_n$$

$$f_i'(x) = Df_{i,x}(1) \in F_i, \qquad 1 \leqslant i \leqslant n.$$

It follows immediately from Proposition 2.5.1 that

$$f'(x) = (f_1'(x), \ldots, f_n'(x)) \in F_1 \times \cdots \times F_n.$$

Remarks

1. We shall sometimes use the alternative notation $\mathrm{d}f/\mathrm{d}x$ for $f'(x)$, especially when f is a map from $\mathbf{R}$ to $\mathbf{R}$.
2. Notice that if $f: U \subset \mathbf{R} \longrightarrow F$ is differentiable at x, then

$$f(x+h) = f(x) + hf'(x) + o(h).$$

Conversely, if there exists $f'(x) \in F$ such that the above equation holds, then f is differentiable at x. In particular, if $F = \mathbf{R}$, $f'(x)$ is the 'classical' derivative of f at x.

Proposition 2.5.1 is of particular importance when $F_1 = \cdots = F_n = \mathbf{R}$ and the product $F_1 \times \cdots \times F_n$ is $\mathbf{R}^n$.

Let us take the standard basis $(e_i)_{i=1}^n$ of $\mathbf{R}^n$. Notice that the inclusion map $I_j: \mathbf{R} \longrightarrow \mathbf{R}^n$ satisfies the property

$$I_j(1) = e_j, \qquad 1 \leqslant j \leqslant n.$$

Consequently, by Proposition 2.5.1 we have

$$f'(x) = Df_x(1) = \sum_{j=1}^{n} I_j \, . \, Df_{j,x}(1) = \sum_{j=1}^{n} f_j'(x) e_j,$$

and so, relative to the standard basis on $\mathbf{R}^n$, we have

$$f'(x) = (f_1'(x), \ldots, f_n'(x)) \in \mathbf{R}^n.$$

Moreover, the matrix of the linear map Df_x, relative to the standard basis on $\mathbf{R}^n$, is easily seen to be given by

$$[Df_x] = \begin{bmatrix} f_1'(x) \\ \vdots \\ f_n'(x) \end{bmatrix}.$$

Summing up the above in a proposition we have

Proposition 2.5.2

Let A be an open subset of $\mathbf{R}$ and let $f: A \longrightarrow \mathbf{R}^n$. If $x \in A$ and we set $f = (f_1, \ldots, f_n)$, f is differentiable at x if and only if each f_i is differentiable at x. The derivatives are related by

$$Df_x = \sum_{j=1}^{n} I_j \, . Df_{j,x},$$

or, equivalently, by

$$f'(x) = (f_1'(x), \ldots, f_n'(x))$$

and the matrix of Df_x, relative to the standard basis on $\mathbf{R}^n$, is given in terms of the derivatives $f_i'(x)$ by

$$[Df_x] = \begin{bmatrix} f_1'(x) \\ \vdots \\ f_n'(x) \end{bmatrix}.$$

Example. Let $r = (x_1, x_2, x_3)$ $\mathbf{R} \longrightarrow \mathbf{R}^3$ be C^1. Then r defines a curve in $\mathbf{R}^3$.

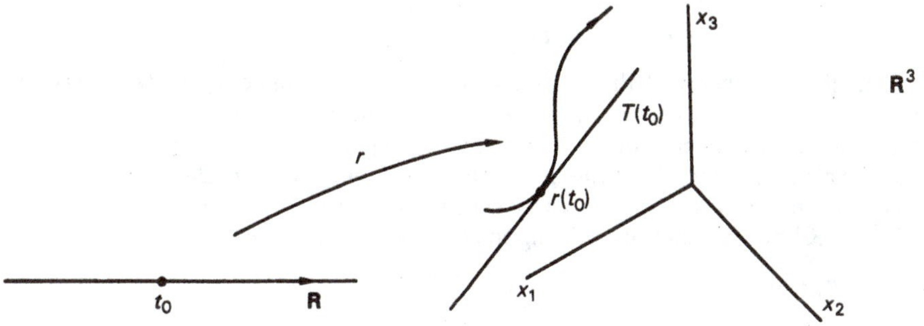

FIG. 2.3

The tangent line $T(t_0)$ to the curve r at $t = t_0$ is given by the affine linear approximation to r at t_0. That is, $T(t_0)$ is given as the image of the map $g: \mathbf{R} \longrightarrow \mathbf{R}^3$ defined by

$$g(t) = c + (x_1'(t_0)t, x_2'(t_0)t, x_3'(t_0)t),$$

where

$$c = (c_1, c_2, c_3)$$

and

$$c_i = x_i(t_0) - x_i'(t_0)t_0, \qquad 1 \leqslant i \leqslant 3.$$

We may think of the curve r as parametrizing the motion of a particle moving in $\mathbf{R}^3$. The velocity at time t_0 of this particle would then be $r'(t_0) \in \mathbf{R}^3$. The x_1, x_2, x_3 components of the velocity are then $x_1'(t_0), x_2'(t_0), x_3'(t_0)$ respectively. The direction of motion of the particle is given by the tangent line $T(t_0)$. Note, however, that if $r'(t_0) = 0$ then $T(t_0)$ degenerates to a point.

Suppose we take the standard inner product $\langle\ ,\ \rangle$ on $\mathbf{R}^3$ and let the mass of the particle be m. Then the (kinetic) energy of the particle at time t is defined by the equation

$$E(t) = \tfrac{1}{2}m\langle r'(t), r'(t)\rangle = \tfrac{1}{2}m(x_1'(t)^2 + x_2'(t)^2 + x_3'(t)^2).$$

Since we assumed r to be C^1, E is clearly a continuous function from $\mathbf{R}$ to $\mathbf{R}$.

Let us assume that the map $r' : \mathbf{R} \longrightarrow \mathbf{R}^3$ is differentiable and investigate the differentiability of the energy function E. We shall denote the derivative of r' by r'', thus $r'' : \mathbf{R} \longrightarrow \mathbf{R}^3$. From Proposition 2.5.2 it follows that x'_1, x'_2 and x'_3 are differentiable and that

$$r''(t) = (x_1''(t), x_2''(t), x_3''(t)) \qquad \text{for all } t \in \mathbf{R}.$$

Now E may be written as the composite $\phi r'$, where ϕ: $\mathbf{R}^3 \longrightarrow \mathbf{R}$ is defined by

$$\phi(x) = \tfrac{1}{2}m\langle x, x\rangle.$$

Since ϕ is differentiable (Section 2.3), it follows from the composite-mapping formula that E is differentiable and that

$$\begin{aligned}\frac{\mathrm{d}E}{\mathrm{d}t} = DE_t(1) &= \tfrac{1}{2}m(D\phi_{r'(t)} \cdot Dr'_t)(1)\\ &= \tfrac{1}{2}m(2\langle r'(t), r''(t)\rangle), \qquad \text{(Section 2.3)}\\ &= m\langle r'(t), r''(t)\rangle.\end{aligned}$$

In particular, the energy of the particle will be *constant* if and only if $r'(t) \perp r''(t)$ for all $t \in \mathbf{R}$. If we let $F(t)$ denote the force acting on the particle at time t, it follows from Newton's law of motion—$F = mr''$—that the energy is constant if and only if F always acts *orthogonally* to the direction of motion of the particle.

Let us conclude this example with a specific illustration of the above. Let $r : \mathbf{R} \longrightarrow \mathbf{R}^3$ be the function defining the helix with axis the x_3-axis of $\mathbf{R}^3$:

$$r(t) = (\cos t, \sin t, t), \qquad \text{for all } t \in \mathbf{R}.$$

Then we have

$$r'(t) = (-\sin t, \cos t, 1), \qquad t \in \mathbf{R}.$$

$$r''(t) = (-\cos t, -\sin t, 0), \qquad t \in \mathbf{R}.$$

and so

$$E(t) = \tfrac{1}{2}m(\sin^2 t + \cos^2 t + 1) = m \qquad \text{and} \qquad \frac{\mathrm{d}E}{\mathrm{d}t} = 0.$$

This last fact is also seen to be true by noting that $\langle r'(t), r''(t)\rangle = 0$ for all $t \in \mathbf{R}$.

In physical terms: If, at time $t = 0$, we have a particle mass m at the point $(1, 0, 0)$ $(= r(0))$ moving with velocity $(0, 1, 1)$ $(= r'(0))$, then, if we apply the force $m(-\cos t, -\sin t, 0)$ $(= r''(t))$ to the particle subsequently, it will trace out the helix and will be at the point $(\cos t, \sin t, t)$ $(= r(t))$ at time t.

Proposition 2.5.3

Let A and B be open subsets of $\mathbf{R}$ and $f : A \longrightarrow \mathbf{R}$ be differentiable at $x \in A$ and satisfy $f(A) \subset B$. Then, if $\phi : B \longrightarrow \mathbf{R}^n$ is differentiable at $f(x)$, the composite

$h = \phi f : A \longrightarrow \mathbf{R}^n$ is differentiable at x and the derivative $h'(x)$, regarded as an element of $\mathbf{R}^n$, is given by

$$h'(x) = (f'(x)\phi_1'(f(x)), \ldots, f'(x)\phi_n'(f(x))).$$

Proof. By the composite-mapping formula, h is differentiable at x and

$$Dh_x = D\phi_{f(x)} \cdot Df_x.$$

Now $h'(x) = Dh_x(1)$ and so

$$\begin{aligned} h'(x) &= D\phi_{f(x)}(Df_x(1)) \\ &= D\phi_{f(x)}(f'(x)1) \\ &= f'(x)D\phi_{f(x)}(1) \qquad (\text{since } D\phi_{f(x)} \in L(\mathbf{R}, \mathbf{R}^n)) \\ &= f'(x)\phi'(f(x)). \quad \blacksquare \end{aligned}$$

Remarks

1. Proposition 2.5.3 can, of course, be deduced from the composite-mapping formula for real-valued functions defined on open subsets of **R**. Indeed, applying the composite-mapping formula to the component functions $\phi_i f$, $1 \leqslant i \leqslant n$, yields the required result.
2. The map f of Proposition 2.5.3 gives a *reparametrization* of the curve defined by ϕ.

Example. Notice that we can have curves in $\mathbf{R}^n$ for which there are discontinuities in the direction of the tangent line $T(t)$. Thus the curve in $\mathbf{R}^2$ defined to be the union of the positive x-axis and the positive y-axis can be parametrized by the following C^1 function $\phi : \mathbf{R} \longrightarrow \mathbf{R}^2$:

$$\phi(t) = \begin{cases} (t^2, 0), & t \leqslant 0, \\ (0, 0), & t \in [0, 1], \\ (0, (t-1)^2), & t \geqslant 1. \end{cases}$$

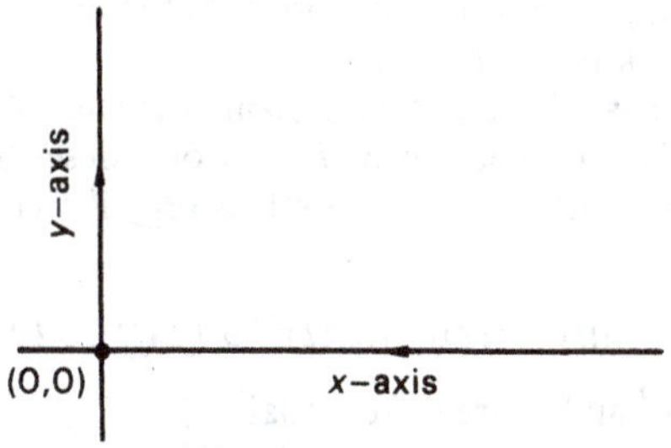

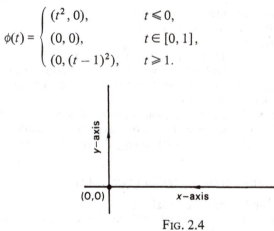

FIG. 2.4

ϕ is easily seen to be C^1 with

$$\phi'(t) = \begin{cases} (2t, 0), & t \leqslant 0 \\ (0, 0), & t \in [0, 1] \\ (0, 2(t-1)), & t \geqslant 1. \end{cases}$$

However, $T(t)$ degenerates to a point for $t \in [0, 1]$ because $\phi'(t) = 0$ there. There clearly can exist no C^1 function $\phi : \mathbf{R} \longrightarrow \mathbf{R}^2$ parametrizing this curve which satisfies the additional condition that $\phi'(t) \neq 0$ for all $t \in \mathbf{R}$. Provided that $\phi'(t) \neq 0, t \in \mathbf{R}$, it is easy to show that the tangent line has no degeneracies or discontinuities in direction. We will return to a more careful study of these points in Chapters 3 and 4.

We conclude this section with a slight generalization of Proposition 2.5.2. Suppose that E is a finite-dimensional vector space with basis $\mathscr{E} = (e_i)_{i=1}^n$, and that $f : A \longrightarrow E$, where A is an open subset of $\mathbf{R}$. We may write f in component form as $(f_1, \ldots, f_n)$, where $f_i : A \longrightarrow \mathbf{R}$, $1 \leqslant i \leqslant n$. That is, the coordinates of $f(x)$, $x \in A$, are $(f_1(x), \ldots, f_n(x))$. If we denote the dual basis of $\mathscr{E}$ by $\mathscr{E}^* = (e_i^*)_{i=1}^n$ the maps f_i may be given explicitly in terms of f by

$$f_i = e_i^* f, \qquad 1 \leqslant i \leqslant n.$$

As in Proposition 2.5.1, it follows from the composite-mapping formula that f is differentiable at x if and only if the component functions are all differentiable at x. The reader may easily verify that $f'(x)$ has coordinates $(f_1'(x), \ldots, f_n'(x))$.

In other words: The coordinates of the derivative of f are the derivatives of the coordinates of f.

Remark. We point out to the reader that all the material of this section applies equally well to differentiable maps from a *complex* normed vector space into a product of complex normed spaces. The only change is that the linear maps are now complex linear rather than real linear. Thus the complex version of Proposition 2.5.2 makes use of the canonical isomorphism between $L(\mathbf{C}, E)$ and E, where $L(\mathbf{C}, E)$ is now the space of *complex* linear maps from $\mathbf{C}$ to E.

Exercises

1. Prove Corollary 2.5.1.1.
2. Find the derivatives of the following functions:
   (a) $f : \mathbf{R} \longrightarrow \mathbf{R}^3, f(t) = (e^t \cos t, e^t \sin t, t^2)$.
   (b) $g : \mathbf{R} \longrightarrow \mathbf{R}^n, g(t) = (1, t, t^2, \ldots, t^{n-1})$.
   Find also the affine linear approximations to f and g at $t = 1$.
3. Consider the curve $r : \mathbf{R} \longrightarrow \mathbf{R}^3$ defined by $r(t) = (at^2, bt^2, ct)$, where $a, b, c \in \mathbf{R}$. Is it possible to find a, b, c so that r represents the motion of a particle moving in $\mathbf{R}^3$ with constant kinetic energy?
4. Let $\phi : \mathbf{R} \longrightarrow \mathbf{R}^3$ be a C^1 map parametrizing the curve C in $\mathbf{R}^3$. Let $f : \mathbf{R} \longrightarrow \mathbf{R}$ be C^1. Show that the parametrization $h = \phi f$ of C describes the motion of a particle on C moving with constant kinetic energy if f satisfies the differential equation

$$f''(t)\langle \phi'(f(t)), \phi'(f(t)) \rangle + f'(t)^2 \langle \phi''(f(t)), \phi'(f(t)) \rangle = 0.$$

   (You may assume ϕ' and f' are differentiable.)

   Show by example that, if there exists $t_0 \in \mathbf{R}$ such that $\phi'(t_0) = 0$, this equation need not have a solution.
5. Let E and F be normed vector spaces and let $f : E \longrightarrow F$ be differentiable. Suppose that f is positively homogeneous of degree k:

$$F(tx) = t^k F(x), \qquad t \geqslant 0, k \geqslant 0.$$

   Show that

$$Df_x(x) = kf(x) \qquad \text{(Euler's theorem).}$$

(*Hint*: Consider the function $f\phi_x : (0, \infty) \longrightarrow F$, where $\phi_x(t) = tx$, $t \in (0, \infty)$ and apply the composite-mapping formula.)

Conversely, suppose that $Df_x(x) = kf(x)$ for all $x \in E$, $k \geqslant 0$. By considering the function $\theta_x = f\phi_x$, find a differential equation for θ_x and deduce that $\theta_x(t)t^{-k}$ is constant for fixed x. Hence show that f is positively homogeneous of degree k.

6. Let U be an open subset of the normed vector space E and let $f, g : U \longrightarrow \mathbf{R}$ be C^1 maps. Prove that the map

$$f \times g : U \times U \longrightarrow \mathbf{R} \times \mathbf{R},$$

defined by $(f \times g)(x, y) = (f(x), g(y))$, is C^1 on $U \times U$ with derivative given by

$$D(f \times g)_{(x,\, y)} = Df_x \times Dg_y : E \times E \longrightarrow \mathbf{R} \times \mathbf{R}.$$

7. Continuing with the assumptions of Exercise 6, show that the function $h : U \longrightarrow \mathbf{R}$ defined by $h(x) = f(x)g(x)$ is C^1 on U with derivative

$$Dh_x = f(x)Dg_x + g(x)Df_x.$$

(*Hint*: Show that $h = \zeta(f \times g)\Delta$, where ζ is scalar multiplication in $\mathbf{R}$ and $\Delta : E \longrightarrow E \times E$ is the diagonal map, then apply the composite-mapping formula.)

2.6 Differentiable maps from products and partial derivatives

In this section we wish to consider functions $f : A \longrightarrow F$, where F is a normed vector space and A is an open subset of the product of normed vector spaces $E_1 \times \cdots \times E_n$.

Definition 2.6.1

Let E_1, E_2, F be normed vector spaces and let A be an open subset of the product normed vector space $E_1 \times E_2$. If $f : A \longrightarrow F$ is continuous and $x = (x_1, x_2) \in A$, we shall say that f is partially differentiable with respect to the 1st variable at x if f is differentiable with respect to the 1st variable at x_1, the second variable being kept fixed and equal to x_2. That is, f is partially differentiable with respect to the first variable at x if there exists $D_1f_x \in L(E_1, F)$ such that

$$f(x_1 + h, x_2) - f(x_1, x_2) - D_1f_x(h) = o(h).$$

D_1f_x is called the *1st partial derivative of f at x*. We may similarly define the 2nd partial derivative of f at x, $D_2f_x \in L(E_2, F)$.

More generally, if $E_1, \ldots, E_n, F$ are normed vector spaces, $A \subset E_1 \times \cdots \times E_n$ is open and $f : A \longrightarrow F$ is continuous, we shall say that f is partially differentiable with respect to the jth variable at $x = (x_1, \ldots, x_n) \in A$ if there exists $D_jf_x \in L(E_j, F)$ such that

$$f(x_1, \ldots, x_j + h, \ldots, x_n) - f(x_1, \ldots, x_j, \ldots, x_n) - D_jf_x(h) = o(h).$$

D_jf_x is called the *jth partial derivative of f at x*.

Remarks

1. It follows from the theory of derivatives that the partial derivatives D_jf_x are unique.
2. If $A \subset E_1 \times \cdots \times E_n$ is a product $A_1 \times \cdots \times A_n$ of open sets, the partial

differentiability of $f: A \longrightarrow F$ with respect to the jth variable at $x = (x_1, \ldots, x_n) \in A$ is equivalent to the differentiability at $x_j \in A_j$ of the map

$$t \longmapsto f(x_1, \ldots, x_{j-1}, t, x_{j+1}, \ldots, x_n), \; t \in A_j.$$

Figure 2.5 illustrates the concept of partial differentiability in the special case when $A = A_1 \times A_2 \subset E_1 \times E_2$ and $f: A_1 \times A_2 \longrightarrow F$ is partially differentiable with respect to the 1st variable at (x_1, x_2).

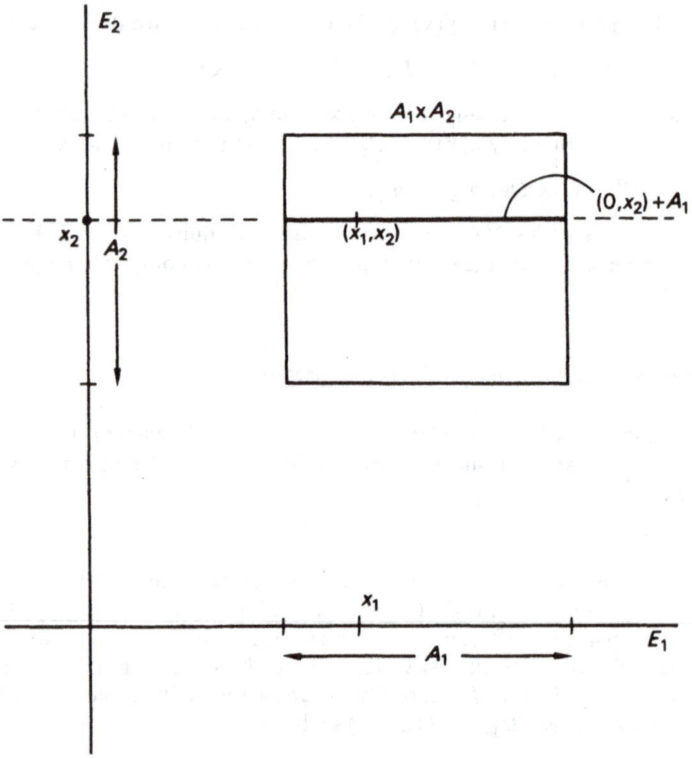

FIG. 2.5

With the notation of the diagram, f is partially differentiable with respect to the 1st variable at (x_1, x_2) if the restriction of f to the segment $(0, x_2) + A_1$ is differentiable at x_1. More precisely, if the map from A_1 to F defined by mapping $t \in A_1$ to $f(t, x_2)$ is differentiable at $t = x_1$.

Example. Let $f: \mathbf{R}^n \longrightarrow \mathbf{R}$ be the function

$$f(x_1, \ldots, x_n) = x_1^3 + \cdots + x_n^3.$$

Fix $y = (y_1, \ldots, y_n) \in \mathbf{R}^n$. The function f is partially differentiable with respect to the jth variable at y if the map from $\mathbf{R}$ to $\mathbf{R}$ defined by

$$x_j \longmapsto y_1^3 + \cdots + y_{j-1}^3 + x_j^3 + y_{j+1}^3 + \cdots + y_n^3$$

is differentiable at $x_j = y_j$. This is certainly the case and, differentiating, we find that for all $t \in \mathbf{R}$ we have

$$D_j f_y(t) = 3y_j^2 t, \qquad 1 \leqslant j \leqslant n.$$

Before stating the next proposition it may be helpful to recall that we have a canonical isomorphism between $L(E_1 \times \cdots \times E_n, F)$ and $\times_{i=1}^n L(E_i, F)$ defined by mapping $A \in L(E_1 \times \cdots \times E_n, F)$ to the n-tuple

$$(A \,.\, I_1, \ldots, A \,.\, I_n) \in \mathop{\times}_{i=1}^{n} L(E_i, F).$$

In particular, we have the relation

$$A(e_1, \ldots, e_n) = \sum_{i=1}^{n} (A \,.\, I_i)(e_i) \text{ for all } (e_1, \ldots, e_n) \in E_1 \times \cdots \times E_n.$$

Proposition 2.6.2

Let $E_1, \ldots, E_n, F$ be normed vector spaces and let A be an open subset of $E_1 \times \cdots \times E_n$. Suppose $f : A \longrightarrow F$. Then, if f is differentiable at $x = (x_1, \ldots, x_n) \in A$, f is partially differentiable at x with respect to each variable and the derivative and partial derivatives are connected by the relations

$$Df_x = \sum_{j=1}^{n} D_j f_x \cdot p_j$$

$$D_j f_x = Df_x \,.\, I_j, \qquad 1 \leqslant j \leqslant n.$$

That is, Df_x corresponds to the point $(D_1 f_x, \ldots, D_n f_x)$ under the canonical isomorphism between $L(E_1 \times \cdots \times E_n, F)$ and $\times_{i=1}^n L(E_i, F)$.

Proof. Let $h = (h_1, \ldots, h_n) \in E_1 \times \cdots \times E_n$. Since f is differentiable at x we have

$$f(x + h) - f(x) - Df_x(h) = o(h).$$

Suppose that $h_2, \ldots, h_n$ are all zero. Then $h = I_1(h_1)$ and $o(h) = o(h_1)$. We may write the above equation in terms of h_1 to obtain

$$f(x_1 + h_1, x_2, \ldots, x_n) - f(x_1, x_2, \ldots, x_n) - Df_x(I_1(h_1)) = o(h_1).$$

But $Df_x(I_1(h_1)) = Df_x \,.\, I_1(h_1)$ and so, by definition, f is partially differentiable with respect to the 1st variable at x and $D_1 f_x = Df_x \,.\, I_1$. Similarly, f is partially differentiable with respect to the remaining variables at x. ■

Corollary 2.6.2.1

Let $E_1, \ldots, E_n, F$ be normed vector spaces and let A be an open subset of $E_1 \times \cdots \times E_n$. Suppose that $f : A \longrightarrow F$ is differentiable at $y = (y_1, \ldots, y_n) \in A$. Then the affine linear approximation to f at y is given in terms of the partial derivatives of f at y by the formula

$$g(x_1, \ldots, x_n) = c + \sum_{j=1}^{n} D_j f_y(x_j),$$

where $c = f(y_1, \ldots, y_n) - \sum_{j=1}^n D_j f_y(y_j)$.

Proof. Immediate from Proposition 2.6.2 and the remarks preceding the statement of that proposition. ∎

Examples

1. Figure 2.6 illustrates Proposition 2.6.2 in the special case when E_1, E_2 and F are of dimension one and $y = 0$. Referring to the diagram, L_1 denotes part of

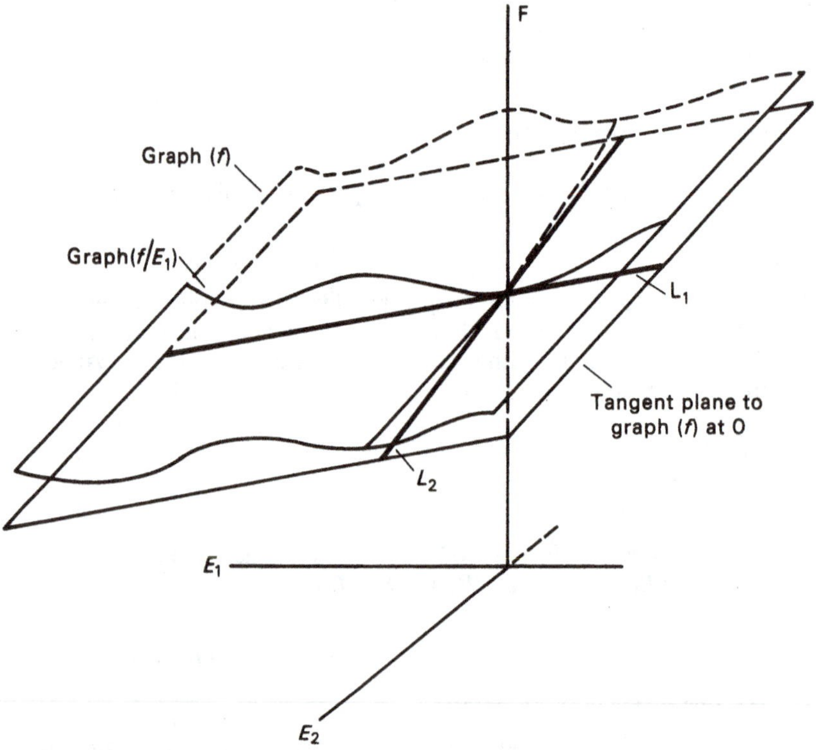

FIG. 2.6

the tangent line to $f|E_1$ at 0. Thus $L_1 \subset E_1 \times F$. Similarly, L_2 denotes the tangent line to $f|E_2$ at 0 and is contained in $E_2 \times F$. If we let T denote the tangent plane to the graph of f at 0 we have the relations

$$T \cap (E_1 \times F) = L_1, \qquad T \cap (E_2 \times F) = L_2.$$

This follows from the corollary to Proposition 2.6.2. Hence the tangent plane determines the tangent lines L_1 and L_2, which in turn correspond to the first and second partial derivatives of f at 0. Thus the geometric interpretation of partial derivatives tells us that to find the partial derivative corresponding to E_1 we just take the intersection of the tangent plane with the subspace $E_1 \times F$ and this then gives us the graph of the partial derivative (up to translation). We leave it to the reader to formulate the appropriate statements in the case when $y \neq 0$.

As we shall soon see, the existence of the tangent lines L_1 and L_2 does not necessarily imply the existence of a tangent plane to the graph of f. That is, the partial differentiability of f need not imply that f is differentiable.

2. Let $f : \mathbf{R}^n \longrightarrow \mathbf{R}$ be the homogeneous cubic polynomial defined by

$$f(x_1, \ldots, x_n) = \sum_{i=1}^{n} x_i^3.$$

It follows from Section 2.3 that f is everywhere differentiable on $\mathbf{R}^n$ and that if $x = (x_1, \ldots, x_n) \in \mathbf{R}^n$ the derivative is given by

$$Df_x(t) = 3 \sum_{i=1}^{n} x_i^2 t_i \qquad \text{for all } t = (t_1, \ldots, t_n) \in \mathbf{R}^n.$$

From Proposition 2.6.2 we see that f is partially differentiable in all variables at every point of $\mathbf{R}^n$ and that for $1 \leqslant j \leqslant n$ we have

$$D_j f_x(s) = Df_x . I_j(s) = 3x_j^2 s, \qquad s \in \mathbf{R}.$$

This formula for $D_j f$ coincides with that obtained by direct computation as in the example following Definition 2.6.1.

3. Let E and F be normed vector spaces and let $T \in L^n(E; F)$. Therefore, T is a map from $\mathsf{X}^n E$ to F. Let $y = (y_1, \ldots, y_n) \in \mathsf{X}^n E$. Since T is continuous and linear in each variable separately, T is partially differentiable with respect to all variables at y and for $1 \leqslant j \leqslant n$ we have

$$D_j T_y(s) = T(y_1, \ldots, y_{j-1}, s, y_{j+1}, \ldots, y_n), \qquad s \in E.$$

Since T is differentiable it follows from Proposition 2.6.2 that

$$DT_y = \sum_{j=1}^{n} D_j T_y \cdot p_j.$$

Equivalently, for all $(e_1, \ldots, e_n) \in \mathsf{X}^n E$ we have

$$DT_y(e_1, \ldots, e_n) = \sum_{j=1}^{n} D_j T_y(e_j),$$

which is the formula for the derivative of T obtained in Section 2.3. Later on in Section 2.8, we shall show how the above argument may be used to obtain the formula for DT without assuming that T is differentiable.

4. In view of Proposition 2.5.1 it is natural to ask whether the converse of Proposition 2.6.2 holds: 'f partially differentiable at y implies f differentiable at y'. Such a statement is false, without further conditions on the partial derivatives of f, as the following example shows.

   Let $f : \mathbf{R}^2 \longrightarrow \mathbf{R}$ be defined by

$$f(x, y) = \begin{cases} \dfrac{xy^3}{x^2 + y^4}, & (x, y) \neq (0, 0) \\ 0, & (x, y) = (0, 0). \end{cases}$$

We remark first that f is continuous at (0, 0). Indeed, set

$$x = r \cos t, \qquad y^2 = r \sin t,$$

where $r \geqslant 0$ and $t \in [0, \pi)$. Then

$$f(x, y) = \frac{(r^2 \cos t \sin t)y}{r^2} = y \cos t \sin t.$$

Hence, $|f(x, y)| \leqslant |y|$. This estimate is clearly sufficient to prove the continuity of f at $(0, 0)$. Since f is obviously continuous away from the origin of $\mathbf{R}^2$, it follows that f is continuous on $\mathbf{R}^2$.

Now $f(x, 0) = f(0, y) = 0$ for all $x, y \in \mathbf{R}$. Therefore, f has partial derivatives

$$D_1 f_{(0,0)} = D_2 f_{(0,0)} = 0.$$

If f were differentiable at $(0, 0)$, it would follow from Corollary 2.6.2.1 that

$$\begin{aligned} f(h, k) &= f(0, 0) + D_1 f_{(0,0)}(h) + D_2 f_{(0,0)}(k) + o(h, k) \\ &= o(h, k). \end{aligned}$$

Let us evaluate $f(h^2, h)$, $h \in \mathbf{R}$. We see that

$$f(h^2, h) = \frac{h^5}{h^4 + h^4} = \frac{h}{2}.$$

On the other hand, for $0 < h < 1$, we have $\|(h^2, h)\| = h$ (product norm) and so

$$\|f(h^2, h)\| / \|(h^2, h)\| = 1/2.$$

But this contradicts $f(h, k) = o(h, k)$. Therefore, f is not differentiable at $(0, 0)$.

Notation. Suppose that A is an open subset of $\mathbf{R}^n$ and that $f : A \longrightarrow F$ is differentiable at $y = (y_1, \ldots, y_n) \in A$. Then $D_j f_y \in L(\mathbf{R}, F)$ and, using the canonical isomorphism between $L(\mathbf{R}, F)$ and F, we shall define the point $f_{;j}(y) \in F$ by

$$f_{;j}(y) = D_j f_y(1), \qquad 1 \leqslant j \leqslant n.$$

If we take the standard basis $(e_i)_{i=1}^n$ of $\mathbf{R}^n$ and note that $I_j(1) = e_j$, $1 \leqslant j \leqslant n$, we may write $f_{;j}$ in terms of the derivative of f at y

$$f_{;j}(y) = Df_y(e_j), \qquad 1 \leqslant j \leqslant n.$$

In particular, Df_y may be identified with $(f_{;1}(y), \ldots, f_{;n}(y)) \in \mathsf{X}^n F$.

Remark. $f_{;j}(y)$ is what is classically called the jth partial derivative of f at y. There are many other notations for partial derivatives. In particular, we mention the notation $\partial f/\partial x_j$ for the jth partial derivative of f. Unfortunately, the notation $\partial f/\partial x_j$ does not clearly exhibit the functional dependence of the partial derivative on the variable x. Notations like $(\partial f/\partial x_j)_{x=y}$ or $(\partial f/\partial x_j)_y$ are not only cumbersome but are also capable of misinterpretation. On occasions, however, when it is not important to emphasize the functional dependence of $f_{;j}(y)$ on y we shall write $\partial f/\partial x_j$ in place of $f_{;j}(y)$.

Examples

1. Let F be a normed vector space and let $T \in L(\mathbf{R}^n, F)$. Denoting the standard basis of $\mathbf{R}^n$ by $(e_i)_{i=1}^n$, we have, for all $y \in \mathbf{R}^n$,

$$T_{;j}(y) = D_j T_y(1) = T(e_j), \qquad 1 \leqslant j \leqslant n.$$

2. Let $f : \mathbf{R}^n \longrightarrow F$ be differentiable at $y = (y_1, \ldots, y_n)$. Then the affine linear approximation to f at y is given by

$$g(x_1, \ldots, x_n) = c + \sum_{j=1}^{n} x_j f_{;j}(y),$$

where $c = f(y_1, \ldots, y_n) - \Sigma_{j=1}^n y_j f_{;j}(y)$.

3. Let $f: \mathbf{R}^n \longrightarrow \mathbf{R}$ be partially differentiable with respect to the 1st variable at $x = (x_1, \ldots, x_n)$. To compute $f_{;1}(x_1, \ldots, x_n)$, all we have to do is to find the derivative at $t = x_1$ of the function

$$t \longmapsto f(t, x_2, x_3, \ldots, x_n).$$

This follows since the partial derivative $f_{;1}(x_1, \ldots, x_n)$ satisfies the equation

$$f(x_1 + h, x_2, \ldots, x_n) - f(x_1, x_2, \ldots, x_n) - hf_{;1}(x_1, x_2, \ldots, x_n) = o(h).$$

(Remember that $D_1 f_x(h) = hf_{;1}(x)$ for all $h \in \mathbf{R}$ since $D_1 f_x \in L(\mathbf{R}, \mathbf{R})$.) As an example, suppose that $f: \mathbf{R}^2 \longrightarrow \mathbf{R}$ is defined by

$$f(x, y) = y^2 e^{2x} + y \sin x$$

and we wish to find $f_{;1}$ at the point (x_0, y_0). This is given as the derivative of the function

$$t \longmapsto y_0^2 e^{2t} + y_0 \sin t$$

at $t = x_0$. Hence

$$f_{;1}(x_0, y_0) = 2y_0^2 e^{2x_0} + y_0 \cos x_0.$$

Similarly,

$$f_{;2}(x_0, y_0) = 2y_0 e^{2x_0} + \sin x_0.$$

4. Let $f: \mathbf{R}^2 \longrightarrow \mathbf{R}$ be differentiable at the origin of $\mathbf{R}^2$. Let the graph of f be as shown in Fig. 2.7. The graph of f restricted to the y-axis is denoted by a thick

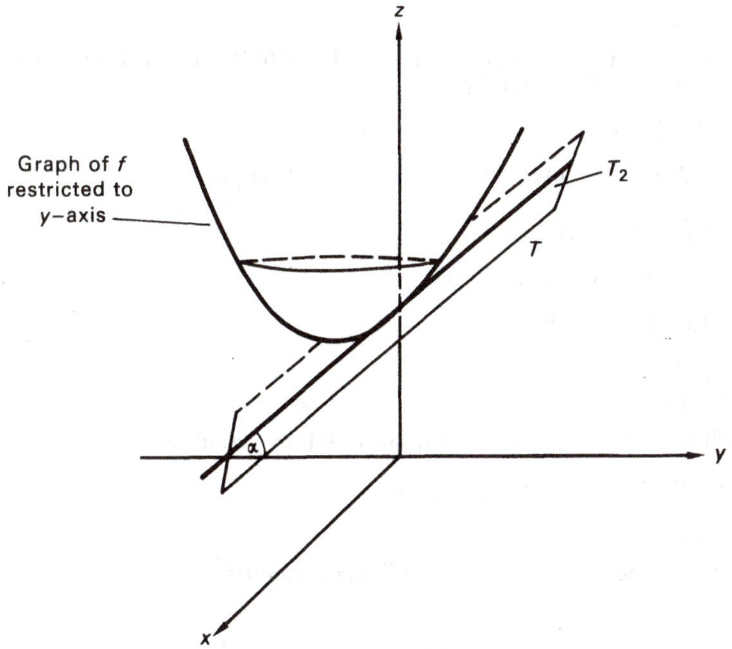

FIG. 2.7

black line. The tangent line to this graph, T_2, is precisely the intersection of the tangent plane T, to the graph of f at 0, with the yz-plane. The gradient $\tan\alpha$ of the line T_2 is equal to $f_{;2}(0)$. To see the truth of these remarks, we note that the equation of the tangent plane to graph(f) at 0 is given by

$$g(x, y) = f(0, 0) + f_{;1}(0)x + f_{;2}(0)y.$$

If we restrict g to the y-axis we obtain the affine linear map $g_2 : \mathbf{R} \longrightarrow \mathbf{R}$ defined by

$$g_2(y) = f(0, 0) + f_{;2}(0)y.$$

This is, therefore, the equation of the line formed by intersecting T with the yz-plane. But if we let f_2 denote the restriction of f to the y-axis we have

$$(f_2 - g_2)(y) = (f - g)(0, y) = o(0, y) = o(y).$$

Therefore, g_2 defines the tangent line to the graph of f, restricted to the y-axis.

A similar argument shows that the intersection of T with the xz-plane gives the tangent line of f restricted to the x-axis. More generally, if f is differentiable at $(x_0, y_0) \in \mathbf{R}^2$, $(x_0, y_0) \neq 0$, all the above continues to hold with the modification that the yz-plane becomes the parallel translation of the yz-plane through (x_0, y_0), the y-axis becomes the translation of the y-axis through (x_0, y_0), etc. In the obvious way, we have a similar interpretation of partial derivatives for maps from $\mathbf{R}^n$ to $\mathbf{R}$, $n > 2$.

The following proposition is very useful in practice.

Proposition 2.6.3

Let A be an open subset of $\mathbf{R}^m$ and suppose

$$f = (f_1, \ldots, f_m) : A \longrightarrow \mathbf{R}^n$$

is differentiable at $y = (y_1, \ldots, y_n) \in A$. Then, relative to the standard bases of $\mathbf{R}^m$ and $\mathbf{R}^n$, the matrix of Df_y is given by

$$[Df_y] = [f_{i;j}(y)],$$

where $f_{i;j}(y) = (f_i)_{;j}(y)$ denotes the ijth component of $[Df_y]$.

Proof. Let $[Df_y] = [a_{ij}]$. Then

$$a_{ij} = i\text{th coordinate of } Df_y(e_j).$$

From Proposition 2.5.1 it follows that

$$Df_y = \sum_{j=1}^{n} I_j \,.\, Df_{j,y}.$$

Hence, the ith coordinate of $Df_y(e_j) = Df_{i,y}(e_j)$. But, by definition,

$$f_{i;j}(y) = Df_{i,y} \,.\, I_j(1) = Df_{i,y}(e_j). \blacksquare$$

Corollary 2.6.3.1

Let $A \subset \mathbf{R}^m$ be open and let $f : A \longrightarrow \mathbf{R}^n$ be differentiable at $y = (y_1, \ldots, y_m) \in A$. Then,

$$Df_y(h_1, \ldots, h_m) = \left(\sum_{j=1}^{m} f_{1;j}(y)h_j, \ldots, \sum_{j=1}^{m} f_{n;j}(y)h_j\right), \qquad (h_1, \ldots, h_m) \in \mathbf{R}^m.$$

Definition 2.6.4

Let $A \subset \mathbf{R}^m$ be open and let $f: A \longrightarrow \mathbf{R}^n$ be differentiable at $y = (y_1, \ldots, y_m) \in A$. Then the matrix $[f_{i;j}(y)]$ is called the *Jacobian matrix* of f at y.

If $m = n$, determinant $([f_{i;j}(y)])$ is called the *Jacobian* of f at y.

Examples

1. The function $f: \mathbf{R}^n \longrightarrow \mathbf{R}^n$ defined by

$$f(x_1, \ldots, x_n) = (x_1^2, \ldots, x_n^2),$$

is partially differentiable in all variables, and the Jacobian matrix of f at $(y_1, \ldots, y_n) \in \mathbf{R}^n$ is the diagonal matrix

$$\begin{bmatrix} 2y_1 & 0 & \cdots & 0 \\ 0 & 2y_2 & \cdots & 0 \\ \vdots & \vdots & & \vdots \\ 0 & 0 & \cdots & 2y_n \end{bmatrix}.$$

The Jacobian of f at y is equal to $2^n y_1 \cdots y_n$.

2. Let us find the partial derivatives of the map $f: \mathbf{R}^3 \longrightarrow \mathbf{R}^2$ defined by

$$f(x, y, z) = (xye^z + x^2 \sin yz, \sin(\pi xy))$$

and hence the affine linear approximation to f at $(1, 1, 0)$. Computing the partial derivatives we find that

$$f_{1;1}(x, y, z) = ye^z + 2x \sin yz,$$

$$f_{1;2}(x, y, z) = xe^z + zx^2 \cos yz,$$

$$f_{1;3}(x, y, z) = xye^z + yx^2 \cos yz,$$

$$f_{2;1}(x, y, z) = \pi y \cos(\pi yx),$$

$$f_{2;2}(x, y, z) = \pi x \cos(\pi yx),$$

$$f_{2;3}(x, y, z) = 0.$$

Hence the partial derivatives at $(1, 1, 0)$ are given by

$$f_{1;1}(1, 1, 0) = 1 \qquad f_{1;2}(1, 1, 0) = 1 \qquad f_{1;3}(1, 1, 0) = 2$$

$$f_{2;1}(1, 1, 0) = -\pi \qquad f_{2;2}(1, 1, 0) = -\pi \qquad f_{2;3}(1, 1, 0) = 0.$$

If we *assume* that f is differentiable at $(1, 1, 0)$ it follows from Corollary 2.6.2.1 that the affine linear approximation to f at $(1, 1, 0)$ is given by the function

$$g(x_1, x_2, x_3) = c + (x_1 + x_2 + 2x_3, -\pi x_1 - \pi x_2),$$

where $c = (3, -2\pi)$.

3. All the material of this section works equally well for differentiable maps between complex vector spaces, except that the derivatives are always complex linear maps. Suppose, for example, that $f: \mathbf{C} \longrightarrow \mathbf{C}$ is continuous and differentiable at $z_0 \in \mathbf{C}$. Thus $Df_{z_0} \in L(\mathbf{C}, \mathbf{C}) \approx \mathbf{C}$. We may make the natural identification of $\mathbf{C}$ with $\mathbf{R}^2$ by mapping $x + iy$ to (x, y) (essentially we 'forget'

the complex structure of the complex plane). With this identification of **C** with $\mathbf{R}^2$, scalar multiplication by i acts on $\mathbf{R}^2$ as follows:

$$i(x,y) = (-y,x), \qquad (x,y) \in \mathbf{R}^2.$$

Now a (real) linear map $A \in L(\mathbf{R}^2, \mathbf{R}^2)$ is complex linear if and only if $Ai = iA$. Relative to the standard basis of $\mathbf{R}^2$, it follows easily that A is complex linear if and only if the matrix of A is of the form

$$\begin{bmatrix} a & -b \\ b & a \end{bmatrix}$$

for some constants $a, b \in \mathbf{R}$. Using our identification of $\mathbf{R}^2$ with **C** and the isomorphism $L(\mathbf{C}, \mathbf{C}) \approx \mathbf{C}$, A corresponds to the point $a + ib \in \mathbf{C}$. Now if $f: \mathbf{R}^2 \longrightarrow \mathbf{R}^2$ is differentiable at (x, y), the matrix of $Df_{(x,y)}$ is given, according to Proposition 2.6.3, by

$$\begin{bmatrix} f_{1;1}(x,y) & f_{1;2}(x,y) \\ f_{2;1}(x,y) & f_{2;2}(x,y) \end{bmatrix}.$$

Hence, $Df_{(x,y)}$ is complex linear if and only if

$$f_{1;1}(x,y) = f_{2;2}(x,y), \qquad f_{2;1}(x,y) = -f_{1;2}(x,y).$$

Making the identification of $\mathbf{R}^2$ with **C**, we may write $f = (f_1, f_2) = u + iv$, where $u = f_1$ is the real part of f and $v = f_2$ is the imaginary part of f. Rewriting the above conditions on the partial derivatives of f_1 and f_2 in terms of u and v we obtain the conditions

$$\partial u/\partial x = \partial v/\partial y \qquad \partial v/\partial x = -\,\partial u/\partial y,$$

which are the Cauchy-Riemann equations. Thus, to say that a function $f: \mathbf{C} \longrightarrow \mathbf{C}$ is complex differentiable at z is equivalent to saying that f satisfies the Cauchy-Riemann equations at z. In other words f is analytic on an open subset of the complex plane if and only if f is (complex) differentiable on the domain.

4. Let $f: \mathbf{R}^2 \longrightarrow \mathbf{R}^2$ be the function defined by

$$f(x,y) = (e^x \sin y, e^{-x} \cos y).$$

Assuming that f is differentiable everywhere on $\mathbf{R}^2$ it follows that the derivative of f at (x, y) is given in matrix form by

$$[Df_{(x,y)}] = \begin{bmatrix} e^x \sin y & e^x \cos y \\ -e^{-x} \cos y & -e^{-x} \sin y \end{bmatrix}$$

and hence the Jacobian determinant of f at (x,y) is $\cos^2 y - \sin^2 y$.

The next proposition describes the composite-mapping formula in terms of partial derivatives.

Proposition 2.6.5

Let A and B be open subsets of $\mathbf{R}^m$ and $\mathbf{R}^n$ respectively and suppose $f: A \longrightarrow \mathbf{R}^n$ is differentiable at $x \in A$, with Jacobian $[f_{i;j}(x)]$ and $g: B \longrightarrow \mathbf{R}^p$ is differentiable at $f(x) \in B$, with Jacobian $[g_{i;j}(f(x))]$. Then the composite $gf : A \cap f^{-1}(B) \longrightarrow \mathbf{R}^p$ is differentiable at x with Jacobian matrix

$$[g_{i;k}(f(x))]\,[f_{k;j}(x)] = \left[\sum_{k=1}^{n} g_{i;k}(f(x)) f_{k;j}(x)\right].$$

Proof. Follows immediately from the composite-mapping formula. ■

Examples

1. Let $f : \mathbf{R} \longrightarrow \mathbf{R}^n$ and $g : \mathbf{R}^n \longrightarrow \mathbf{R}$ be continuous functions, differentiable at t_0 and $f(t_0)$ respectively. Then the composite $p = gf : \mathbf{R} \longrightarrow \mathbf{R}$ is differentiable at t_0 and

$$p'(t_0) = \sum_{i=1}^{n} g_{;i}(f(t_0)) f_i'(t_0),$$

or, in classical partial-derivative notation,

$$\frac{dp}{dt} = \sum_{i=1}^{n} \frac{df_i}{dt}\frac{\partial g}{\partial x_i}.$$

2. Let $g : \mathbf{R} \longrightarrow \mathbf{R}^n$ and $f : \mathbf{R} \times \mathbf{R}^n \longrightarrow \mathbf{R}$ be continuous functions, differentiable at at t_0 and $(t_0, g(t_0))$ respectively. Then the function $V : \mathbf{R} \longrightarrow \mathbf{R}$ defined by $V(t) = f(t, f(t))$, $t \in \mathbf{R}$, is differentiable at t_0 and

$$\frac{dV}{dt} = \frac{\partial f}{\partial t} + \sum_{i=1}^{n} \frac{\partial f}{\partial x_i}\frac{dg_i}{dt},$$

where coordinates in $\mathbf{R} \times \mathbf{R}^n$ are written $(t, (x_1, \ldots, x_n))$. To see the truth of this formula, observe that V may be written as the composite $f(I, g)$, where

$$(I, g) : \mathbf{R} \longrightarrow \mathbf{R} \times \mathbf{R}^n$$

is defined by $(I, g)(t) = (t, g(t))$, $t \in \mathbf{R}$. From Proposition 2.5.2 it follows that (I, g) is differentiable at t_0 with derivative (I, Dg_{t_0}). From the composite-mapping formula it follows that V is differentiable at t_0 with derivative given by

$$\begin{aligned} DV_{t_0} &= Df_{(t_0, g(t_0))} \cdot (I, Dg_{t_0}) \\ &= D_1 f_{(t_0, g(t_0))} \cdot I + D_2 f_{(t_0, g(t_0))} \cdot Dg_{t_0}. \end{aligned}$$

Hence $V'(t_0) = DV_{t_0}(1)$ is given by

$$V'(t_0) = f_{;1}(t_0, g(t_0)) + D_2 f_{(t_0, g(t_0))}(g'(t_0)).$$

Writing this expression in classical partial-derivative notation gives the formula

$$\frac{dV}{dt} = \frac{\partial f}{\partial t} + \sum_{i=1}^{n} \frac{\partial f}{\partial x_i}\frac{dg_i}{dt}.$$

This formula is of some practical importance. Suppose, for example, that P is a particle moving in $\mathbf{R}^3$. The position and velocity of P may be described by a point in $\mathbf{R}^6$. Let the motion of P be given by a function $\phi : \mathbf{R} \longrightarrow \mathbf{R}^6$ and write ϕ in component form as $\phi = (x_1, x_2, x_3, x_1', x_2', x_3')$, where $x_1, \ldots, x_3'$ are regarded as functions from $\mathbf{R}$ to $\mathbf{R}$. The coordinates $x_1(t), x_2(t), x_3(t)$ give the position (in $\mathbf{R}^3$) of P at time t and the coordinates $x_1'(t), x_2'(t), x_3'(t)$ give the corresponding components of the velocity of P at time t.

Suppose that we are given a fixed 'energy' function V which depends on position, velocity and time: $V\colon \mathbf{R}\times\mathbf{R}^6 \longrightarrow \mathbf{R}$.

The rate of change of energy along the trajectory of P determined by ϕ is then given as the time derivative of $V(t, \phi(t))$. Thus,

$$\frac{\mathrm{d}V}{\mathrm{d}t} = \frac{\partial V}{\partial t} + \sum_{i=1}^{3}\left(\frac{\partial V}{\partial x_i}\frac{\mathrm{d}x_i}{\mathrm{d}t} + \frac{\partial V}{\partial x_i'}\frac{\mathrm{d}x_i'}{\mathrm{d}t}\right)$$

$$= \frac{\partial V}{\mathrm{d}t} + \sum_{i=1}^{3}\left(\frac{\partial V}{\partial x_i}x_i' + \frac{\partial V}{\partial x_i'}x_i''\right).$$

Taking the standard inner product on $\mathbf{R}^6$ we may equivalently write this formula as

$$\frac{\mathrm{d}V}{\mathrm{d}t} = \frac{\partial V}{\mathrm{d}t} + \langle V_{;2}(t, \phi(t)), \phi'(t)\rangle.$$

3. An important application of the material in Example 2 is to the stability theory of ordinary differential equations. An example will suffice to illustrate the methods: Consider the second-order linear differential equation

$$\frac{\mathrm{d}^2x}{\mathrm{d}t^2} + \frac{\mathrm{d}x}{\mathrm{d}t} + x = 0.$$

The general theory of ordinary differential equations tells us that corresponding to any initial condition $(x_0, x_0') \in \mathbf{R}^2$ there exists a unique solution $x(t)$ of the differential equation satisfying the conditions: $x(0) = x_0$; $x'(0) = x_0'$. Now $(x(t), x'(t))$ defines a curve in $\mathbf{R}^2$. Let $V\colon \mathbf{R}^+ \times \mathbf{R}^2 \longrightarrow \mathbf{R}$ be the 'energy' function defined by

$$V(t, (x, y)) = (1 + 1/t)x^2 + y^2, \qquad (x, y) \in \mathbf{R}^2, \quad t \in \mathbf{R}^+.$$

It follows from Example 2 that the time derivative of V along the trajectory $(x(t), x'(t))$ is given by

$$\frac{\mathrm{d}V}{\mathrm{d}t} = -x^2/t^2 + 2(1 + 1/t)xx' + 2x'x''.$$

Since x satisfies the differential equation $x'' + x' + x = 0$, it follows that $x'' = -x' - x$ and the formula for $\mathrm{d}V/\mathrm{d}t$ may be simplified to

$$\frac{\mathrm{d}V}{\mathrm{d}t} = -\frac{x^2}{t^2} + \frac{2xx'}{t} - 2x'^2, \qquad t > 0.$$

Recall that a quadratic form $ax^2 + 2bxy + cy^2$ is positive definite if and only if $b^2 < ac$. It follows that the form

$$\frac{x^2}{t^2} - \frac{2xy}{t} + 2y^2$$

is positive definite and hence that $\mathrm{d}V/\mathrm{d}t$ is less than zero for all $t > 0$. Hence $V(t, x(t), x'(t))$ is a decreasing function. Since V is positive for all $t > 0$, it follows that $\lim_{t\to\infty} V(t, x(t), x'(t))$ exists. It follows easily that this limit must be zero. Since

$$V(t, x(t), x'(t)) = (1 + 1/t)x(t)^2 + x'(t)^2$$

it follows that $\lim_{t\to\infty} x(t) = 0$. In physical terms: The origin is a stable equilibrium point for the differential equation $x'' + x' + x = 0$. Notice that we obtained this result without having to use an explicit formula for the solution $x(t)$. The method we use works equally well for non-linear equations provided we can find suitable energy functions. For more details we refer the interested reader to [12].

Suppose that E is a normed vector space and that u is a fixed vector of unit length in E. We denote the line in E which passes through u and the origin of E by L_u. Thus,

$$L_u = \{ku: k \in K\}.$$

If $x \in E$, we let $L_{u,x}$ denote the affine line

$$x + L_u = \{x + ku : k \in K\}.$$

$L_{u,x}$ is thus the image of the continuous affine linear map

$$\phi_{u,x} : K \longrightarrow E,$$

defined by

$$\phi_{u,x}(k) = x + ku, \qquad k \in K.$$

Suppose that A is an open subset of E. Then $\phi_{u,x}^{-1}(A)$ is an open subset of K (since $\phi_{u,x}$ is continuous). We set

$$A_{u,x} = \phi_{u,x}^{-1}(A)$$

Diagrammatically, this is shown in Fig. 2.8.

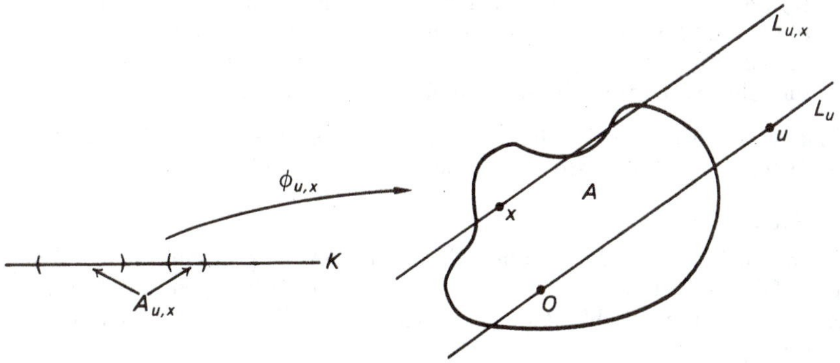

FIG. 2.8

Definition 2.6.6

Let E and F be normed vector spaces and let A be an open subset of E. Suppose that u is a vector (of unit length) in E. A function $f : A \longrightarrow F$ is said to be differentiable in direction u at $x \in A$ if the map

$$f_{u,x} = f\phi_{u,x} : A_{u,x} \subset K \longrightarrow F$$

is differentiable at 0.

We denote the derivative of $f_{u,x}$ at 0 by $D_u f_x \in L(K, E)$. We set $f_{;u}(x) = D_u f_x(1) \in F$. $f_{;u}(x)$ is called the *directional derivative* of f at x in direction u.

Remark. For f to be differentiable in direction u at x means essentially that f, restricted to the line $L_{u,x}$ through x, is differentiable at x. The following proposition makes this point clearer.

Proposition 2.6.7

Let E and F be normed vector spaces and let A be an open subset of E. Suppose u is a vector (of unit length) in E. A function $f : A \longrightarrow F$ is differentiable in direction u at $x \in A$ if and only if there exists a point $X_u \in F$ such that

$$\lim_{t \to 0} \left\| \frac{f(x+tu) - f(x)}{t} - X_u \right\| = 0.$$

If this limit exists, $X_u = f_{;u}(x)$, the directional derivative of f at x in direction u.

Proof. Left to the exercises. ∎

Examples

1. If $f : A \longrightarrow F$ is differentiable at x, then f is differentiable in all directions at x. Indeed, we claim that the derivative of f at x in direction u is just $Df_x(u)$. This follows since, with the notation of Definition 2.6.6, $f_{u,x} = f\phi_{u,x}$ and so, by the composite-mapping formula, $f_{u,x}$ is differentiable at $k = 0$ with derivative given by
$$Df_{u,x} = Df_x \,.\, D(\phi_{u,x})_0 = Df_x \,.\, \phi_{u,0}.$$
Evaluating at $k = 1$, we obtain
$$f_{;u}(x) = Df_x \,.\, \phi_{u,0}(1) = Df_x(u).$$
2. If $f : A \longrightarrow F$ is directionally differentiable in all directions at x it does not follow that f is differentiable at x. Indeed the example we used to show that partial differentiability at a point does not imply differentiability provides a suitable counterexample. We leave details to the exercises.

Finally, in this section, we show how directional derivatives may be used to obtain a matrix form for the derivative of a function between finite-dimensional vector spaces.

Proposition 2.6.8

Let E and F be finite-dimensional vector spaces over K and suppose that $\mathscr{E} = (e_\alpha)_{\alpha=1}^m$ and $\mathscr{F} = (f_\beta)_{\beta=1}^n$ are bases for E and F respectively. If A is an open subset of E and $g : A \longrightarrow F$ is differentiable at $x \in A$, the matrix of Dg_x relative to the bases $\mathscr{E}$ and $\mathscr{F}$ is given by

$$[Dg_x] = [g_{i;e_j}(x)],$$

where g is written in component form as $(g_1, \ldots, g_m)$ and $g_{i;e_j}(x)$ is the directional derivative of g_i in direction e_j.

Proof. The components g_i of g are the functions $g_i : A \longrightarrow K$, $1 \leqslant i \leqslant n$, satisfying the condition

$$g(x) = \sum_{i=1}^{n} g_i(x) f_i.$$

It follows from Section 2.5 that the functions g_i are all differentiable at x and hence, from the preceding examples, differentiable in all directions at x.

If a_{ij} denotes the ijth component of $[Dg_x]$, relative to the bases $\mathscr{E}$ and $\mathscr{F}$, we have

$$\begin{aligned} a_{ij} &= i\text{th component of } Dg_x(e_j) \\ &= i\text{th component of } (Dg_{1,x}(e_j), \ldots, Dg_{n,x}(e_j)) \\ &= Dg_{i,x}(e_j) = g_{i\,;e_j}(x). \quad \blacksquare \end{aligned}$$

Remarks

1. Proposition 2.6.3 is just a special case of the above proposition where we have taken the standard bases on $\mathbf{R}^m$ and $\mathbf{R}^n$.
2. In Proposition 2.6.8 we have not stipulated that the basis vectors are of unit length. Of course, what really matters in the definition of directional derivatives is that the vector is non-zero.

Exercises

1. Let $P : \mathbf{R}^n \longrightarrow \mathbf{R}$ be the pth-order homogeneous polynomial defined by
$$P(x_1, \ldots, x_n) = x_1^p + \cdots + x_n^p + x_1^{p-1}\, x_n.$$
Find D_jP_x, $1 \leqslant j \leqslant n$, and hence DP_x, $x \in \mathbf{R}^n$.
2. Let $f : \mathbf{R} \longrightarrow \mathbf{R}^3$ be defined by
$$f(t) = (e^t \sin t, e^{2t} \cos t, t^3)$$
and $g : \mathbf{R}^3 \longrightarrow \mathbf{R}^2$ be defined by
$$g(x_1, x_2, x_3) = (x_1x_3 + 1, x_2x_3 - x_1^2).$$
Assuming that f and g are differentiable on $\mathbf{R}$ and $\mathbf{R}^3$ respectively, find $(gf)'(t)$, $t \in \mathbf{R}$.
3. Let $f: \mathbf{R}^2 \longrightarrow \mathbf{R}^3$ and $g : \mathbf{R}^3 \longrightarrow \mathbf{R}^4$ be respectively defined by
$$f(x_1, x_2) = (x_1 \cos x_2, e^{x_1}x_2, x_1x_2), \qquad (x_1, x_2) \in \mathbf{R}^2,$$
$$g(x_1, x_2, x_3) = (x_1 + x_2, x_3x_2, x_3^2 \sin x_1, x_1x_2x_3), \quad (x_1, x_2, x_3) \in \mathbf{R}^3.$$
Find Df_x and Dg_y in matrix form and hence $D(gf)_{(1,0)}$.
4. Let $f: \mathbf{C}^n \longrightarrow \mathbf{C}$ be continuous. Suppose that f is differentiable as a map between real vector spaces at $z = (z_1, \ldots, z_n) \in \mathbf{C}^n$. Find a condition on the partial derivatives of f at z in order that Df_z be a complex linear map. (*Hint*: identify $\mathbf{C}^n$ with $\mathbf{R}^{2n}$.)
5. Consider the second-order non-linear differential equation
$$x'' + f(x)x' + g(x) = 0,$$
where f and g are continuous functions from $\mathbf{R}$ to $\mathbf{R}$ which satisfy the following conditions:

(a) $xg(x) > 0$, for all $x \neq 0$.
(b) $f(x) > 0$, for all $x \neq 0$.
(c) If $K(x) = \int_0^x g(t)\, dt$, then $K(x) \longrightarrow \infty$ as $|x| \longrightarrow \infty$.

Show, by considering the energy function $V(x, y) = K(x) + y^2/2$, that all solutions of the differential equation are bounded for $t \geqslant 0$, and that the origin is a stable equilibrium point of the system.

6. Prove Proposition 2.6.7.
7. Find a function on $\mathbf{R}^2$ that is directionally differentiable in all directions at (0, 0) but not differentiable there.
8. Let $f : \mathbf{R}^3 \longrightarrow \mathbf{R}$ be defined by

$$f(x,y,z) = x^2 + y^4 + z^2.$$

Find the directional derivative of f in direction $(1/\sqrt{2}, 1/\sqrt{2}, 0)$ at the point (1, 1, 1).

9. Find the affine linear approximation to the function

$$f(x,y,z) = (x \cos(2z - y) + e^{x-y}, z \cos(x + y + z)),$$

at the point (0, 0, 0). (You may assume that f is differentiable on $\mathbf{R}^3$.)

2.7 The mean-value theorem

In this section we prove one of the most important and useful results in analysis: the mean-value theorem. The mean-value theorem, when generalized to functions on normed vector spaces, no longer takes the form of an equality but is an inequality. This does not, however, diminish its usefulness.

Theorem 2.7.1 (Mean-value theorem)

Let $[a, b]$ be a closed interval of the real line and E be a normed vector space. Suppose the function $f: [a, b] \longrightarrow E$ satisfies the following conditions:

1. f is continuous on $[a, b]$.
2. f is differentiable at each point of (a, b).

Then,

$$\|f(b) - f(a)\| \leqslant (b - a) \sup_{x \in (a, b)} \|Df_x\|.$$

Proof. We may suppose that $f(b) \neq f(a)$, since otherwise the result is trivial. Let $\phi \in E^*$ and consider the composite

$$\phi f: [a, b] \longrightarrow \mathbf{R}.$$

Then, by the composite-mapping formula, ϕf is differentiable at all points of (a, b), and we may apply the classical mean-value theorem to obtain, for some $y \in (a, b)$,

$$\phi f(b) - \phi f(a) = (b - a)(\phi f)'(y).$$

If we notice that $|(\phi f)'(x)| = \|D(\phi f)_x\|$, it follows that

$$|\phi f(b) - \phi f(a)| \leqslant (b - a)\|D(\phi f)_y\| \leqslant (b - a) \sup_{x \in (a,b)} \|D(\phi f)_x\|.$$

Again, from the composite-mapping formula, it follows that

$$D(\phi f)_x = D\phi_{f(x)} \cdot Df_x, \qquad x \in (a, b)$$
$$= \phi \, . Df_x.$$

Since $\|\phi . Df_x\| \leqslant \|\phi\| \|Df_x\|$, it follows that we may rewrite the above inequality in the form

$$|\phi f(b) - \phi f(a)| \leqslant (b-a)\|\phi\| \sup_{x\in(a,b)} \|Df_x\| \qquad \text{for all } \phi \in E^*.$$

Consider the line L through the origin of E and the point $f(b) - f(a)$. We may define a continuous linear map

$$\tilde{\psi}: L \longrightarrow \mathbf{R},$$

by

$$\tilde{\psi}(t(f(b) - f(a)) = t\|f(b) - f(a)\|.$$

By the Hahn-Banach extension theorem, $\tilde{\psi}$ extends to an element $\psi \in E^*$ satisfying

(a) $\psi | L = \tilde{\psi}$
(b) $\|\tilde{\psi}\| = \|\psi\|$.

But $\|\tilde{\psi}\| = 1$ and $\tilde{\psi}(f(b) - f(a)) = \|f(b) - f(a)\|$. Hence substituting $\phi = \psi$ in the above inequality yields the required result. ■

Let x, y be points in the normed vector space E. We let $[x, x+y]$ denote the closed line segment in E defined by

$$[x, x+y] = \{x + \lambda y : 0 \leqslant \lambda \leqslant 1\}.$$

Corollary 2.7.1.1

Let E, F be normed vector spaces and A be an open subset of E. Suppose $[x, x+y]$ is a line segment of E contained in A. Then, if $f: A \longrightarrow F$ is continuous and differentiable on $[x, x+y]$, we have

$$\|f(x+y) - f(x)\| \leqslant \|y\| \sup_{0<\theta<1} \|Df_{x+\theta y}\|.$$

Proof. We define the map $\phi: [0, 1] \longrightarrow E$ by $\phi(\theta) = x + \theta y$. Then ϕ is continuous on $[0, 1]$ and differentiable on $(0, 1)$. We let

$$G: [0, 1] \longrightarrow F$$

denote the composite $G = f\phi$. From Theorem 2.7.1 it follows that

$$\|G(1) - G(0)\| \leqslant \sup_{0<\theta<1} \|DG_\theta\|.$$

From the composite-mapping formula, we have

$$\begin{aligned} DG_\theta(t) &= Df_{\phi(\theta)} . D\phi_\theta(t), \qquad t \in \mathbf{R}. \\ &= Df_{x+\theta y}(ty). \end{aligned}$$

Hence,

$$\begin{aligned} \|DG_\theta\| &= \sup_{|t|\leqslant 1} \|Df_{x+\theta y}(ty)\| \\ &= \sup_{|t|\leqslant 1} |t| \|Df_{x+\theta y}(y)\| = \|Df_{x+\theta y}(y)\| \leqslant \|y\| \|Df_{x+\theta y}\|. \end{aligned}$$

Observing that $G(1) - G(0) = f(x+y) - f(x)$ and substituting in the above inequality for $\|G(1) - G(0)\|$ yields the result. ■

Remark. In general we do not have equality in Theorem 2.7.1. The reason for this is that the tangent plane to the graph of f need not be parallel to the chord defined by $(a, f(a))$ and $(b, f(b))$. The following examples make this point clearer.

1. Define $f: \mathbf{R} \longrightarrow \mathbf{R}^2$ by $f(x) = (x^2, x^3)$. Then $f'(x) = (2x, 3x^2)$ and $f(1) - f(-1) = (0, 2) \neq 2f'(x)$ for any $x \in (-1, +1)$. (See Fig. 2.9.)

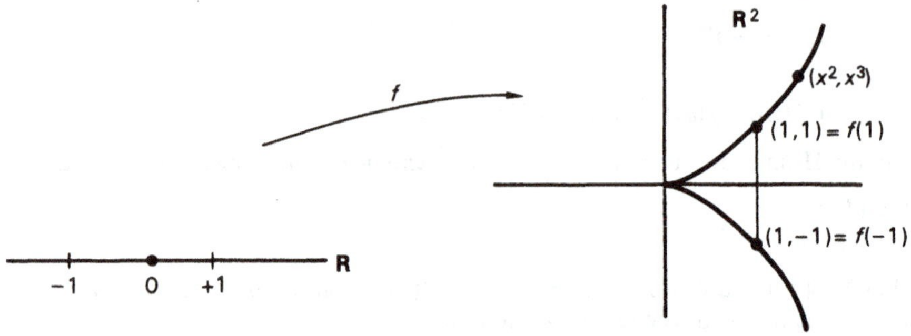

FIG. 2.9

2. Define the helix in $\mathbf{R}^3$ by $g(t) = (\cos t, \sin t, t)$. Then $g(0) = (1, 0, 0)$ and $g(2\pi) = (1, 0, 2\pi)$. We leave it to the reader to verify that the tangent line to $g(t)$ is never parallel to the line joining $g(0)$ and $g(2\pi)$.

Theorem 2.7.2

Let E and F be normed vector spaces and let A be an open connected subset of E. Then, if $f: A \longrightarrow F$ is differentiable on A with derivative everywhere zero on A, f is constant on A.

Proof. Fix $x_0 \in A$ and let $A_0 = \{x \in A : f(x) = f(x_0)\}$. Then A_0 is certainly a closed non-empty subset of A. On the other hand, let $y \in A_0$ and pick $r > 0$ such that $B_r(y) \subset A$. If $z \in B_r(y)$, it follows from Corollary 2.7.1.1 that

$$\| f(z) - f(y) \| \leqslant \| z - y \| 0 = 0.$$

Hence $f(z) = f(y) = f(x_0)$ for all $z \in B_r(y)$ and so A_0 is also open. Since A was assumed connected it follows that $A_0 = A$ and that f is constant on A. ■

Exercises

1. Show that the mean-value theorem can be written in the form of an equality for real-valued functions defined on open subsets of normed vector spaces.
2. Prove the mean-value theorem for functions taking values in a Hilbert space, by using the Riesz representation theorem as opposed to the Hahn–Banach theorem.
3. Let E, F be normed vector spaces and let A be an open subset of E. Suppose that $f: A \longrightarrow F$ satisfies the following conditions: For all $x \in A$, there exists $u(x) \in L(E, F)$ such that

$$\lim_{t \to 0} \frac{f(x + ty) - f(x)}{t} = u(x)(y)$$

and the map $x \longmapsto u(x)$ is continuous.

By applying the mean-value theorem to the map

$$t \longmapsto f(x+ty) - u(x)(ty),$$

prove that f is C^1 on A and that $Df_x = u(x)$ for all $x \in A$.

4. In this example we outline an alternative proof of the mean-value theorem which does not use the Hahn–Banach theorem. With the assumptions and notation of Theorem 2.7.1, let us set

$$M = \sup_{x \in (a,b)} \| Df_x \|.$$

Without loss of generality we may assume that $M \neq \infty$. Let $\epsilon > 0$ and let $\xi \in [a,b]$ denote the supremum of points $t \in [a,b]$ such that for all $s \in [a,t)$ we have

$$\| f(a) - f(s) \| \leqslant (s-a)(M+\epsilon) + \epsilon. \tag{A}$$

Show
(a) $\xi > a$.
(b) $\xi = b$: Show that if we assume $\xi < b$ then we may derive a contradiction by using the differentiability of f at ξ to express $f(\xi + h)$ in terms of $f(\xi - h)$ and $Df_\xi, h > 0$, together with eq. (A).
(c) Using (b) deduce the mean-value theorem.

2.8 Differentiation and partial differentiation

In Section 2.6 we showed that whilst differentiability implied partial differentiability the converse was not necessarily true. In this section we wish to give sufficient conditions for the converse to hold. The reader should note the importance of the mean-value theorem in these arguments.

Theorem 2.8.1

Let E_1, E_2, F be normed vector spaces and A be an open subset of $E_1 \times E_2$. Suppose $f: A \longrightarrow F$ is continuous and that there exist open sets A_1 and A_2 of E_1 and E_2 respectively such that

1. $A_1 \times A_2 \subset A$.
2. f is partially differentiable with respect to the first variable at $(a_1, a_2) \in A_1 \times A_2$.
3. f is partially differentiable with respect to the second variable throughout $A_1 \times A_2$.
4. The map

$$D_2 f; A_1 \times A_2 \longrightarrow L(E_2, F); a \longmapsto D_2 f_a$$

is continuous at (a_1, a_2).

Then f is differentiable at (a_1, a_2) with derivative given by

$$Df_{(a_1, a_2)} = D_1 f_{(a_1, a_2)} \cdot p_1 + D_2 f_{(a_1, a_2)} \cdot p_2.$$

Proof. Let $(h, k) \in E_1 \times E_2$; then

$$\begin{aligned} f(a_1 + h, a_2 + k) - f(a_1, a_2) &= (f(a_1 + h, a_2 + k) - f(a_1 + h, a_2)) \\ &\quad + (f(a_1 + h, a_2) - f(a_1, a_2)). \end{aligned}$$

Let us first consider the second term on the right-hand side of the above equality. Since f is partially differentiable at (a_1, a_2) with respect to the first variable, we have

$$f(a_1 + h, a_2) - f(a_1, a_2) = D_1 f_{(a_1, a_2)}(h) + o(h).$$

Since $\|h\| \leqslant \|(h, k)\|$ and $o(h)$ is *independent of* k it follows that $o(h) = o(h, k)$ and so

$$f(a_1 + h, a_2) - f(a_1, a_2) = D_1 f_{(a_1, a_2)}(h) + o(h, k). \tag{A}$$

The analysis of the term $f(a_1 + h, a_2 + k) - f(a_1 + h, a_2)$ is not so straightforward. We clearly would like to prove that

$$f(a_1 + h, a_2 + k) - f(a_1 + h, a_2) - D_2 f_{(a_1, a_2)}(k) = o(h, k).$$

Let us therefore define, for fixed h, the function g by

$$g(k) = f(a_1 + h, a_2 + k) - f(a_1 + h, a_2) - D_2 f_{(a_1, a_2)}(k).$$

Since f is partially differentiable with respect to the second variable throughout $A_1 \times A_2$, it follows that g is differentiable at k whenever $(a_1 + h, a_2 + k) \in A_1 \times A_2$ with derivative

$$Dg_k = D_2 f_{(a_1 + h, a_2 + k)} - D_2 f_{(a_1, a_2)}.$$

Now $D_2 f$ is assumed continuous at (a_1, a_2) and so, given $\epsilon > 0$, $\exists r > 0$ such that

$$\| Dg_k \| = \| D_2 f_{(a_1 + h, a_2 + k)} - D_2 f_{(a_1, a_2)} \| \leqslant \epsilon, \qquad \|(h, k)\| \leqslant r.$$

We now apply the mean-value theorem to the function g to obtain

$$\|g(k) - g(0)\| \leqslant \|k\| \sup_{0 < \theta < 1} \|Dg_{\theta k}\|.$$

Since $g(0) = 0$, it follows that

$$\|g(k)\| \leqslant \|k\| \sup_{0 < \theta < 1} \|Dg_{\theta k}\|.$$

But if $\|(h, k)\| \leqslant r$, it follows that $\|Dg_k\| \leqslant \epsilon$, and so we obtain the estimate

$$\|g(k)\| \leqslant \epsilon \|k\|, \qquad \|(h, k)\| \leqslant r.$$

Since $\|k\| \leqslant \|(h, k)\|$ it follows that

$$\|g(k)\| \leqslant \epsilon \|(h, k)\|, \qquad \|(h, k)\| \leqslant r.$$

Since $\epsilon > 0$ was arbitrary it follows that $g(k) = o(h, k)$ and that

$$f(a_1 + h, a_2 + k) - f(a_1 + h, a_2) = D_2 f_{(a_1, a_2)}(k) + o(h, k). \tag{B}$$

Combining eqs. (A) and (B) it follows immediately that f is differentiable at (a_1, a_2) and that the derivative of f is given by

$$Df_{(a_1, a_2)} = D_1 f_{(a_1, a_2)} \cdot p_1 + D_2 f_{(a_1, a_2)} \cdot p_2. \quad ■$$

Definition 2.8.2

Let $E_1, \ldots, E_n, F$ be normed vector spaces and let A be an open subset of the product $E_1 \times \cdots \times E_n$. Suppose that $f: A \longrightarrow F$ is partially differentiable with respect to the jth variable at every point of A. Then we say that f is partially differentiable with respect to the jth variable on A. If the map

$$D_j f : A \longrightarrow L(E_j, F),$$

defined by mapping $a \in A$ to $D_j f_a$, is continuous, we say that f is continuously partially differentiable with respect to the jth variable on A. If f is continuously partially differentiable with respect to *all* variables on A we say that f is continuously partially differentiable on A.

Before stating the next theorem we wish to introduce some new notation.

Notation. Recall that if f and g are functions we write their composite as fg. If, on the other hand, A and B are linear maps we write their composite as $A \cdot B$. We now wish to generalize this 'dot' notation. Suppose, then, that

$$f : A \longrightarrow L(F, G),$$

$$g : A \longrightarrow L(G, H),$$

where E, F, G, H are normed vector spaces and A is an open subset of E. We let $g \cdot f : A \longrightarrow L(F, H)$ denote the function defined by

$$(g \cdot f)(x) = g(x) \cdot f(x), \qquad x \in A.$$

We leave it as an easy exercise for the reader to verify that, if f and g are continuous, so is $g \cdot f$ (cf. the proof of the C^1 version of the composite-mapping formula). In applications, either f or g is often a constant function. Suppose, for example, that $f(x) = T \in L(F, G)$ for all $x \in A$. Then $g \cdot T$ is the function whose value at $x \in A$ $g(x) \cdot T \in L(F, H)$.

Theorem 2.8.3

Let E_1, E_2, F be normed vector spaces and let A be an open subset of $E_1 \times E_2$. Let $f : A \longrightarrow F$ be continuous. The following are equivalent:

1. f is continuously differentiable on A.
2. f *is* continuously partially differentiable on A.

If either (1) or (2) holds, the derivative and partial derivatives of f are related by the formulae

$$D_j f = Df \cdot I_j, \qquad j = 1, 2,$$

$$Df = D_1 f \cdot p_1 + D_2 f \cdot p_2.$$

Proof

(1) implies (2). From Proposition 2.6.2, f is partially differentiable with respect to either variable on A and

$$D_j f = Df \cdot I_j, \qquad j = 1, 2.$$

Since Df is assumed continuous and $I_j \in L(E_j, F), j = 1, 2$, it follows immediately from the remarks preceding the statement of the theorem that $D_j f$ is continuous, $j = 1, 2$.

(2) implies (1). From Theorem 2.8.1, f is differentiable on A and

$$Df = D_1 f . p_1 + D_2 f . p_2.$$

Since $D_j f$ is assumed continuous, $j = 1, 2$, and the projection maps p_1 and p_2 are continuous, it follows that Df is the sum of continuous maps and is therefore continuous. ■

Corollary 2.8.3.1

Let $E_1, \ldots, E_n, F$ be normed vector spaces and let A be an open subset of $E_1 \times \cdots \times E_n$. Suppose $f : A \longrightarrow F$ is continuous. Then f is continuously differentiable on A if and only if f is continuously partially differentiable on A. The derivatives and partial derivatives of f are related by the formulae

$$D_j f = Df . I_j, \qquad 1 \leqslant j \leqslant n$$

$$Df = \sum_{j=1}^{n} D_j f . p_j.$$

Proof. Follows inductively from Theorem 2.8.3. ■

Corollary 2.8.3.1 is particularly useful when interpreted in terms of Jacobian matrices.

Proposition 2.8.4

Let A be an open subset of $\mathbf{R}^m$ and suppose that $f = (f_1, \ldots, f_m) : A \longrightarrow \mathbf{R}^n$ is continuously partially differentiable in all variables on A. Then f is C^1 on A and the matrix of the derivative of f at $y \in A$ is given in terms of the partial derivatives at y by

$$[Df_y] = [f_{i;j}(y)].$$

Proof. Immediate from Corollary 2.8.3.1 and Proposition 2.6.3. ■

Examples

1. Let $f : \mathbf{R}^3 \longrightarrow \mathbf{R}^3$ be the function defined by

$$f(x_1, x_2, x_3) = (x_1 e^{x_2}, x_3^2 + x_3^4 x_1, \log(1 + x_1^2)), \qquad (x_1, x_2, x_3) \in \mathbf{R}^3.$$

The component functions of f are immediately seen to be continuously partially differentiable on $\mathbf{R}^3$ and hence the derivative of f at $(y_1, y_2, y_3) \in \mathbf{R}^3$ is given by

$$[Df_{(y_1, y_2, y_3)}] = \begin{bmatrix} e^{y_2} & y_1 e^{y_2} & 0 \\ y_3^4 & 0 & 2y_3 + 4y_3^3 y_1 \\ 2y_1/(1 + y_1^2) & 0 & 0 \end{bmatrix}.$$

The reader should note that a *direct* verification that f is differentiable on $\mathbf{R}^3$, without the use of Proposition 2.8.4, would involve considerable computation.

2. Let $T \in L(E_1, \ldots, E_n; F)$, where $E_1, \ldots, E_n, F$ are normed vector spaces. Then T is certainly continuously partially differentiable on $E_1 \times \cdots \times E_n$ with partial derivatives given by

$$D_j T_{(y_1, \ldots, y_n)}(x) = T(y_1, \ldots, y_{j-1}, x, y_{j+1}, \ldots, y_n), \qquad x \in E_j, \quad 1 \leqslant j \leqslant n$$

This is immediate since T is continuous and linear in the jth variable.

Hence, by Corollary 2.8.3.1, T is differentiable on $E_1 \times \cdots \times E_n$, with derivative given by

$$DT_{(y_1, \ldots, y_n)}(x_1, \ldots, x_n) = \sum_{j=1}^{n} D_j T_{(y_1, \ldots, y_n)}(x_j),$$

which is the formula which we obtained for the derivative in Section 2.3.

Notation. Before stating the next proposition we wish to generalize our 'dot' notation somewhat further. Suppose, therefore, that E, F_1, F_2, G are normed vector spaces and we are given a *fixed* element $\phi \in L(F_1, F_2; G)$. Then we shall write

$$\phi(f_1, f_2) = f_1 . f_2 \qquad \text{for all } (f_1, f_2) \in F_1 \times F_2.$$

If A is an open subset of E and $f: A \longrightarrow F_1, g: A \longrightarrow F_2$, then we shall let $f.g: A \longrightarrow G$ denote the function defined by

$$(f.g)(x) = f(x).g(x), \qquad x \in A.$$

Recall that $L^0(E; F)$ is defined to be F. For $p, q \geqslant 0$, ϕ induces a continuous bilinear map

$$L^p(E; F_1) \times L^q(E; F_2) \longrightarrow L^{p+q}(E; G),$$

denoted by

$$(S, T) \longmapsto S.T, \qquad S \in L^p(E; F_1), \quad T \in L^q(E; F_2).$$

according to the rule

$$(S.T)(e_1, \ldots, e_{p+q}) = S(e_1, \ldots, e_p).T(e_{p+1}, \ldots, e_{p+q})$$

for all

$$(e_1, \ldots, e_{p+q}) \in \times^{p+q} E.$$

In particular, if $p = q = 0$, this composition is precisely ϕ. Other important special cases are when $p = 0, q = 1$, in which case we have a composition

$$F_1 \times L(E, F_2) \longrightarrow L(E, G)$$

defined by

$$(f.T)(e) = f.T(e), \qquad f \in F_1, \quad T \in L(E, F_2), \quad e \in E,$$

and when $p = 1, q = 0$, in which case we have a composition

$$L(E, F_1) \times F_2 \longrightarrow L(E, G)$$

defined by

$$(S.f)(e) = S(e).f, \qquad S \in L(E, F_1), \quad f \in F_2, \quad e \in E.$$

If $f: A \longrightarrow L^p(E; F_1)$ and $g: A \longrightarrow L^q(E; F_2)$, then we let $f.g: A \longrightarrow L^{p+q}(E, G)$ denote the function defined by

$$(f.g)(x) = f(x).g(x), \qquad x \in A.$$

Let us look at some characteristic examples of compositions ϕ that we shall meet in the sequel.

Examples

1. The *evaluation map* ev: $L(E,F) \times E \longrightarrow F$ is defined by

$$\mathrm{ev}\,(T,e) = T(e), \qquad T \in L(E,F), \quad e \in E.$$

Suppose that A is an open subset of the normed vector space G and that $f: A \longrightarrow L(E,F), g: A \longrightarrow E$. Then

$$f.g: A \longrightarrow F$$

is defined by

$$(f.g)(x) = \mathrm{ev}\,(f(x), g(x)) = f(x)(g(x))$$

A particularly important case of this composition occurs when g is a constant function. For example, suppose that $g(x) = e$ for all $x \in A$. We shall write

$$f.e: A \longrightarrow F$$

to denote the function whose value at $x \in A$ is $f(x)(e)$.

2. Composition of linear maps: $L(F,G) \times L(E,F) \longrightarrow L(E,G); (S,T) \longmapsto S.T.$ We have already met this composition in, for example, the expression of the derivative of a function in terms of the partial derivatives and projection maps.

The following proposition provides the natural generalization of the 'product' rule for differentiation.

Proposition 2.8.5

Let E, F_1, F_2, G be normed vector spaces and let $\phi \in L(F_1, F_2; G)$. Suppose that A is an open subset of E and that $f: A \longrightarrow F_1$ and $g: A \longrightarrow F_2$ are differentiable at x. Then $f.g: A \longrightarrow G$ is differentiable at x with derivative given by

$$D(f.g)_x = f(x).Dg_x + Df_x.g(x),$$

where $f(x).Dg_x$ is formed from the composition $F_1 \times L(E, F_2) \longrightarrow L(E,G)$ and $Df_x.g(x)$ is formed from the composition $L(E,F_1) \times F_2 \longrightarrow L(E,G)$.

Proof. We first remark that $f.g$ may be written as the composite

$$\phi(f,g),$$

where $(f,g): A \longrightarrow F_1 \times F_2$ is the map $x \longmapsto (f(x), g(x))$. By Proposition 2.5.1, (f,g) is differentiable at x and

$$D(f,g)_x = (Df_x, Dg_x).$$

Since ϕ is a continuous bilinear map, ϕ is C^1 on $F_1 \times F_2$ and

$$D\phi_{(g_1,g_2)}(f_1,f_2) = \phi(g_1,f_2) + \phi(f_1,g_2),$$

where $(g_1,g_2), (f_1,f_2) \in F_1 \times F_2$.

It follows from the composite-mapping formula that $f.g$ is differentiable at x and that

$$D(f.g)_x = D\phi_{(f(x),g(x))}.(Df_x, Dg_x).$$

Let $e \in E$. Then

$$D(f.g)_x(e) = D\phi_{(f(x),g(x))}(Df_x(e), Dg_x(e))$$
$$= \phi(f(x), Dg_x(e)) + \phi(Df_x(e), g(x)).$$

Since this relation holds for all $e \in E$, it follows, by the definition of 'dot' that

$$D(f.g)_x = f(x).Dg_x + Df_x.g(x). \quad \blacksquare$$

Remark. If, with the notation of Proposition 2.8.5, f and g are C^1 on A, so is $f.g$ and $D(f.g) = f.Dg + Df.g$. Notice, however, that the two compositions on the right-hand side of this equality are not the same, though they are, of course, both induced from ϕ.

Let E, F, G be normed vector spaces and let $r \geqslant 1$. We shall use abbreviated notation for the evaluation map

$$\mathrm{ev}_r : L^r(E; F) \times \mathsf{X}^r E \longrightarrow F$$

and write

$$\mathrm{ev}_r(T, (e_1, \ldots, e_r)) = T.(e_1, \ldots, e_r),$$

where $T \in L^r(E; F)$ and $(e_1, \ldots, e_r) \in \mathsf{X}^r E$.

Notice that ev_r is an $(r+1)$-linear map ($\mathrm{ev}_r \in L(L^r(E;F), E, \ldots, E; F)$). We have the following lemma which will be of great importance in the sequel, especially in our studies of higher derivatives.

Lemma 2.8.6 (Evaluation lemma)

Let E, F, G be normed vector spaces and let U be an open subset of G. Suppose that $f: U \longrightarrow L^r(E; F)$ is differentiable at $x \in U$. Then, if $(e_1, \ldots, e_r)$ is a fixed point of $\mathsf{X}^r E$, the map $f.(e_1, \ldots, e_r): U \longrightarrow F$ is differentiable at x and

$$D(f.(e_1, \ldots, e_r))_x = Df_x.(e_1, \ldots, e_r).$$

In other words: *Differentiation commutes with evaluation at a fixed point.*

Proof. Since f is differentiable at x we have the following equation in $L^r(E; F)$:

$$f(x+g) = f(x) + Df_x(g) + o(g), \qquad g \in U - x.$$

Evaluating this equation at the point $(e_1, \ldots, e_r)$ yields

$$(f.(e_1, \ldots, e_r))(x+g) = (f.(e_1, \ldots, e_r))(x) + (Df_x(g).(e_1, \ldots, e_r)) + o(g).$$

It follows immediately, by the definition of derivative, that $f.(e_1, \ldots, e_r)$ is differentiable at x with derivative $Df_x.(e_1, \ldots, e_r) \in L(G, F)$. $\blacksquare$

Example. The evaluation lemma can be useful in computing derivatives. For example, suppose that $f: \mathbf{R} \longrightarrow L(\mathbf{R}^n, \mathbf{R}^n)$ is a C^1 map. Relative to the standard basis $(e_i)_{i=1}^n$ of $\mathbf{R}^n$, we may write f in matrix form $f = [a_{ij}]$, where $a_{ij}: \mathbf{R} \longrightarrow \mathbf{R}$, $1 \leqslant i, j \leqslant n$. Let

$f'(t) = [b_{ij}]$, $t \in \mathbf{R}$. Then

$$\begin{aligned} b_{ij}(t) &= i\text{th component of } f'(t) \,.\, e_j \\ &= i\text{th component of } (f \,.\, e_j)'(t), \\ &= a'_{ij}(t). \end{aligned}$$

Hence $f' = [a'_{ij}]$.

Exercises

1. Comment on the proof of the following 'theorem'

Theorem

Let A be an open subset of $E_1 \times E_2$ and suppose $f: A \longrightarrow F$ is continuous. Assume $D_2 f_{(x,\, b)}$ exists as a continuous function of x, $x \in B_d(a)$, $d > 0$, and also that $D_1 f_{(a,\, b)}$ exists. Then f is differentiable at (a, b).

Proof. Let $(h, k) \in E_1 \times E_2$. Then

$$f(a + h, b + k) - f(a, b) = f(a + h, b + k) - f(a + h, b) + f(a + h, b) - f(a, b).$$

Now

$$f(a + h, b) - f(a, b) = D_1 f_{(a,\, b)}(h) + o(h)$$

and

$$f(a + h, b + k) - f(a + h, b) = D_2 f_{(a+h,\, b)}(k) + o(k).$$

Given $\epsilon > 0$, there exists d', $0 < d' \leqslant d$, such that

$$\| D_2 f_{(a+h, b)} - D_2 f_{(a,\, b)} \| < \epsilon,$$

for $\| h \| \leqslant d'$. Hence,

$$\| D_2 f_{(a+h,\, b)}(k) - D_2 f_{(a,\, b)}(k) \| \leqslant \epsilon \| k \| \leqslant \epsilon \| (h, k) \|,$$

for $\| (h, k) \| \leqslant d'$. Therefore,

$$D_2 f_{(a+h,\, b)}(k) = D_2 f_{(a,\, b)}(k) + o(h, k)$$

and so

$$f(a + h, b + k) = f(a, b) + D_1 f_{(a,\, b)}(h) + D_2 f_{(a,\, b)}(k) + o(h, k) + o(h) + o(k).$$

But $\| h \| \leqslant \| (h, k) \|$ and $\| k \| \leqslant \| (h, k) \|$, therefore $o(h) = o(h, k)$ and $o(k) = o(h, k)$ and so f is differentiable at (a, b). ■

Produce a counterexample to show that the theorem is false.

2. Show that the function $f: \mathbf{R}^n \setminus \{0\} \longrightarrow \mathbf{R}$, defined by

$$f(x_1, \ldots, x_n) = 1/(x_1^2 + \cdots + x_n^2)^{1/2},$$

is C^1 and find $[Df_x]$, $x \in \mathbf{R}^n$, relative to the standard basis on $\mathbf{R}^n$.

3. Let $f: \mathbf{R} \longrightarrow L(\mathbf{R}^n, \mathbf{R}^n)$ and $g: \mathbf{R} \longrightarrow L(\mathbf{R}^n, \mathbf{R})$ be the maps respectively defined by:

$$f(t) = \begin{bmatrix} t & 0 & \dots & 0 \\ 0 & t^2 & \dots & 0 \\ \vdots & \vdots & & \vdots \\ 0 & 0 & \dots & t^n \end{bmatrix}, \quad t \in \mathbf{R} \text{ and } g(t) = [t^2, t^4, \dots, t^{2n}], \quad t \in \mathbf{R},$$

where the matrices are given relative to the standard bases on $\mathbf{R}^n$. Using Proposition 2.8.5, find $(g \,.\, f)'(t)$.

4. Prove Proposition 2.8.5 directly in the special case when $\phi: L(E, F) \times E \longrightarrow F$ is the evaluation map $\phi(A, x) = A(x)$.

2.9 Higher derivatives

Definition 2.9.1

Let E and F be normed vector spaces and let A be an open subset of E. Suppose that f is a C^1 function on A. We say that f is *twice differentiable* at the point $x \in A$, if the map

$$Df: A \longrightarrow L(E, F)$$

is differentiable at x. We set $D(Df)_x = D^2 f_x$ and call $D^2 f_x$ the *second derivative* of f at x.

With the notation and assumptions of Definition 2.9.1 suppose that f is twice differentiable at x. Then $D^2 f_x \in L(E, L(E, F))$. Using the canonical isomorphism $L(E, L(E, F)) \approx L^2(E; F)$, we shall almost always regard $D^2 f_x$ as defining a point of $L^2(E; F)$.

Notice that the differentiability of Df at x implies that we have the following equation in $L(E, F)$:

$$Df_{x+h} = Df_x + D^2 f_x(h) + o(h),$$

where $o(h), D^2 f_x(h) \in L(E, F)$. Thus, if $k \in E$, we have

$$Df_{x+h}(k) = Df_x(h) + D^2 f_x(h)[k] + o(h)(k),$$

or, regarding $D^2 f_x$ as an element of $L^2(E; F)$, the equation

$$Df_{x+h}(k) = Df_x(k) + D^2 f_x(h, k) + o(h)(k).$$

Now since $o(h) \in L(E, F)$, it follows that $\| o(h)(k) \| \leqslant \| o(h) \| \, \| k \|$, and so $o(h)(k) = o(h, k)$. We have finally, therefore, the equation

$$Df_{x+h}(k) = Df_x(k) + D^2 f_x(h, k) + o(h, k).$$

Definition 2.9.2

Let E, F be normed vector spaces and let A be an open subset of E. If $f: A \longrightarrow F$ is twice differentiable at every point of A we say f is twice differentiable on A. If, moreover, the map

$$D^2 f: A \longrightarrow L^2(E; F),$$

which maps $x \in A$ to $D^2 f_x$, is continuous we say that f is twice continuously differentiable on A or just C^2 on A.

Proceeding inductively, if f is $(p-1)$ times continuously differentiable on A, we say that f is p times differentiable at the point $x_0 \in A$ if the map

$$D^{p-1} f : A \longrightarrow L^{p-1}(E;F),$$

which maps $x \in A$ to $D^{p-1} f_x$, is differentiable at x_0. We denote the pth derivative of f at x by $D^p f_x$. If f is p times differentiable at every point of A we say that f is p times differentiable on A. If, in addition, the map

$$D^p f : A \longrightarrow L^p(E;F)$$

is continuous, we say that f is p times continuously differentiable on A or just C^p (on A). If f is C^p on A for $p = 1, 2, \ldots$ we say that f is C^∞ (on A) or infinitely differentiable on A.

Remark. In the above definition we make use of the identification between $L^p(E;F)$ and $L(E, L^{p-1}(E;F))$ given by Theorem 1.6.5.

Examples

1. Let $u \in L(E,F)$. Then u is C^1 on E and $Du : E \longrightarrow L(E,F)$ is the constant map, $Du = u$. Consequently, $D^2 u_x = 0$ for all $x \in E$. More generally

$$D^p u = 0, \qquad p \geqslant 2.$$

2. Let $T \in L_s^n(E;F)$ and let $P \in H^n(E;F)$ denote the polynomial associated with T. Then P is C^1 on E and, as we have previously shown, the derivative of P at $x \in E$ is given by

$$DP_x(y) = nT(x^{n-1}y) \qquad \text{for all } y \in E.$$

We claim that $DP \in H^{n-1}(E; L(E,F))$. Indeed, if we define $T_1 \in L_s^{n-1}(E; L(E,F)$ by

$$T_1(x_1, \ldots, x_{n-1})(y) = T(x_1, \ldots, x_{n-1}, y)$$
$$\text{for all } x_1, \ldots, x_{n-1}, \quad y \in E,$$

we have

$$DP = n(T_1 \,.\, \Delta).$$

Hence $D^2 P$ is a homogeneous polynomial of degree $n-2$ lying in the space $H^{n-2}(E; L_s^2(E;F))$ and characterized by

$$D^2 P_x(z) = n(n-1)T_1(x^{n-2}z) \qquad \text{for all } x, z \in E.$$

If we make the identification of $L(E, L(E,F))$ with $L^2(E;F)$ it follows that for all $x \in E$ we have

$$D^2 P_x(z, y) = n(n-1)T(x^{n-2}zy)$$

for all $z, y \in E$.

Proceeding inductively, it follows that $D^r P \in H^{n-r}(E; L_s^r(E;F))$. In particular,

$$D^r P = 0, \qquad r \geqslant (\text{degree of } P) + 1.$$

In our study of higher derivatives we shall follow a similar pattern to that used in the analysis of derivatives and first examine maps into products.

Suppose $E, F_1, \ldots, F_n$ are normed vector spaces. We have a natural isomorphism

$$L^p(E; F_1 \times \cdots \times F_n) \approx \mathop{\times}_{j=1}^{n} L^p(E; F_j), \qquad p \geqslant 1,$$

defined by mapping

$$A \in L^p(E; F_1 \times \cdots \times F_n)$$

to

$$(p_1 . A, \ldots, p_n . A) \in \mathop{\times}_{j=1}^{n} L^p(E; F_j).$$

Proposition 2.9.3

Let $E, F_1, \ldots, F_n$ be normed vector spaces and let A be an open subset of E. Suppose

$$f = (f_1, \ldots, f_n): A \longrightarrow F_1 \times \cdots \times F_n.$$

Then, f is p times differentiable at $x \in A$ if and only if each component function f_i, $1 \leqslant i \leqslant n$, is p times differentiable at x.

The derivatives of f and the components of f are related by

$$D^p f_x = \sum_{j=1}^{n} I_j . D^p f_{j,x}; D^p f_{j,x} = p_j . D^p f_x, \qquad 1 \leqslant j \leqslant n,$$

or, in terms of the isomorphism between $L^p(E; F_1 \times \cdots \times F_n)$ and $\times_{j=1}^{n} L(E; F_j)$, by

$$D^p f_x = (D^p f_{1,x}, \ldots, D^p f_{n,x}).$$

Proof. We proceed by induction on p. The result is certainly true for $p = 1$ by Proposition 2.5.1. Assume the result is true for $p - 1$ and suppose f is p times differentiable at $x \in A$. Then f is C^{p-1} on A and

$$D^{p-1} f_j = p_j . D^{p-1} f, \qquad 1 \leqslant j \leqslant n,$$

by the inductive hypothesis. Differentiating the above equality once, it follows from Proposition 2.8.5 that f_j is p times differentiable at x and that

$$D^p f_{j,x} = p_j . D^p f_x, \qquad 1 \leqslant j \leqslant n.$$

The reverse implication follows similarly. ■

Suppose now, with the notation and assumptions of Proposition 2.9.3, that A is an open subset of the real line $\mathbf{R}$.

Suppose

$$f = (f_1, \ldots, f_n) : A \longrightarrow F_1 \times \cdots \times F_n$$

is twice differentiable at x. Then

$$D^2 f_x \in L^2(\mathbf{R}; F_1 \times \cdots \times F_n).$$

Noting that $D^2 f_x$ is uniquely determined by its value at the point $(1, 1) \in \mathbf{R}^2$, we set

$$f''(x) = D^2 f_x(1, 1), \qquad \text{and} \qquad f_j''(x) = D^2 f_{j,x}(1, 1), \qquad 1 \leqslant j \leqslant n.$$

It follows immediately from Proposition 2.9.3 that

$$f''(x) = (f_1''(x), \ldots, f_n''(x)) \in F_1 \times \cdots \times F_n.$$

More generally, if f is p times differentiable at x, we set

$$f^{(p)}(x) = D^p f_x(1, \ldots, 1),$$

and

$$f_j^{(p)}(x) = D^p f_{j,x}(1, \ldots, 1), \qquad 1 \leqslant j \leqslant n.$$

It follows again from Proposition 2.9.3 that

$$f^{(p)}(x) = (f_1^{(p)}(x), \ldots, f_n^{(p)}(x)) \in F_1 \times \cdots \times F_n.$$

The case when $F_1 = F_2 = \cdots = F_n = \mathbf{R}$ is particularly important.

Proposition 2.9.4

Let A be an open subset of $\mathbf{R}$ and suppose $f = (f_1, \ldots, f_n) : A \longrightarrow \mathbf{R}^n$. Then f is p times differentiable at $x \in A$ if and only if each component function is p times differentiable at x.

The derivatives of f and the components of f are related by

$$f^{(p)}(x) = (f_1^{(p)}(x), \ldots, f_n^{(p)}(x))$$

or, in classical derivative notation, by

$$\frac{\mathrm{d}^p f}{\mathrm{d}x^p} = \left(\frac{\mathrm{d}^p f_1}{\mathrm{d}x^p}, \ldots, \frac{\mathrm{d}^p f_n}{\mathrm{d}x^p}\right).$$

Example. Let $f: \mathbf{R} \longrightarrow \mathbf{R}^3$ be defined by

$$f(t) = (e^t, e^{2t}, e^{-t});$$

then f is C^∞ on $\mathbf{R}$ and

$$f^{(p)}(t) = (e^t, 2^p e^{2t}, (-1)^p e^{-t}), \qquad t \in \mathbf{R}, \quad p \geqslant 1.$$

Suppose that A is an open subset of $\mathbf{R}$ and that $f: A \longrightarrow F$ is $r + s$ times differentiable at x. We claim that $f^{(r)}: A \longrightarrow F$ is s times differentiable at x and that

$$(f^{(r)})^{(s)} = f^{(r+s)}.$$

Since

$$f^{(p)}(x) = D^p f_x(1, 1, \ldots, 1), \qquad 1 \leqslant p \leqslant r + s,$$

this result follows by an easy induction from the next proposition.

Proposition 2.9.5

Let E and F be normed vector spaces and let A be an open subset of E. Suppose that $f: A \longrightarrow F$ is p times differentiable at x and let $(x_2, \ldots, x_p) \in \mathsf{X}^{p-1}E$. Let

$$D^{p-1}f.(x_2, \ldots, x_p): A \longrightarrow F$$

be the map defined by

$$(D^{p-1}f.(x_2, \ldots, x_p))(y) = D^{p-1}f_y(x_2, \ldots, x_p), \qquad y \in A.$$

Then $D^{p-1}f.(x_2, \ldots, x_p)$ is differentiable at x with derivative given by

$$D(D^{p-1}f.(x_2, \ldots, x_p))_x(x_1) = D^p f_x(x_1, \ldots, x_p), \qquad x_1 \in E.$$

Proof. An immediate consequence of the evaluation lemma (Lemma 2.8.6) applied to the function $D^{p-1}f$. ■

Before beginning our study of higher derivatives of maps from products we shall prove a fundamental theorem about the symmetry of higher derivatives.

Theorem 2.9.6

Let E and F be normed vector spaces and let A be an open subset of E. If $f: A \longrightarrow F$ is twice differentiable at x, then

$$D^2 f_x \in L_s^2(E; F).$$

Proof. We must prove that $D^2 f_x(h, k) = D^2 f_x(k, h)$ for all $h, k \in E$. Choose $d > 0$ such that $\bar{B}_{2d}(x) \subset A$ and assume in all that follows that $(h, k) \in \bar{B}_d(x)$, that is, $\|(h, k)\| \leqslant d$.

Consider the map $S: \bar{B}_d(0) \times \bar{B}_d(0) \longrightarrow F$ defined by

$$S(h, k) = f(x + h + k) - f(x + h) - f(x + k) + f(x).$$

Clearly S is symmetric: $S(h, k) = S(k, h)$ for all $(h, k) \in \bar{B}_d(0) \times \bar{B}_d(0)$. We shall show that

$$S(h, k) - D^2 f_x(h, k) = \gamma(h, k),$$

where $\gamma(h, k)/\|(h, k)\|^2 \longrightarrow 0$ as $(h, k) \longrightarrow 0$. It will then follow easily from the symmetry of S and the continuous bilinearity of $D^2 f_x$ that $D^2 f_x$ is symmetric.

We define $g: [0, 1] \longrightarrow F$ by

$$g(t) = f(x + h + tk) - f(x + tk) - tD^2 f_x(h, k).$$

Notice that

$$g(1) - g(0) = S(h, k) - D^2 f_x(h, k).$$

g is clearly continuous on $[0, 1]$ and differentiable on $(0, 1)$ and so it follows from the mean-value theorem that

$$\|g(1) - g(0)\| \leqslant \sup_{0<t<1} \|Dg_t\| = \sup_{0<t<1} \|g'(t)\|.$$

Computing $g'(t)$ we find

$$\begin{aligned} g'(t) &= Df_{(x+tk+h)}(k) - Df_{(x+tk)}(k) - D^2 f_x(h, k) \\ &= (Df_{(x+tk+h)}(k) - Df_x(k)) - (Df_{(x+tk)}(k) - Df_x(k)) - D^2 f_x(h, k). \end{aligned}$$

Now Df is differentiable at x. Hence, given $\epsilon > 0$, $\exists d_1, 0 < d_1 \leqslant d$, such that, if $\|m\| \leqslant 2d_1$,

$$Df_{x+m} = Df_x + D^2 f_x(m) + r(m),$$

where $\|r(m)\| \leqslant \epsilon \|m\|$. Evaluating this equation in $L(E, F)$ at the point $k \in E$ it follows that

$$Df_{x+m}(k) = Df_x(k) + D^2 f_x(m, k) + r(m, k),$$

where

$$\|r(m,k)\| = \|r(m)(k)\| \leqslant \epsilon \|m\| \|k\|, \qquad \|(m,k)\| \leqslant d_1.$$

Substituting in our expression for $g'(t)$ we have

$$g'(t) = D^2f_x(tk+h,k) - D^2f_x(tk,k) - D^2f_x(h,k) + r(tk+h,k) - r(tk,k).$$

The first three terms on the right-hand side cancel, since D^2f_x is bilinear. Hence,

$$g'(t) = r(tk+h,k) - r(tk,k),$$

and so

$$\begin{aligned} \|g'(t)\| &\leqslant \|r(tk+h,k)\| + \|r(tk,k)\| \\ &\leqslant \epsilon \|k\|(\|tk+h\| + \|tk\|), \qquad \text{for } \|(h,k)\| \leqslant d_1 \\ &\leqslant 2\epsilon \|k\|(\|h\| + \|k\|). \end{aligned}$$

Substituting this estimate for $g'(t)$ in our estimate of $\|g(1) - g(0)\|$ it follows that

$$\|g(1) - g(0)\| \leqslant 2\epsilon \|k\|(\|h\| + \|k\|), \qquad \|(h,k)\| \leqslant d_1.$$

That is

$$\|S(h,k) - D^2f_x(h,k)\| \leqslant 2\epsilon \|k\|(\|h\| + \|k\|), \qquad \|(h,k)\| \leqslant d_1.$$

Since S is symmetric, it follows from the triangle inequality that

$$\|D^2f_x(h,k) - D^2f_x(k,h)\| \leqslant 2\epsilon(\|h\| + \|k\|)^2, \qquad \|(h,k)\| \leqslant d_1.$$

Since $\|h\|, \|k\| \leqslant \|(h,k)\|$, we arrive at the final estimate

$$\|D^2f_x(h,k) - D^2f_x(k,h)\| \leqslant 8\epsilon \|(h,k)\|^2, \qquad \|(h,k)\| \leqslant d_1.$$

But the bilinearity of D^2f_x implies that this estimate holds for all $(h,k) \in E \times E$. Since $\epsilon > 0$ was arbitrary, it follows that

$$D^2f_x(h,k) = D^2f_x(k,h) \qquad \text{for all } h,k \in E. \ \blacksquare$$

Corollary 2.9.6.1

Let E and F be normed vector spaces and let A be an open subset of E. Then, if $f: A \longrightarrow F$ is p times differentiable at x,

$$D^pf_x \in L_s^p(E;F), \qquad p \geqslant 1.$$

Proof. We prove the result by induction on p. Thus let us assume that the result is true for derivatives of order less than or equal to $p-1$. Let $(e_1, \ldots, e_p) \in \mathsf{X}^p E$. Then, from Proposition 2.9.5, the map $D^{p-1}f.(e_2, \ldots, e_p): A \longrightarrow F$ is differentiable at x with derivative given by

$$D(D^{p-1}f.(e_2, \ldots, e_p))_x(e_1) = D^pf_x(e_1, \ldots, e_p).$$

It follows, from the inductive hypothesis, that D^pf_x is symmetric in the last $p-1$ variables. On the other hand, consider

$$D^{p-2}f.(e_3, \ldots, e_p): A \longrightarrow F.$$

Again by Proposition 2.9.5 this map is twice differentiable at x and

$$D^2(D^{p-2}f.(e_3, \ldots, e_p))_x(e_1, e_2) = D^pf_x(e_1, \ldots, e_p).$$

But by Theorem 2.9.6 it follows that $D^p f_x$ must be symmetric in the first two variables. Combining this with the fact that $D^p f_x$ is symmetric in the last $p-1$ variables it follows that $D^p f_x$ is symmetric. ■

Definition 2.9.7

Let $E_1, \ldots, E_n, F$ be normed vector spaces and let A be an open subset of $E_1 \times \cdots \times E_n$. Suppose that $f: A \longrightarrow F$ is C^1 on A and that the map

$$D_j f: A \longrightarrow L(E_j, F)$$

is differentiable at x. Then we set

$$D_{ij} f_x = D_i(D_j f)_x, \qquad 1 \leqslant i \leqslant n.$$

Notice that

$$D_{ij} f_x \in L(E_i, L(E_j, F)) \approx L(E_i, E_j; F).$$

If $D_j f$ is differentiable on A, we let

$$D_{ij} f: A \longrightarrow L(E_i, E_j; F)$$

denote the map which takes value $D_{ij} f_x$ at x.

Proposition 2.9.8

Let $E_1, \ldots, E_n, F$ be normed vector spaces and let A be an open subset of $E_1 \times \cdots \times E_n$. If the map $f: A \longrightarrow F$ is twice differentiable at x, then the partial-derivative maps

$$D_j f: A \longrightarrow L(E_j, F), \qquad 1 \leqslant j \leqslant n,$$

are differentiable at x and

$$D_{ij} f_x = D_{ji} f_x, \qquad 1 \leqslant i, j \leqslant n,$$

in the sense that $D_{ij} f_x(h, k) = D_{ji} f_x(k, h)$ for all $(h, k) \in E_i \times E_j$.

Moreover, we have the following relations between $D^2 f_x$ and $D_{ij} f_x$:

$$D_{ij} f_x = D^2 f_x \cdot (I_i \times I_j), \qquad 1 \leqslant i, j \leqslant n, \tag{1}$$

$$D^2 f_x = \sum_{i,j=1}^{n} D_{ij} f_x \cdot (p_i \times p_j). \tag{2}$$

Proof. We first remark that it is sufficient to prove eq. (1). Indeed, (1) obviously implies (2) (by substitution in the right-hand side of (2)). That $D_{ij} f_x = D_{ji} f_x$ is an immediate consequence of (1) together with Theorem 2.9.6.

Since $D_j f = Df . I_j$, it follows from the composite-mapping formula, or Proposition 2.8.5, that $D_j f$ is differentiable at x. Let $(h, k) \in E_i \times E_j$. We shall follow the notation of Proposition 2.9.5 and the evaluation lemma (Lemma 2.8.6). We have

$$\begin{aligned} D_{ij} f_x(h, k) &= D(D_j f)_x(I_i(h), k) && \text{(definition of } D_{ij}) \\ &= D(D_j f . k)_x(I_i(h)) && \text{(evaluation lemma)} \\ &= D(Df . I_j(k))(I_i(h)) && \text{(definition of } D_j) \\ &= D^2 f_x(I_i(h), I_j(k)) && \text{(Proposition 2.9.5).} \end{aligned}$$

Since this equality holds for all $(h, k) \in E_i \times E_j$, (1) is proved. ■

Notation. Let $E_1, \ldots, E_n, F$ be normed vector spaces and let A be an open subset of $E_1 \times \cdots \times E_n$. Suppose that $f: A \longrightarrow F$ and $j_1, \ldots, j_p$ are integers satisfying $1 \leqslant j_1, \ldots, j_p \leqslant n$. Then by

$$D_{j_1 \cdots j_p} f_x, \qquad x \in A,$$

is meant

$$D_{j_1}(D_{j_2}(\cdots(D_{j_p} f)\cdots))_x,$$

assuming that f, and its derivatives, are sufficiently differentiable. Also we let

$$D_{j_1 \cdots j_p} f: A \longrightarrow L(E_{j_1}, \ldots, E_{j_p}; F)$$

denote the corresponding derivative function.

If $i_1, \ldots, i_n$ are integers satisfying $i_1, \ldots, i_n \geqslant 0$, we let

$$D_1^{i_1} \cdots D_n^{i_n} f_x, \qquad x \in A,$$

denote

$$\underbrace{D_1(\cdots D_1}_{i_1}(D_2 \cdots \underbrace{(D_n f)\cdots)}_{i_n}\cdots)_x,$$

it being understood that if $i_s = 0$ no differentiations are performed with respect to the sth variable.

Proposition 2.9.9

Let $E_1, \ldots, E_n, F$ be normed vector spaces and let A be an open subset of $E_1 \times \cdots \times E_n$. Suppose that $f: A \longrightarrow F$ is p times differentiable at x. Then, if $j_1, \ldots, j_p$ are integers satisfying $1 \leqslant j_1, \ldots, j_p \leqslant n$, the derivatives $D_{j_1 \cdots j_p} f_x$ all exist and, in addition, we have

1. $D_{j_1 \cdots j_p} f_x$ is symmetric in the j's
$$D_{j_{\sigma(1)} \cdots j_{\sigma(p)}} f_x = D_{j_1 \cdots j_p} f_x \qquad \text{for all } \sigma \in S_p$$
in the sense of Proposition 2.9.8.
2. $D_{j_1 \cdots j_p} f_x = D^p f_x . (I_{j_1} \times \cdots \times I_{j_p})$.
3. $$D^p f_x = \sum_{j_1, \ldots, j_p = 1}^{n} D_{j_1 \cdots j_p} f_x . (p_{j_1} \times \cdots \times p_{j_p}).$$
4. $$D^p f_x = \sum_{|m| = p} \frac{p!}{m_1! \cdots m_n!} D_1^{m_1} \cdots D_n^{m_n} f_x . (p_1^{m_1} \times \cdots \times p_n^{m_n}).$$

Proof. As in the proof of Proposition 2.9.8 it is sufficient to show that the derivative $D_{j_1 \cdots j_p} f_x$ exists and that (2) holds. Indeed (2) implies (3), and (2), together with Corollary 2.9.6.1, implies (1). (4) follows from (1) and (3) just as in the proof of Lemma 1.7.8.

The proof of (2) goes by induction on p and uses the evaluation lemma. The details are similar to those of Proposition 2.9.8 and we leave them to the reader. ■

Examples

1. Let $T \in L_s^n(E; F)$ and let $P \in H^n(E; F)$ denote the corresponding polynomial. Then,

$$D^r P \in H^{n-r}(E; L_s^r(E; F))$$

and is characterized by

$$D^r P_x(e_1, \ldots, e_r) = n(n-1) \cdots (n-r+1) T(x^{n-r} e_1 e_2 \cdots e_r),$$
$$(e_1, \ldots, e_r) \in \mathrm{X}^r E.$$

2. Let $T \in L(E_1, \ldots, E_n; F)$. Since T is linear in each variable separately, $D_{ii}T = 0$, $1 \leqslant i \leqslant n$. It follows that $D_{j_1 \ldots j_p} T = 0$ if any two of the j's are equal. In particular, $D^{n+1} T = 0$. We also have, from Proposition 2.9.9,

$$D^p T = \sum_{\substack{|m| = p \\ m_i = 0, 1}} \frac{p!}{m_1! \cdots m_n!} D_1^{m_1} \cdots D_n^{m_n} T, \qquad p \geqslant 1.$$

Suppose that A is an open subset of $\mathbf{R}^n$ and that F is a normed vector space. Let $f: A \longrightarrow F$ be p times differentiable at $x \in A$. Then, for $1 \leqslant j_1, \ldots, j_p \leqslant n$, we shall define the elements $f_{;j_1 \cdots j_p}(x) \in F$ by

$$f_{;j_1 \cdots j_p}(x) = D_{j_1 \ldots j_p} f_x(1, 1, \ldots, 1),$$

where we notice that

$$D_{j_1 \cdots j_p} f_x \in L(\mathbf{R}, \mathbf{R}, \ldots, \mathbf{R}; F) = L^p(\mathbf{R}; F).$$

If we let $(e_j)_{j=1}^n$ denote the standard basis of $\mathbf{R}^n$ it follows from condition (2) of Proposition 2.9.9 that

$$f_{;j_1 \cdots j_p}(x) = D^p f_x(e_{j_1}, \ldots, e_{j_p}).$$

This is immediate since $I_j(1) = e_j$, $1 \leqslant j \leqslant n$.

It follows from Proposition 2.9.5 that

$$f_{;j_1 \cdots j_p}(x) = (f_{;j_2 \cdots j_p})_{;j_1}(x)$$
$$= (\ldots (f_{;j_p})_{;j_{p-1}} \cdots)_{;j_1}(x).$$

Consequently, the evaluation of $f_{;j_1 \ldots j_p}(x)$ reduces to the successive computation of partial derivatives.

The collection $\{f_{;j_1 \ldots j_p}(x) : 1 \leqslant j_1, \ldots, j_p \leqslant n\}$ is called the set of pth-order partial derivatives at x. Any vector $f_{;j_1 \ldots j_p}(x)$ is called a pth-order partial derivative of f at x.

We remark the alternative classical notation for the pth-order partial derivative $f_{;j_1 \cdots j_p}(x)$

$$\frac{\partial^p f}{\partial x_{j_1} \cdots \partial x_{j_p}}.$$

Thus, when $A \subset \mathbf{R}^n$ and $F = \mathbf{R}$, Proposition 2.9.8 reduces to the well-known symmetry condition

$$\frac{\partial^2 f}{\partial x_i \partial x_j} = \frac{\partial^2 f}{\partial x_j \partial x_i}, \qquad 1 \leqslant i, j \leqslant n.$$

Finally, suppose $m = (m_1, \ldots, m_n) \in \mathbf{N}^n$ is a multi-index with $|m| = p$. Then by $f_{;m}(x)$ is meant the pth partial derivative of f at x corresponding to differentiation with respect to x_1, m_1 times, . . ., differentiation with respect to x_n, m_n times. In classical notation

$$f_{;m}(x) = \frac{\partial^p f}{\partial x_1^{m_1} \cdots \partial x_n^{m_n}}.$$

The next proposition is just a restatement of Proposition 2.9.9 in the above notation.

Proposition 2.9.10

Let F be a normed vector space and let A be an open subset of $\mathbf{R}^n$. Suppose $f: A \longrightarrow F$ is p times differentiable at y. Then $f_{;j_1 \cdots j_p}(y) \in F$ is a symmetric function of the j's and we have

$$D^p f_y = \sum_{j_1, \ldots, j_p = 1}^{n} f_{;j_1 \cdots j_p}(y)(p_{j_1} \times \cdots \times p_{j_p})$$

$$= \sum_{|m| = p} \frac{p!}{m_1! \cdots m_n!} f_{;m}(y)\,(p_1^{m_1} \times \cdots \times p_n^{m_n}).$$

Examples

1. Let A be an open subset of $\mathbf{R}^2$ and let $f: A \longrightarrow \mathbf{R}$ be twice differentiable on A. Then $f_{;12} = f_{;21}$ on A. For example, if $f(x, y) = x \cos y + y^2 \cos x$, f is twice differentiable on $\mathbf{R}^2$ and the second partial derivatives of f are

$$f_{;11}(x, y) = -y^2 \cos x; f_{;22}(x, y) = -x \cos y + 2 \cos x,$$

$$f_{;12}(x, y) = -\sin y - 2y \sin x = f_{;21}(x, y).$$

$D^2 f_{(x,y)}$ is given in terms of the partial derivatives $f_{;ij}(x, y)$ by the formula

$$D^2 f_{(x,y)} = \sum_{i,j=1}^{n} f_{;ij}(x, y)(p_i \times p_j).$$

Thus, if $(h_1, k_1), (h_2, k_2) \in \mathbf{R}^2$, we have

$$D^2 f_{(x,y)}((h_1, k_1), (h_2, k_2)) = \sum_{i,j=1} h_i k_j f_{;ij}(x, y).$$

2. Let $f: \mathbf{R}^3 \longrightarrow \mathbf{R}$ be the C^∞ function defined by

$$f(x, y, z) = xe^y + (z - x) \cos y + z \sin x.$$

We shall find the set of second-order partial derivatives of f at $(0, 0, 0)$. We have

$$f_{;11}(x, y, z) = -z \sin x; f_{;22}(x, y, z) = xe^y - (z - x) \cos y; f_{;33}(x, y, z) =$$

$$f_{;12}(x, y, z) = e^y + \sin y; f_{;13}(x, y, z) = \cos x; f_{;23}(x, y, z) = -\sin y.$$

Substituting, we find that

$$f_{;11}(0, 0, 0) = f_{;22}(0, 0, 0) = f_{;33}(0, 0, 0) = f_{;23}(0, 0, 0) = 0$$

and

$$f_{;12}(0, 0, 0) = f_{;13}(0, 0, 0) = 1.$$

3. Let E be a normed vector space, A an open subset of E and suppose that $f: A \longrightarrow \mathbf{R}$ is twice differentiable on A. Then $Df: A \longrightarrow L(E, \mathbf{R}) = E^*$ and so

$$D^2 f: A \longrightarrow L(E, E^*).$$

Suppose that $E = \mathbf{R}^n$ and that $\mathbf{R}^n$ is given the standard basis $(e_j)_{j=1}^n$. Let us find, for $x \in A$, the matrix of $D^2 f_x$ relative to the given basis on $\mathbf{R}^n$ and the corresponding dual basis of $\mathbf{R}^{n*}$. Set $[D^2 f_x] = [a_{ij}]$. We have

$$\begin{aligned} a_{ij} &= i\text{th component of } D^2 f_x(e_j) \\ &= (D^2 f_x(e_j))(e_i) \qquad \text{(definition of dual basis)} \\ &= D^2 f_x(e_j, e_i) \qquad \text{(using } L(E, E^*) \approx L^2(E; \mathbf{R})\text{)} \\ &= f_{;ji}(x). \end{aligned}$$

Hence $[D^2 f_x] = [f_{;ji}(x)]$ and therefore $[D^2 f_x]$ is a symmetric matrix.

Exercises

1. Prove Proposition 2.9.5.
2. Let $f: \mathbf{R} \longrightarrow \mathbf{R}^3$ be defined by $f(t) = (t^6, e^{2t}, t^3)$. Show that f is C^∞ and find $f^{(p)}(t)$ for all $p \geqslant 1$. Hence find $D^p f_t$, $t \in \mathbf{R}$ and $p \geqslant 1$.
3. Complete the proof of Proposition 2.9.9.
4. Let $E_1, \ldots, E_n, F$ be normed vector spaces and let A be an open subset of $E_1 \times \cdots \times E_n$. Suppose $f: A \longrightarrow F$ is p times differentiable at x. If $e = (e_1, \ldots, e_n) \in E_1 \times \cdots \times E_n$, find an expression for $D^p f_x(e^p)$ in terms of $D_1^{m_1} \cdots D_n^{m_n} f_x$, where $m_1 + \cdots + m_n = p$.
5. Find the set of third-order partial derivatives at $(0, 0)$ of the function $f(x, y) = e^x y + \sin(x^2 + y)$.
6. Let $f = (f_1, f_2): \mathbf{R}^2 \longrightarrow \mathbf{R}^2$ be defined by $f(x, y) = (x \sin(x + y), ye^x \sin x)$. Find the second-order partial derivatives $f_{1;11}(0, 0), f_{2;12}(0, 1), f_{2;11}(1, 0)$.
7. Let A be an open subset of $\mathbf{R}^2$ and suppose that $f: A \longrightarrow \mathbf{R}$ is C^1 and that $D_{21} f$ exists and is continuous on A. By considering the function

$$g(t) = f(x + t, y + k) - f(x + t, y) - D_{21} f_{(x, y)}(tk), \qquad t \in [0, 1],$$

show that $D_{12} f$ exists on A and is equal to $D_{21} f$. Let

$$f(x, y) = xy(x^2 - y^2)/(x^2 + y^2), \qquad (x, y) \neq (0, 0),$$

$$f(0, 0) = 0.$$

Verify that f is C^1 on $\mathbf{R}^2$ and that the derivatives $D_{11} f, D_{12} f, D_{21} f$ and $D_{22} f$ all exist on $\mathbf{R}^2$ but $D_{21} f \neq D_{12} f$ at $(0, 0)$.
8. Outline the generalization of directional derivatives to derivatives of higher order.

2.10 Leibniz' theorem and the general composite-mapping formula

Let us recall that the classical Leibniz theorem states that

$$(f.g)^{(p)}(t) = \sum_{r=0}^{p} \binom{p}{r} f^{(r)}(t) g^{(p-r)}(t),$$

where f and g are real-valued p times differentiable functions on an open interval of the real line. Here 'dot' denotes multiplication of functions and $\binom{p}{r}$ the binomial coefficient $p!/r!\,(p-r)!$.

Suppose that E, F_1, F_2, G are normed vector spaces and $\phi \in L(F_1, F_2; G)$ is a fixed bilinear map. Then, if $f : A \longrightarrow F_1, g : A \longrightarrow F_2$, where A is an open subset of E, we may define

$$f.g : A \longrightarrow G.$$

If f and g are differentiable on A, then Proposition 2.8.5 gives a Leibniz theorem for the case $p = 1$

$$D(f.g) = Df.g + f.Dg.$$

Using Proposition 2.8.5 inductively we might hope to achieve a Leibniz theorem, for the case when f and g are p times differentiable, in the form

$$D^p(f.g) = \sum_{r=0}^{p} \binom{p}{r} D^r f.D^{p-r}g.$$

However, inspection of the above formula shows that the terms on the right-hand side are not symmetric forms (unlesss $r = 0$ or p) and so the above formula is not true as stated. The problem arises because the dot products in $Df.g + f.Dg$ are induced from the different compositions

$$L(E, F_1) \times F_2 \longrightarrow G \qquad \text{and} \qquad F_1 \times L(E, F_2) \longrightarrow G.$$

Thus, we are not entitled to combine the two $Df.Dg$ terms arising from differentiation of $Df.g + f.Dg$ without further investigation.

What we shall do is to define a 'symmetric' dot product associated with ϕ.

Definition 2.10.1

Let E, F_1, F_2, G be normed vector spaces and suppose we are given a fixed element $\phi \in L(F_1, F_2; G)$. Then, for $p, q \geqslant 0$, we define the symmetric dot product

$$\odot : L_s^p(E; F_1) \times L_s^q(E; F_2) \longrightarrow L_s^{p+q}(E; G)$$

by the formula

$$(A \odot B)(e_1, \ldots, e_{p+q}) = \frac{1}{(p+q)!} \sum_{\sigma \in S_{p+q}} A(e_{\sigma(1)}, \ldots,$$

$$e_{\sigma(p)}).B(e_{\sigma(p+1)}, \ldots, e_{\sigma(p+q)}),$$

where $A \in L_s^p(E; F_1), B \in L_s^q(E; F_2)$ and $(e_1, \ldots, e_{p+q}) \in \mathsf{X}^{p+q} E$.

Remarks. Notice that $A \odot B$ is certainly a *symmetric* G-valued $(p+q)$-linear map. Observe that, if $p = q = 1$, then

$$A \odot B(e, f) = \tfrac{1}{2}(A(e).B(f) + A(f).B(e)), \qquad A \in L(E, F_1), \quad B \in L(E, F_2).$$

If, for example, $A \in F_1 (= L_s^0(E; F_1))$ and $B \in L^q(E; F_2)$, then $A \odot B$ is defined by

$$A \odot B(e_1, \ldots, e_q) = A.B(e_1, \ldots, e_q)), \qquad (e_1, \ldots, e_q) \in \mathsf{X}^q E.$$

We may now state the general version of Leibniz' theorem.

Theorem 2.10.2

Let E, F_1, F_2, G be normed vector spaces and let ϕ be a fixed element of $L(F_1, F_2; G)$. Suppose that A is an open subset of E and that $f: A \longrightarrow F_1$ and $g: A \longrightarrow F_2$ are both p times differentiable at x. Then

$$f.g : A \longrightarrow G$$

is p times differentiable at x with derivative given by

$$D^p(f.g)_x = \sum_{r=0}^{p} \binom{p}{r} D^r f_x \odot D^{p-r} g_x.$$

Proof. We prove by induction on p. The theorem is certainly true for $p = 1$ by Proposition 2.8.5 and the preceding remarks. Suppose we have proved it for $p - 1$. Thus we have

$$D^{p-1}(f.g) = \sum_{r=0}^{p-1} \binom{p-1}{r} D^r f \odot D^{p-r-1} g.$$

It follows easily from the composite-mapping formula (or Proposition 2.8.5) that $f.g$ is p times differentiable at x. Fix $(e_2, \ldots, e_p) \in \mathsf{X}^{p-1}E$. We have, with the notation of Proposition 2.9.5,

$$D^{p-1}(f.g).(e_2, \ldots, e_p) = \sum_{r=0}^{p-1} \binom{p-1}{r} (D^r f \odot D^{p-r-1} g).(e_2, \ldots, e_p).$$

Differentiating once, it follows from Proposition 2.9.5 that

$$D^p(f.g)_x(e_1, \ldots, e_p) = \sum_{r=0}^{p-1} \binom{p-1}{r} D((D^r f \odot D^{p-r-1} g).(e_2, \ldots, e_p))_x(e_1).$$

Expanding the terms on the right-hand side and using the identity

$$\binom{p-1}{r-1} + \binom{p-1}{r} = \binom{p}{r}$$

yields the result. We leave the details to the exercises. ■

Example. Let F_1, F_2, G be normed vector spaces and let $\phi \in L(F_1, F_2; G)$. Suppose that A is an open subset of the real line **R** and that $f: A \longrightarrow F_1$ and $g: A \longrightarrow F_2$ are p times differentiable at x. Then $f.g: A \longrightarrow G$ is p times differentiable at x with derivative given by

$$(f.g)^{(p)}(x) = \sum_{r=0}^{p} \binom{p}{r} f^{(r)}(x).g^{(p-r)}(x).$$

In the remainder of this section we wish to consider the composite-mapping formula for p times differentiable maps. The following theorem is of great importance.

Theorem 2.10.3 (Composite-mapping theorem)

Let E, F and G be normed vector spaces and let A and B be open subsets of E and F respectively. Suppose that $f: A \longrightarrow F$ and $g: B \longrightarrow G$ are p times differentiable at x and $f(x)$ respectively. Then the composite

$$h = gf : A \cap f^{-1}(B) \longrightarrow G$$

is p times differentiable at x.

Proof. We prove the result by induction. By Theorem 2.4.1 the result is true for $p = 1$. Suppose it is true for $p - 1$. Now f and g are C^{p-1} on A and B respectively. We must therefore prove

(a) $gf = h$ is C^{p-1} on $A \cap f^{-1}(B)$.
(b) $gf = h$ is p times differentiable at x.

Without loss of generality we may assume $f(A) \subset B$ for, if not, replace A by $A \cap f^{-1}(B)$.

Let us consider the map

$$(Dg)f : A \longrightarrow L(F, G); x \longmapsto Dg_{f(x)}.$$

Since f is C^{p-1} on A and Dg is $(p - 1)$ times differentiable at x, it follows, by the inductive hypothesis, that $(Dg)f$ is $(p - 1)$ times differentiable at x.

Now $Dh = (Dg)f . Df = \phi((Dg)f, Df)$, where $\phi : L(F, G) \times L(E, F) \longrightarrow L(E, F)$ is just composition of linear maps.

Since ϕ is a continuous bilinear map, ϕ is C^∞. Hence, by the inductive hypothesis, Dh is $(p - 1)$ times differentiable at x. It follows that Dh is C^{p-2} on A and so h is C^{p-1} on A. Similarly, Dh is $p - 1$ times differentiable at x and so h is p times differentiable at x. ■

Essentially the same proof yields the C^p version of the composite-mapping theorem.

Theorem 2.10.4

Let E, F and G be normed vector spaces and let A and B be open subsets of E and F respectively. Suppose that $f : A \longrightarrow F$ and $g : B \longrightarrow G$ are C^p on A and B respectively. Then the composite

$$h = gf : A \cap f^{-1}(B) \longrightarrow G$$

is C^p.

In the above we have not obtained an explicit formula for the pth derivative of the composite of two functions. In practice, the only important case is a formula for the *first* derivative of the composite. However, as an application of Leibniz' theorem, we give the general case.

Theorem 2.10.5

Let E, F, G be normed vector spaces and let A and B be open subsets of E and F respectively. Suppose $f : A \longrightarrow F$ and $g : B \longrightarrow G$ are p times differentiable at x and $f(x)$ respectively. Then $h = gf : A \cap f^{-1}(B) \longrightarrow G$ is p times differentiable at x with derivative given by the formula

$$D^p h_x = \sum_{q=1}^{k} \left(\sum_{m_1 + \cdots + m_q = k} a_k(m_1, \ldots, m_q) D^q g_{f(x)}(D^{m_1} f_x \odot \cdots \odot D^{m_q} f_x) \right),$$

where $a_k(m_1, \ldots, m_q)$ are the positive integers determined inductively from the relations

$$a_{k+1}(k + 1) = 1, \qquad k = 0, 1, 2, \ldots$$

$$a_{k+1}(m_1, \ldots, m_j, k + 1 - q) = \binom{k}{q} a_k(m_1, \ldots, m_j).$$

Proof. By induction on p using Leibniz' theorem. We leave details to the exercises. ■

Example. Let F and G be normed vector spaces and suppose A and B are open subsets of $\mathbf{R}$ and F respectively. Let $f: A \longrightarrow F$ and $g : B \longrightarrow G$ be p times differentiable at x and $f(x)$ respectively. Then the second derivative of gf is given by

$$(gf)''(x) = D^2 g_{f(x)}(f'(x), f'(x)) + Dg_{f(x)}(f''(x)).$$

Exercises

1. Complete the proof of Theorem 2.10.2.
2. Let $(H, \langle\ ,\ \rangle)$ be an inner product space. Suppose $f: A \longrightarrow H$ and $g: A \longrightarrow H$ are C^p, where A is an open subset of $\mathbf{R}$. Setting $(f.g)(x) = \langle f(x), g(x)\rangle, x \in A$, find

   (a) $(f.g)''(x), x \in A$,
   (b) $(f.g)^{(3)}(x), x \in A$,

   in terms of the derivatives of f and g.
3. Complete the proof of Theorem 2.10.5.
4. Let $f : \mathbf{R} \longrightarrow \mathbf{R}^3$ and $g : \mathbf{R}^3 \longrightarrow \mathbf{R}^2$ be defined by

$$f(t) = (\sin t, e^t \cos t, t^4)$$

$$g(x_1, x_2, x_3) = (e^{x_1 x_2}, \cos(x_1^2 + x_2)).$$

   Find $(gf)''(0)$.

2.11 Taylor's theorem

In this section our main objective will be to show that we can use higher derivatives to define polynomial approximations to a function.

Theorem 2.11.1 (Taylor's theorem)

Let E, F be normed vector spaces and let A be an open subset of E. Suppose that $f: A \longrightarrow F$ is p times differentiable at x.

Then, if we define the remainder term $R(h)$ by

$$R(h) = f(x+h) - \left(f(x) + \cdots + \frac{1}{r!} D^r f_x(h^r) + \cdots + \frac{1}{p!} D^p f_x(h^p)\right),$$

$\|R(h)\| / \|h\|^p \longrightarrow 0$ as $h \longrightarrow 0$.

In other words, the function

$$G(h) = f(x) + \cdots + \frac{1}{p!} D^p f_x(h^p)$$

provides an approximation to f at x of order p.

Proof. We proceed by induction. The result is certainly true for $p = 1$ by the very definition of derivative. Suppose it is true for $p - 1$. Fix $d > 0$ such that $B_d(x) \subset A$ and assume in all that follows that $h \in B_d(0)$ and so $x + h \in B_d(x)$.

Define $S : B_d(0) \longrightarrow F$ by

$$S(k) = f(x+k) - f(x).$$

Then S is certainly p times differentiable at 0 and we have

$$D^r S_h = D^r f_{x+h}, \qquad r = 1, 2, \ldots, p.$$

Substituting in the defining equation for $R(h)$ it follows that

$$S(k) = DS_0(k) + \cdots + \frac{1}{r!} D^r S_0(k^r) + \cdots + \frac{1}{p!} D^p S_0(k^p) + R(k).$$

Differentiating with respect to k and setting $DS_h = g(h)$ gives

$$g(h) = g(0) + \cdots + \frac{1}{(r-1)!} D^{r-1} g_0(h^{r-1}) + \cdots + \frac{1}{(p-1)!} D^{p-1} g_0(h^{p-1}) + DR$$

Hence, by the inductive hypothesis, given $\epsilon > 0$, there exists $\delta > 0$ such that if $\| h \| \leqslant \delta$ $\| DR_h \| \leqslant \epsilon \| h \|^{p-1}$.

But, by the mean-value theorem,

$$\| R(h) \| = \| R(h) - R(0) \| \leqslant \| h \| \sup_{0 < t < 1} \| DR_{th} \|.$$

Since

$$\| DR_{th} \| \leqslant \epsilon \| th \|^{p-1} \leqslant \epsilon \| h \|^{p-1}, \qquad \| h \| \leqslant \delta,$$

it follows that

$$\| R(h) \| \leqslant \epsilon \| h \|^p, \qquad \| h \| \leqslant \delta.$$

Since $\epsilon > 0$ was arbitrary, the result follows. ∎

Corollary 2.11.1.1

Let A be an open subset of $\mathbf{R}^n$ and suppose that $f: A \longrightarrow \mathbf{R}$ is p times differentiable at x; then

$$f(x+h) = f(x) + \sum_{r=1}^{p} \left(\sum_{|m|=r} \frac{r1}{m!} f_{;m}(x) h^m \right) + R(h),$$

where $h \in \mathbf{R}^n$ and $R(h)/\| h \|^p \longrightarrow 0$ as $h \longrightarrow 0$.

Proof. An immediate consequence of the formula for $D^r f_x$ given in Proposition 2.9.10. ∎

Remark. In classical partial derivative notation, the expression for $f(x+h)$ given by Corollary 2.11.1.1 is

$$f(x) + \sum_{r=1}^{p} \left(\sum_{m_1 + \cdots + m_n = r} \frac{r1}{m_1! \cdots m_n!} (\partial^r f / \partial_{x_1}^{m_1} \ldots \partial_{x_n}^{m_n}) h_1^{m_1} \cdots h_n^{m_n} \right) + R(h),$$

where $h = (h_1, \ldots, h_n) \in \mathbf{R}^n$.

Example. We find the quadratic approximation to the function $f: \mathbf{R}^3 \longrightarrow \mathbf{R}$ defined by

$$f(x_1, x_2, x_3) = e^{x_1} \cos x_2 \sin x_3,$$

at $x = 0$. Listing the relevant derivatives at $x = 0$, we have

$$f(0) = 0$$

$$f_{;1}(0) = f_{;2}(0) = 0; f_{;3}(0) = 1$$

$$f_{;12}(0) = f_{;23}(0) = f_{;11}(0) = f_{;22}(0) = f_{;33}(0) = 0; f_{;13}(0) = 1.$$

It follows from Corollary 2.11.1.1 that the quadratic approximation to f at 0 is given by

$$g(x_1, x_2, x_3) = x_3 + x_1 x_3.$$

In the remainder of this section we wish to prove some other versions of Taylor's theorem with different conditions on the differentiability and alternative remainder terms.

Theorem 2.11.2 (Scalar-valued Taylor's theorem)

Let A be an open subset of the normed vector space E and suppose $f: A \longrightarrow \mathbf{R}$ is p times differentiable on the line segment $[x, x + h] \subset A$. Then there exists a point $\theta \in (0, 1)$, such that

$$f(x + h) = f(x) + Df_x(h) + \cdots + \frac{1}{r!} D^r f_x(h^r) + \cdots + \frac{1}{p!} D^p f_{x+\theta h}(h^p).$$

Proof. We define ϕ $[0, 1] \longrightarrow A \subset E$ by $\phi(t) = x + th$, $t \in [0, 1]$. Then, if we set $F = f\phi: [0, 1] \longrightarrow \mathbf{R}$, F is p times differentiable on $(0, 1)$ and

$$F^r(t) = D^r f_{x+th}(h^r), \qquad r = 0, 1, \ldots, p.$$

Applying the classical Taylor's theorem to F gives the result. ■

Corollary 2.11.2.1

Let A be an open subset of $\mathbf{R}^n$ and suppose $f: A \longrightarrow \mathbf{R}$ is p times differentiable on $[x, x + h] \subset A$. Then, $\exists\ \theta \in (0, 1)$ such that

$$f(x + h) = f(x) + \sum_{r=1}^{p-1} \left(\sum_{|m| = r} \frac{1}{m!} f_{;m}(x) h^m \right) + \sum_{|m| = p} \frac{1}{m!} f_{;m}(x + \theta h) h^m.$$

Remark. Just as we cannot, in general, obtain equality in the mean-value theorem, we cannot obtain a version of Theorem 2.11.2 for maps into an arbitrary normed vector space.

Finally we wish to obtain an explicit form for the remainder term in Theorem 2.11.1 in the case when f is p times differentiable in a neighbourhood of x. However, we first need to develop a little of the theory of integration of normed vector-space-valued functions.

Thus, let $f: [a, b] \longrightarrow E$ be a continuous function on the closed interval $[a, b]$ taking values in the normed vector space E. We let $\hat{E}$ denote the completion of E. We shall mimic the theory of Riemann integration and form the partial sums

$$S_N(f) = \sum_{n=1}^{N} (f(x_{n+1}) - f(x_n))(x_{n+1} - x_n),$$

where $a = x_1 < \cdots < x_{N+1} = b$. Set $\delta = \max |x_{n+1} - x_n|$. Since $[a, b]$ is compact, f is uniformly continuous on $[a, b]$ and thus the limit

$$\lim_{\delta \to 0} S_N(f)$$

exists, as a point in $\hat{E}$, and is independent of the choice of partitions of $[a, b]$. We define this limit to be the integral of f on $[a, b]$ and we denote it by

$$\int_a^b f(s)\, ds.$$

Notice that, in general, $\int_a^b f(s)\, ds \notin E$ (see the exercises at the end of this section). We remark that if $\phi \in E^* = \hat{E}^*$.

$$\phi\left(\int_a^b f(s)\, ds\right) = \int_a^b \phi f(s)\, ds.$$

This is an immediate consequence of the fact that $\phi S_N(f) = S_N(\phi f)$.

We may now state our final version of Taylor's theorem.

Theorem 2.11.3

Let E and F be normed vector spaces and let A be an open subset of E. Then, if $f: A \longrightarrow F$ is C^p on A and $[x, x+h] \subset A$, we have

$$f(x+h) = f(x) + \cdots + \frac{1}{(p-1)!} D^{p-1} f_x(h^{p-1}) + \int_0^1 \frac{(1-s)^{p-1}}{(p-1)!} D^p f_{x+sh}(h^p) ds.$$

Proof. Let us first prove the special case when A is an open subset of $\mathbf{R}$ containing the unit interval $[0, 1]$ and $f: A \longrightarrow \mathbf{R}$ is C^p. Suppose that $g: A \longrightarrow \mathbf{R}$ is also C^p. It follows easily from the product rule for derivatives that

$$f(t) g^{(p)}(t) - (-1)^p f^{(p)}(t) g(t) = \sum_{j=0}^{p-1} (-1)^j f^{(j)}(t) g^{(p-j-1)}(t), \qquad t \in [0, 1].$$

Taking $g(t) = (1-t)^{p-1}/(p-1)!$ and integrating the above expression between 0 and 1 gives the theorem for this special case.

Continuing our assumption that $E = \mathbf{R}$, let us suppose that f takes values in the normed vector space F. Let $\phi \in F^*$. Then, by what we have already proved, the theorem holds for the map $\phi f: [0, 1] \longrightarrow \mathbf{R}$. Noticing that

$$D^r(\phi f) = \phi . D^r f, \qquad r = 0, 1, \ldots, p$$

and that

$$\int_0^1 \frac{(1-s)^{p-1}}{(p-1)!} \phi . D^p f_{x+sh}(h^p)\, ds = \phi\left(\int_0^1 \frac{(1-s)^{p-1}}{(p-1)!} D^p f_{x+sh}(h^p)\, ds\right),$$

it follows that

$$\left[f(x+h) - \left(f(x) + \sum_{r=1}^{p-1} \frac{1}{r!} D^r f_x(h^r)\right.\right.$$
$$\left.\left. + \int_0^1 \frac{(1-s)^{p-1}}{(p-1)!} D^p f_{x+sh}(h^p)\, ds\right)\right] \in \operatorname{Ker}(\phi).$$

Since $\phi \in F^*$ was arbitrary, it follows by the Hahn-Banach theorem that the theorem is true in this case also.

For the general case when $f : A \subset E \longrightarrow F$ we define γ: $[0, 1] \longrightarrow E$ by $\gamma(t) = x + th$ and apply the preceding result to the composition γf, just as in Theorem 2.11.2. ■

Remark. Notice that the integral term

$$\int_0^1 \frac{(1-s)^{p-1}}{(p-1)!} D^p f_{x+sh}(h^p)\, ds$$

in Theorem 2.11.3 lies in E and not just $\hat{E}$.

Exercises

1. Evaluate the cubic approximation given by Taylor's theorem at $(0, 0)$ to the function $f : \mathbf{R}^2 \longrightarrow \mathbf{R}^3$ defined by

$$f(x, y) = (\sin(x + y^2), ye^x, e^{(x^2 + y)}).$$

2. Find the quadratic approximation given by Taylor's theorem to the functions

   (a) $f : \mathbf{R}^3 \longrightarrow \mathbf{R}; (x_1, x_2, x_3) \longmapsto e^{\cos(x_1 + x_2)} \sin(x_2^2 + x_3)$,

   (b) $f : \mathbf{R}^3 \longrightarrow \mathbf{R}; (x_1, x_2, x_3) \longmapsto \log(1 + x_1^2 + x_2^2) \log(1 + x_1^2 + x_3^2)$.

3. Show that Theorem 2.11.3 implies that $\|R(h)\|/\|h\|^p \longrightarrow 0$ as $h \longrightarrow 0$, where $R(h)$ is as defined in the statement of Theorem 2.11.1.
4. Let E be an incomplete normed vector space. Construct a continuous function $f: [0, 1] \longrightarrow E$ such that $\int_0^1 f(s)\, ds \notin E$.

   (*Hint*: Let x be a point in $\hat{E}$ but not in E and choose a sequence $(x_n)_{n=2}^{\infty}$ of points of E such that $\Sigma_{n=2}^{\infty} x_n = x$ and $\lim_{n\to\infty} n^2 x_n = 0$. Define a map $\tilde{f}:[0, 1] \longrightarrow E$ by requiring that $\tilde{f} \mid [1/n, 1/(n+1)) = n(n+1)x_n$ and $\tilde{f}(1) = 0$. Show that $\tilde{f}$ is continuous at 0 and that $\int_0^1 \tilde{f}(s)\, ds = x$. Modify f to a continuous function $f:[0, 1] \longrightarrow E$ such that $\int_0^1 f(s)\, ds = x$.)

2.12 Extrema

In this section we wish to show how the classical theory of maxima and minima of differentiable functions generalizes to functions of more than one variable. We start with a definition.

Definition 2.12.1

Let X be a subset of the normed vector space E and suppose that $f : X \longrightarrow \mathbf{R}$. A point $x_0 \in X$ is said to be a *local maximum* for f if there exists a neighbourhood V of x_0 in E such that

$$f(x) \leqslant f(x_0) \qquad \text{for all } x \in X \cap V.$$

x_0 is said to be a *strict* local maximum if we can choose V such that

$$f(x) < f(x_0) \qquad \text{for all } x \in (X \setminus \{x_0\}) \cap V.$$

x_0 is said to be an *absolute* maximum for f on X if

$$f(x) \leqslant f(x_0) \qquad \text{for all } x \in X.$$

x_0 is said to be a strict absolute maximum for f if

$$f(x) < f(x_0) \qquad \text{for all } x \in X \setminus \{x_0\}.$$

The various notions of local and absolute minima for f are defined similarly.

Remark. In practice we shall be considering differentiable functions defined on open subsets of normed vector spaces. In general, there need not exist any maxima or minima for functions defined on open sets. Of course if, in Definition 2.12.1, X were *compact* and f continuous, then f would have absolute maxima and minima on X. In practice extrema of a function defined on an open set often occur on the boundary of the set. We shall look at one or two examples later, but remark in passing the maximum modulus principle for analytic functions, which states that the maximum value of the modulus of an analytic function always occurs on the boundary of the domain on which the function is defined.

Proposition 2.12.2

Let U be an open subset of the normed vector space E and suppose that f is differentiable on U. Then a necessary condition for the point $x_0 \in U$ to be a local maximum or minimum for f is that

$$Df_{x_0} = 0.$$

Proof. Suppose that x_0 is a local maximum for f. Then there exists a neighbourhood V of x_0 contained in U such that

$$f(x) \leqslant f(x_0), \qquad x \in V.$$

Since f is differentiable at x_0,

$$f(x_0 + h) = f(x_0) + Df_{x_0}(h) + o(h), \qquad x + h \in V.$$

Therefore,

$$f(x_0 + h) - f(x_0) = Df_{x_0}(h) + o(h).$$

Hence,

$$Df_{x_0}(h) + o(h) \leqslant 0, \qquad h \in V \setminus \{x_0\}.$$

Replacing h by $-h$ in the above inequality, it follows easily that

$$Df_{x_0}(h) = o(h).$$

Hence, by Proposition 2.2.1, $Df_{x_0} = 0$. ■

Remarks

1. Geometrically, the condition $Df_{x_0} = 0$ means that the tangent plane to the graph of f at x_0 is parallel to the subspace E of $E \times \mathbf{R}$. For example, consider the paraboloid defined by $f(x, y) = x^2 + y^2 + 2$. The graph of f is illustrated in Fig. 2.10.

   It can be seen that the minimum value of f must occur where the tangent plane is parallel to the xy-plane.
2. The condition Df_{x_0} is certainly not sufficient for the point x_0 to be a maximum or minimum value of the function f. For example the function on $\mathbf{R}^2$ defined by $f(x, y) = x^2 - y^2$ has zero derivative at the origin. However, the origin is

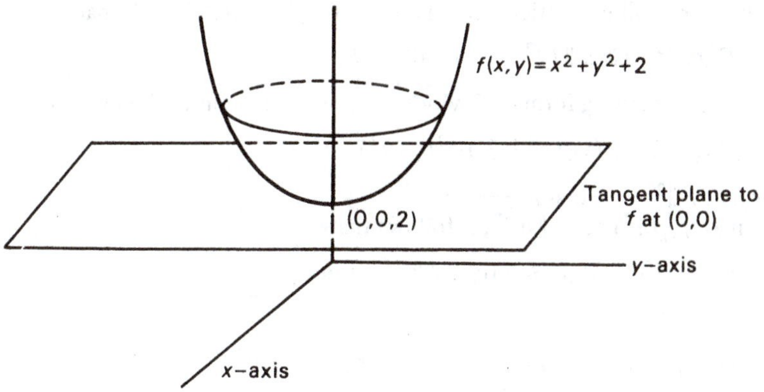

FIG. 2.10

certainly not a maximum or minimum for the function. By restricting the function to the x-axis and the y-axis the reader may easily verify that the graph of f has a 'saddle' at the origin.

Definition 2.12.3

Let U be an open subset of the normed vector space E and suppose that $f: U \longrightarrow \mathbf{R}$ is differentiable on U. If $Df_x = 0$, $x \in U$, x is called a *critical point* of the function f. The value of f at x is called the *critical value* of f at x.

Example. Let us find the set of critical points and values of the function $f: \mathbf{R}^3 \longrightarrow \mathbf{R}$ defined by

$$f(x, y, z) = x^4 + z^4 - 2(x^2 + z^2) + y^2.$$

Since f is certainly C^1, the critical points for f occur at those points where all three partial derivatives of f vanish. That is, we must solve the set of equations:

$$f_{;1}(x, y, z) = 4x^3 - 4x = 0,$$

$$f_{;2}(x, y, z) = 2y = 0,$$

$$f_{;3}(x, y, z) = 4z^3 - 4z = 0.$$

The set of solutions of these equations is easily seen to be:

$$(0, 0, 0), (1, 0, 0), (-1, 0, 0), (0, 0, 1), (0, 0, -1),$$

$$(1, 0, 1), (-1, 0, 1), (1, 0, -1), (-1, 0, -1),$$

and the corresponding critical values are $0, -1, -1, -1, -1, -2, -2, -2, -2$ respectively.

Proposition 2.12.4

Let U be an open subset of the normed vector space E. Suppose that $f: U \longrightarrow \mathbf{R}$ is differentiable on U and twice differentiable at x_0 with $Df_{x_0} = 0$. Then a sufficient condition for the point x_0 to be a strict local maximum for f is that there exist $c > 0$ such that

$$D^2 f_{x_0}(t, t) \leqslant -c \,\|t\|^2 \qquad \text{for all } t \in E.$$

Similarly, x_0 will be a strict local minimum for f if $\exists\ d > 0$ such that

$$D^2 f_{x_0}(t, t) \geqslant d \, \| t \|^2 \qquad \text{for all } t \in E.$$

Proof. From the strong form of Taylor's theorem (Theorem 2.11.1) we have

$$f(x_0 + h) = f(x_0) + (1/2!) D^2 f_{x_0}(h^2) + r(h),$$

where $r(h)/\| h \|^2 \longrightarrow 0$ as $h \longrightarrow 0$.

Now if $D^2 f_{x_0}(h^2) \leqslant -c \, \| h \|^2$, it follows that

$$f(x_0 + h) - f(x_0) \leqslant -(c/2) \, \| h \|^2 + r(h),$$

and so, for $h \neq 0$,

$$(f(x_0 + h) - f(x_0))/\| h \|^2 \leqslant -c/2 + r(h)/\| h \|^2.$$

Choose $d > 0$, such that $| r(h)/\| h \|^2 | \leqslant c/4$ for all $h \in B_d(0)$. It follows immediately that

$$f(x_0 + h) - f(x_0) \leqslant -(c/4) \, \| h \|^2, \qquad h \in B_d(0).$$

Hence x_0 is a strict local maximum for f. ■

Remark. If $Df_{x_0}, D^2 f_{x_0}, \ldots, D^{2r-1} f_{x_0}$ had all been zero, then the same proof as above would have shown that a sufficient condition that x_0 be a local maximum for f would have been the existence of $c > 0$ such that

$$D^{2r} f_{x_0}(h^{2r}) \leqslant -c \, \| h \|^{2r} \qquad \text{for all } h \in E.$$

Example. In the previous example we computed the critical points of the function $f : \mathbf{R}^3 \longrightarrow \mathbf{R}$ defined by

$$f(x, y, z) = x^4 + z^4 - 2(x^2 + z^2) + y^2.$$

Let us find which of these critical points are strict local maxima or minima.

f is clearly C^2 on $\mathbf{R}^3$ and the second-order partial derivatives are given by

$$f_{;11}(x, y, z) = 12x^2 - 4,$$
$$f_{;22}(x, y, z) = 2,$$
$$f_{;33}(x, y, z) = 12z^2 - 4,$$

all other partial derivatives of order two vanishing.

It follows from Proposition 2.12.4 that a critical point (x, y, z) will be a local maximum (minimum) for f if $f_{;11}(x\ y, z), f_{;22}(x, y, z)$ and $f_{;33}(x, y, z)$ are strictly negative (respectively, strictly positive). It follows that no critical point is a strict local maximum for f and that $(1, 0, 1), (-1, 0, 1), (1, 0, -1), (-1, 0, -1)$ are all strict local minima for f with corresponding minimum value -2. Since f is certainly bounded below it follows that -2 is actually the absolute minimum value of f on $\mathbf{R}^3$. Inspection shows that there is no local maximum value of f on $\mathbf{R}^3$. Indeed, at any one of the critical points of f, perturbation of y increases the value of f.

If E is a finite-dimensional vector space the condition on the second derivative for a strict local extrema can be weakened slightly as the following proposition shows.

Proposition 2.12.5

Let E be a finite-dimensional normed vector space and let $A \in L_s^2(E; \mathbf{R})$. Then if

$A(t, t) > 0$ for all $t \neq 0$, there exists $c > 0$ such that

$$A(t, t) \geqslant c \| t \|^2 \qquad \text{for all } t \in E.$$

Proof. Let $c = \inf_{\| t \| = 1} A(t, t)$. Since E is locally compact, and hence the unit disc in E is compact, it follows that $c > 0$. The proposition now follows by an argument similar to that of Theorem 1.5.7. ∎

Let us recall from Section 1.16 that if E is a finite-dimensional normed vector space and $A \in L_s^2(E; \mathbf{R})$, then A defines a quadratic form q on E by the relation

$$q(x) = A(x, x), \qquad x \in E.$$

q is said to be *non-degenerate* if $q(x) = 0$ if and only if $x = 0$. If $q(x) > 0$ for all $x \neq 0$, q is said to be *positive definite.* By abuse of language we shall also use the terms non-degenerate and positive definite to describe A. If $-q$ is positive definite we shall say that q is *negative definite*

If $\mathscr{E} = (e_i)_{i=1}^n$ is any basis for E, we may define an $n \times n$ symmetric matrix $[a_{ij}]$ by

$$a_{ij} = A(e_i, e_j), \qquad 1 \leqslant i, j \leqslant n.$$

The characteristic roots of this matrix are real and it follows from Theorem 1.16.12 that q is a non-degenerate quadratic form if and only if all the characteristic roots of $[a_{ij}]$ are non-zero. If all the characteristic roots of $[a_{ij}]$ are strictly positive, then q is positive definite. It also follows from Theorem 1.16.12 that it is possible to find a basis $\mathscr{E}$ for E such that, relative to $\mathscr{E}, q$ has the form

$$q(x_1, \ldots, x_n) = \sum_{j=1}^{p} x_i^2 - \left(\sum_{i=p+1}^{p+m} x_i^2 \right), \qquad (x_1, \ldots, x_n) \in E.$$

We shall now apply the above results to critical points of functions defined on open subsets of finite-dimensional vector spaces.

Proposition 2.12.6

Let U be an open subset of the finite-dimensional normed vector space E. Suppose that

1. $f : U \longrightarrow \mathbf{R}$ is C^1.
2. $x \in U$ is a critical point for f.
3. f is twice differentiable at x and $D^2 f_x$ is non-degenerate.

Then x is a strict local maximum for f if and only if $D^2 f_x$ is negative definite. x is a strict local minimum for f if and only if $D^2 f_x$ is positive definite.

In particular, if $(e_i)_{i=1}^n$ is any basis for E, x is a strict local maximum (minimum) if and only if the matrix $[D^2 f_x(e_i, e_j)]$ has all characteristic roots strictly negative (respectively, strictly positive).

Proof. The 'if' part of the proposition follows immediately from Proposition 2.12.4 and the above remarks. For the 'only if' part we proceed as follows. Suppose that $D^2 f_x$ is non-degenerate but not negative definite. We may choose a basis $(e_i)_{i=1}^n$ for E such that the Taylor expansion for f at $x = (x_1, \ldots, x_n)$ takes the form

$$f(x_1 + h_1, \ldots, x_n + h_n) = f(x_1, \ldots, x_n) + \tfrac{1}{2} \left(\sum_{i=1}^{p} h_i^2 - \sum_{i=p+1}^{n} h_i^2 \right) + R_2(h),$$

where $h = (h_1, \ldots, h_n)$, $R_2(h)/\| h \|^2 \longrightarrow 0$ as $h \longrightarrow 0$ and p is strictly positive.

Let E_p denote the space spanned by the first p coordinates of E. If we restrict f to $U \cap (x + E_p)$, we see from Proposition 2.12.4 that x cannot be a strict local maximum for f. Similarly, if $D^2 f_x$ is not positive definite, x cannot be a strict local minimum for f. ■

Remark. Notice that we require $D^2 f_x$ to be non-degenerate in Proposition 2.12.6. It is certainly not true in general that $D^2 f_x$ must be negative definite in order that x be a local maximum for f.

Definition 2.12.7

Let U be an open subset of the finite-dimensional normed vector space E. Suppose that

1. $f : U \longrightarrow \mathbf{R}$ is C^1.
2. x is a critical point for f and f is twice differentiable at x.

x is said to be a *non-degenerate critical point* for f if $D^2 f_x$ is non-degenerate.

If x is a non-degenerate critical point for f, we define the index of the critical point to be the integer m defined in Definition 1.16.11.

Examples

1. Let $\mathbf{R}^3$ have the usual inner product. Find the point of the plane $2x - 2y - z = 5$ which is closest to the point $(1, 1, 1)$. First we remark that if the point minimizes distance it also minimizes distance squared. We define $d^2 : \mathbf{R}^3 \longrightarrow \mathbf{R}$ by

$$d^2(x, y, z) = (x - 1)^2 + (y - 1)^2 + (z - 1)^2.$$

We must find the minimum of d^2 subject to the constraint that $2x - 2y - z = 5$. Any point on the plane $2x - 2y - z = 5$ is given uniquely in terms of x and y by $(x, y, 2x - 2y - 5)$. It is therefore sufficient to find the minimum value of the function $F : \mathbf{R}^2 \longrightarrow \mathbf{R}$ defined by

$$F(x, y) = (x - 1)^2 + (y - 1)^2 + (2x - 2y - 6)^2.$$

F is clearly C^∞ on $\mathbf{R}^2$ and so we have

$$[DF_{(x,y)}] = \left[\frac{\partial F}{\partial x}; \frac{\partial F}{\partial y}\right] = [10x - 26 - 8y, 10y - 8x + 22].$$

$$[D^2 F_{(x,y)}] = \begin{bmatrix} 10 & -8 \\ -8 & 10 \end{bmatrix}$$

Solving $DF_{(x,y)} = 0$, we find that f has just one critical point $(7/3, -1/3)$. Since $D^2 F_{(x,y)}$ is positive definite on the whole of $\mathbf{R}^2$, it follows that $(7/3, -1/3)$ is a strict local minimum for F. This point must be the absolute minimum for F since we know that F does have a minimum (Section 1.12 or 'geometry') and at that minimum $DF = 0$.

Substituting for z, we find that the required point in the plane is given as $(7/3, -1/3, -2/3)$.

2. Show that all critical points of the function $f : \mathbf{R}^3 \longrightarrow \mathbf{R}$ defined by

$$f(x, y\ z) = x^3 + x^2(y + z) - (x^2 + y^2 + z^2)/2$$

are non-degenerate. In each case find the corresponding index. Computing the partial derivatives of f we find that

$$f_{;1}(x,y,z) = x(3x + (y+z) - 1),$$
$$f_{;2}(x\ y,z) = x^2 - y,$$
$$f_{;3}(x,y,z) = x^2 - z.$$

Solving the equations $f_{;1} = f_{;2} = f_{;3} = 0$, we find that f has the two critical points $(0, 0, 0)$ and $(-4/3, 16/9, 16/9)$. We now compute the second derivatives of f:

$$f_{;11}(x,y,z) = 6x + (y+z) - 1; f_{;22}(x,y,z) = f_{;33}(x,y,z) = -1$$
$$f_{;12}(x,y,z) = 2x; f_{;13}(x,y,z) = 2x; f_{;23}(x,y,z) = 0.$$

It follows that $D^2f_{(0,0,0)} = -I$ and so $(0, 0, 0)$ is a strict maximum for f and is of index 3.

We find that the matrix of D^2f at the second critical point is

$$\begin{bmatrix} -49/9 & -8/3 & -8/3 \\ -8/3 & -1 & 0 \\ -8/3 & 0 & -1 \end{bmatrix}.$$

By inspection we see that -1 is a characteristic root of this matrix. Since the determinant of the matrix is easily checked to be positive it follows that one of the remaining characteristic roots must be positive and the other negative and that no characteristic root is zero. Hence the critical point $(-4/3, 16/9, 16/9)$ is non-degenerate and of index 1.

Exercises

1. Find the critical points of the following functions,

   (a) $f: \mathbf{R}^2 \longrightarrow \mathbf{R}; (x,y) \longmapsto x^4 + y^4 + x + y,$

   (b) $g: \mathbf{R}^2 \longrightarrow \mathbf{R}; (x,y) \longmapsto x^4 + y^4 + x^2 - y^2.$

   In each case state whether the critical points are non-degenerate and, where applicable, find the index of the critical point.
2. Let $f: \mathbf{R}^2 \longrightarrow \mathbf{R}$ be C^1. Need f have any critical points? What about if f is bounded above or below?
3. Let B be a rectangular box without a lid of length x, width y and height z. Suppose that the area of the sides and the base of the box is fixed and equal to A. Show that the largest volume of B is attained when $x = y = (A/3)^{1/2}$ and $z = x/2$.
4. Let $f: l^2 \longrightarrow \mathbf{R}$ be defined by

   $$f(x_1, \ldots) = x_1^4 + x_2^4 - (\|(x_1 \ldots)\|_2^2).$$

   Find the critical points of f and ascertain whether or not they are non-degenerate.
5. Show, by means of examples, that if x is a degenerate critical point, then x can be a strictly local maximum or minimum.

2.13 Curves in $\mathbf{R}^n$

In this section we wish to develop a little of the theory of differentiable curves in $\mathbf{R}^n$. We shall not, however, prove any general theorems asserting that this or that subset of $\mathbf{R}^n$ is a smooth curve. Instead, we shall emphasize properties of smooth curves and we shall make our definitions sufficiently strong so that we can establish interesting general properties. In Chapter 4 we shall undertake a more serious study of 'differentiable subsets' of $\mathbf{R}^n$ which will, in particular, give us ways of constructing many smooth curves.

First we need to discuss a technical point: Suppose that $\phi: [a, b] \longrightarrow \mathbf{R}^n$ is continuous and differentiable on (a, b). How may we define differentiability of ϕ at the end points of $[a, b]$? We shall in fact say that ϕ is differentiable at a, with (right-handed) derivative $\phi'(a) \in \mathbf{R}^n$, if

$$\phi(a + h) - \phi(a) = h\phi'(a) + o(h) \qquad \text{for all } h \geqslant 0.$$

We similarly define the (left-handed) derivative of ϕ at b.

If both $\phi'(a)$ and $\phi'(b)$ exist, we shall say that ϕ is differentiable on $[a, b]$. If $\phi': [a, b] \longrightarrow \mathbf{R}$ is continuous, we shall say that ϕ is C^1 on $[a, b]$. Similar definitions may be given for higher derivatives of ϕ on $[a, b]$.

Remarks

1. It can easily be shown that if $\phi: [a, b] \longrightarrow \mathbf{R}^n$ is differentiable on $[a, b]$, then there exists an open interval $(c, d) \supset [a, b]$ and a differentiable function $\Phi: (c, d) \longrightarrow \mathbf{R}^n$ such that $\Phi \mid [a, b] = \phi$.
2. The extension of the notion of differentiability to closed subsets of $\mathbf{R}^m$, $m > 1$, is much less trivial. By remark (1), one might say a function defined on a closed subset X of $\mathbf{R}^m$ is differentiable if it is the restriction of a differentiable function defined on an open neighbourhood of X. Whilst this definition is perfectly valid, it has the drawback of not defining differentiability directly in terms of the values of the function on X. In particular, it gives no idea of when such an extension can exist. An important theorem of Whitney (the Whitney extension theorem, see [2] for a proof) gives a set of necessary and sufficient conditions on a function in order that it have an extension to an open set. In general, however, the question of extension is also related to the geometry of the boundary of X.

Definition 2.13.1

Suppose that $\phi: [a, b] \longrightarrow \mathbf{R}^n$ is differentiable. Set $\phi([a, b]) = C$. Then ϕ is said to define a differentiable parametrization of the curve C. By abuse of language, we shall often refer to ϕ as a differentiable curve.

Definition 2.13.2

Suppose $\phi: [a, b] \longrightarrow \mathbf{R}^n$ is a differentiable curve.

1. If ϕ is injective, we shall say that ϕ defines an *arc*.
2. If $\phi(a) = \phi(b)$ and $\phi \mid [a, b)$ is injective, we shall say that ϕ defines a *simple closed curve.*
3. If ϕ is C^1 and $\phi' \neq 0$ on $[a, b]$ then ϕ is said to define a *smooth curve.* If, in addition, ϕ is injective, we shall say that ϕ defines a smooth arc.
4. Suppose that ϕ is a smooth curve such that $\phi \mid [a, b)$ is injective and $\phi(a) = \phi(b)$, $\phi'(a) = \phi'(b)$, then we shall say that ϕ defines a *smooth simple closed curve.*

Remark. In practice we shall be only interested in smooth arcs and curves. Without the condition on the non-vanishing of the derivative of ϕ, the image of ϕ need not have tangents at every point as the following examples illustrate.

Examples

1. Define $\phi: [-1, +1] \longrightarrow \mathbf{R}^2$ by

$$\phi(t) = (t^2, t^3), \qquad t \in [-1, +1].$$

ϕ is certainly injective and C^∞ but $\phi'(0) = 0$. Hence ϕ defines an arc but not a smooth arc. The image of ϕ in $\mathbf{R}^2$ is shown below and the reader should observe the discontinuity in the tangent to the curve at the origin (Fig. 2.11).

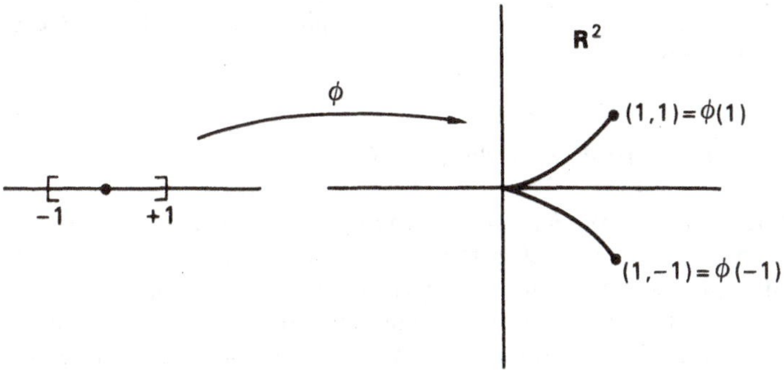

FIG. 2.11

2. Define $\phi: [0, 4\pi] \longrightarrow \mathbf{R}^2$ by $\phi(t) = (\sin(t/2), \sin t)$. ϕ is a smooth curve since $\phi'(t) = (\frac{1}{2}\cos(t/2), \cos t) \neq 0$ for all $t \in \mathbf{R}$. However, $\phi(0) = \phi(2\pi) = \phi(4\pi) = 0$ and so ϕ has self-intersections. The image of ϕ is shown in Fig. 2.12.

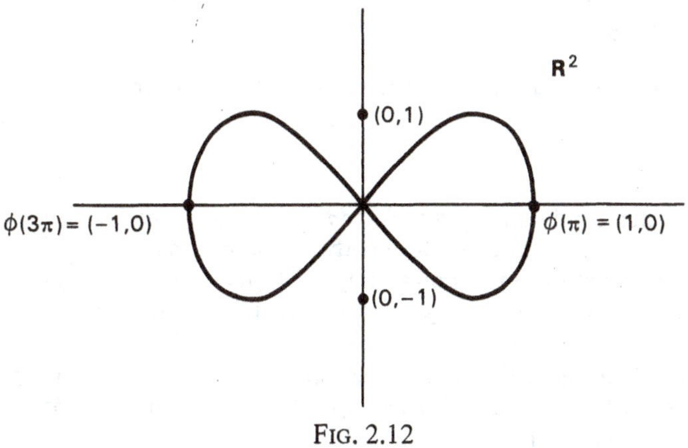

FIG. 2.12

3. Define $\phi: [0, 2\pi] \longrightarrow \mathbf{R}^2$ by $\phi(t) = (a\cos t, b\sin t)$, where $ab > 0$. Then ϕ is easily checked to be a smooth simple closed curve whose image in $\mathbf{R}^2$ is the ellipse $x^2/a^2 + y^2/b^2 = 1$.

Remark. No particular significance should be attached to the fact that a curve is defined as the image of a closed interval. The use of closed, rather than open intervals, is just a technical convenience. For example, it enables us easily to define the end points of an arc and also make use of the fact that curves are compact subsets of $\mathbf{R}^n$. We can include non-compact curves under Definitions 2.13.1 and 2.13.2 if we allow the real line $\mathbf{R}$ and the half-infinite intervals $(-\infty, 0]$ and $[0, \infty)$ as closed intervals. That is, if we replace 'closed interval' in the definitions by 'connected closed subset of $\mathbf{R}$'. It should be noticed that every open interval of $\mathbf{R}$ can be put in differentiable 1:1 correspondence with $\mathbf{R}$ and so this extension of the definition of a curve really includes the case of open and closed intervals.

Definition 2.13.3

Let $\phi: [a, b] \longrightarrow \mathbf{R}^n$ be a differentiable parametrization of the curve C and suppose that $\gamma: [c, d] \longrightarrow [a, b]$ is a homeomorphism which is C^1 with C^1 inverse. Then the function

$$\phi\gamma: [c, d] \longrightarrow \mathbf{R}^n$$

is said to be a *reparametrization* of the curve C.

Remark. It follows from the *inverse-function theorem*, proved in Chapter 3, that γ satisfies the conditions of the definition if γ is a smooth curve with image $[a, b]$. In particular, it is sufficient to assume that $\gamma'(t) \neq 0$ for all $t \in [c, d]$ as this condition is sufficient to ensure the existence of a differentiable inverse for γ. We shall assume this result in the remainder of this section. We shall also generally assume that the reparametrization does not change the direction associated with the parametrization of the curve. That is, the initial and terminal points of the curve remain the same or, equivalently, $\gamma'(t) > 0$.

Example. Let $\phi: [a, b] \longrightarrow \mathbf{R}^n$ be a C^2 parametrization of the curve C. Then we claim that we can find a reparametrization $\psi: [c, d] \longrightarrow \mathbf{R}^n$ of C such that $\| \psi'(t) \|_2 = 1$ for all $t \in [c, d]$. Indeed, let us define $\tau: [a, b] \longrightarrow \mathbf{R}$ by

$$\tau(t) = 1/\| \phi'(t) \|_2.$$

τ is C^1 since $\phi'(t) \neq 0$ on $[a, b]$. We consider the first-order differential equation

$$\frac{\mathrm{d}f}{\mathrm{d}t} = \tau(f(t))$$

subject to the initial condition $f(a) = a$. Since τ is C^1 (and hence Lipschitz) it follows by the existence theory for ordinary differential equations that we may find a unique solution f for this equation defined on an interval $[c, d]$, where $c = a = f(a)$ and $f(d) = b$ (we use the fact that, since $[a, b]$ is compact, $\exists\ \alpha > 0$ such that $\tau(t) \geqslant \alpha$ for all $t \in [a, b]$).

Since $\tau > 0$, it follows by the above remark that $f: [c, d] \longrightarrow [a, b]$ is a C^1 homeomorphism with C^1 inverse. Consider the reparametrization $\psi = \phi f$ of C. By the composite-mapping formula,

$$\psi'(t) = \phi'(f(t))f'(t) = (1/\tau(f(t))f'(t) = 1.$$

Hence ψ defines the required reparametrization of C. It can also be shown that, up to a constant, ψ is unique and, in particular, that the length of the interval $[c, d]$ is an invariant of the curve C. We shall soon see that this constant is naturally associated with the length of the curve.

Suppose that $\phi:[a,b] \longrightarrow \mathbf{R}^n$ is a smooth curve, possibly with self-intersections. Let $t \in [a,b]$ and suppose that $[t, t+\delta t] \subset [a,b]$, where δt is supposed small. Then,

$$\phi(t+\delta t) = \phi(t) + (\delta t)\phi'(t) + o(\delta t).$$

In particular,

$$\phi(t+\delta t) - \phi(t) = (\delta t)\phi'(t) + o(\delta t)$$

and so the length of the chord $[\phi(t), \phi(t+\delta t)]$ is approximately equal to $\delta t \|\phi'(t)\|_2$ (see Fig. 2.13).

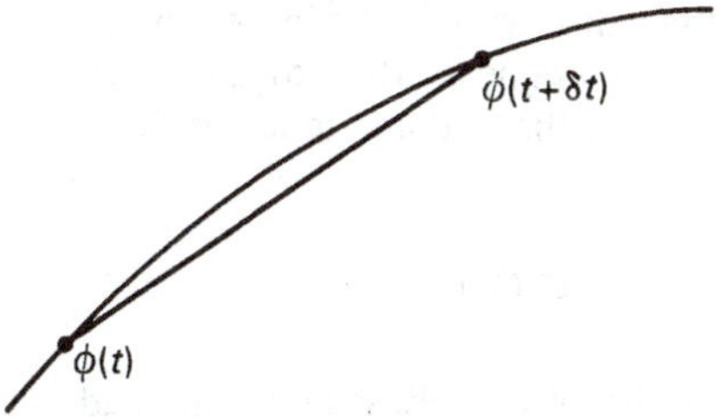

FIG. 2.13

On the other hand, one would expect that a good approximation for the length of the segment of curve between $\phi(t)$ and $\phi(t+\delta t)$ would be given by the length of the chord $[\phi(t), \phi(t+\delta t)]$. That is, an approximation to the length of the segment should be given by $\delta t \|\phi'(t)\|_2$. More generally, suppose that we take a partition

$$a = t_0 < t_1 < \cdots < t_n = b$$

of $[a,b]$, then we might expect an approximation to the length of the curve to be given by

$$\sum_{i=0}^{n-1} \|\phi'(t_i)\|_2 (t_{i+1} - t_i).$$

This is, of course, just an approximating sum for the Riemann integral

$$\int_a^b \|\phi'(t)\|_2 \, dt.$$

However, we have not shown that the length of the curve defined by ϕ is this integral: Indeed we have not even *defined* what is meant by the length of a curve in $\mathbf{R}^n$ which is not a straight line. What we have done though, is to indicate that any reasonable definition of length should lead to the above integral as a formula for the length of a *smooth curve.* Motivated by these remarks we make the following definition of the length of a smooth curve.

Definition 2.13.4

Let $\phi: [a,b] \longrightarrow \mathbf{R}^n$ be a smooth curve. We define the length of ϕ, $s(\phi)$, to be the integral

$$s(\phi) = \int_a^b \|\phi'(t)\|_2 \, dt.$$

Remark. Notice that, so far, we have defined the length of a curve in terms of a parametrization rather than intrinsically.

Examples

1. Let $\phi: [0, 2\pi] \longrightarrow \mathbf{R}^2$ be the usual parametrization of the circle of radius r and centre 0: $\phi(t) = (r \cos t, r \sin t)$. Then ϕ defines a simple smooth closed curve and $\| \phi'(t) \|_2 = r$ for all $t \in [0, 2\pi]$. Hence
$$s(\phi) = \int_0^{2\pi} r \, dt = 2\pi r,$$
which coincides with the usual formula for the circumference of a circle of radius r.
2. Let us parametrize the ellipse $x^2/a^2 + y^2/b^2 = 1$ by $(a \cos t, b \sin t)$, $t \in [0, 2\pi]$. Then this gives a smooth parametrization of the ellipse as a simple closed curve. The reader may verify that the circumference of the ellipse is given by the formula
$$4a \int_0^{\pi/2} \sqrt{(1 - e^2 \sin^2 t)} \, dt,$$
where e is the eccentricity of the ellipse and is defined in terms of a and b by $e = a^{-1}\sqrt{(a^2 - b^2)}$, if $a > b$.

Remark. The integral occurring in Example 2 is an example of an *elliptic integral of the second kind*, the general form of which is given by
$$E(k, \phi) = \int_0^{\phi} \sqrt{(1 - k^2 \sin^2 t)} \, dt, \qquad 0 < k < 1.$$

The integral cannot be evaluated in terms of standard integrals. Values of $E(k, \phi)$ for various values of k and ϕ are given in tables. Problems in curve length often lead to elliptic integrals.

A priori, our definition of the length of a smooth curve C depends on the parametrization of C. In fact the length of C is independent of the parametrization used to define it. The full proof of this statement will have to wait until Chapter 4. The following proposition shows, however, that a reparametrization of C does not change its length.

Proposition 2.13.5

Let $\phi: [a, b] \longrightarrow \mathbf{R}^n$ be a C^1 parametrization of the smooth curve C. Suppose that $f: [c, d] \longrightarrow [a, b]$ defines a reparametrization $\phi f: [c, d] \longrightarrow \mathbf{R}^n$ of C. Then $s(\phi) = s(\phi f)$.

Proof. We are assuming that f is C^1 with a C^1 inverse. By definition,
$$\begin{aligned} s(\phi f) &= \int_c^d \| (\phi f)'(t) \|_2 \, dt \\ &= \int_{f^{-1}(a)}^{f^{-1}(b)} | f'(t) | \, \| \phi'(f(t)) \|_2 \, dt \\ &= \int_a^b \| \phi'(t) \|_2 \, dt \text{ (by change-of-variables formula)} \\ &= s(\phi). \end{aligned}$$

Notice that to apply the change-of-variables formula, we needed the fact that $|f'(t)| \neq 0$ on $[c, d]$. ■

Remark. What Proposition 2.13.5 does not show is that if ϕ_1 and ϕ_2 are C^1 parametrizations of the curve C then $s(\phi_1) = s(\phi_2)$. We shall in fact prove this in Chapter 4 by showing that ϕ_2 can always be regarded as a reparametrization of ϕ_1. In any case, for the rest of this section, we shall assume that the length of a smooth curve is uniquely defined, independent of parametrization. We shall denote the length of the curve C by $S(C)$, or just S, and omit any reference to the parametrization.

Suppose that $\phi: [a, b] \longrightarrow \mathbf{R}^n$ is a parametrization of the smooth curve C and that C has length S. We define a function $L: [a, b] \longrightarrow [0, S]$ by

$$L(t) = \int_a^t \|\phi'(u)\|_2 \, du, \qquad t \in [a, b].$$

Remark. L measures the distance travelled by a particle moving along C, with velocity ϕ', after time t.

By the fundamental theorem of the integral calculus we have

$$L'(t) = \|\phi'(t)\|_2, \qquad t \in [a, b]. \tag{A}$$

In particular, $L' > 0$ on $[a, b]$ and so L is a strictly increasing function. Hence L defines a bijection between $[a, b]$ and $[0, S]$. By the inverse-function theorem, L has a C^1 inverse $P: [0, L] \longrightarrow [a, b]$. Thus

$$LP = \text{identity on } [0, S]$$

$$PL = \text{identity on } [a, b].$$

If $x \in [0, S]$, $P(x)$ is uniquely determined by the condition

$$x = \int_a^{P(x)} \|\phi'(u)\|_2 \, du. \tag{B}$$

Indeed,

$$\int_a^{P(x)} \|\phi'(u)\|_2 \, du = L(P(x)) = x.$$

Remark. P answers the question: How long do we have to travel at velocity ϕ' to cover distance x?

The derivative of P is given by the usual rules of the differential calculus in terms of the derivative of L. In fact, since LP = identity, $L'(P(x))P'(x) = 1$, and so

$$P'(x) = 1/L'(P(x)), \qquad x \in [0, S].$$

From formula (A) above it follows that

$$P'(x) = 1/\|\phi'(P(x))\|_2, \qquad x \in [0, S]. \tag{C}$$

Proposition 2.13.6

Let $\phi: [a, b] \longrightarrow \mathbf{R}^n$ be a parametrization of the smooth curve C. Then, with the above notation, the function $\Phi = \phi P: [0, S] \longrightarrow \mathbf{R}^n$ defines a reparametrization of C satisfying the additional condition

$$\|\Phi'(x)\|_2 = 1 \qquad \text{for all } x \in [0, S].$$

We also have

$$\phi'(t) = \| \phi'(t) \|_2 \Phi'(L(t)), \qquad t \in [a, b].$$

Proof. By the composite-mapping formula

$$\Phi'(x) = \phi'(P(x))P'(x)$$
$$= \phi'(P(x))/\| \phi'(P(x)) \|_2, \qquad \text{(by formula (C))}$$

Hence $\| \Phi'(x) \|_2 = 1$. ■

Remarks

1. The parametrization Φ of the curve C corresponds to the motion of a particle along C at unit velocity. After time x the particle will have moved a distance x along C.
2. As we have previously indicated the parametrization Φ of C is unique. Indeed, this is an easy consequence of the fact that the length of C is unique.
3. The reader should notice that the parametrization Φ of C depends on the inner product. If we change the inner product on $\mathbf{R}^n$ then both the length of C and the parametrization Φ may change.

Notation. We shall call the parametrization of C given by Proposition 2.13.6 the *canonical parametrization* of C. Now $\Phi' : [0, S] \longrightarrow \mathbf{R}^n$ and we shall set $T = \Phi'$. $T(x)$ is then a vector of unit length for all $x \in [0, S]$, and we shall call $T(x)$ the *unit tangent* to C at x (strictly $\Phi(x)$).

It is useful to regard $T(x)$ as actually *being* tangent to C at x. To do this we introduce the 'arrow' notation. The arrow $\Phi(x) \longrightarrow T(x)$ in Fig. 2.14 denotes the line of unit length with end points $\Phi(x)$ and $\Phi(x) + T(x)$. That is, the line is the translation through $\Phi(x)$ of the line joining 0 to $T(x)$.

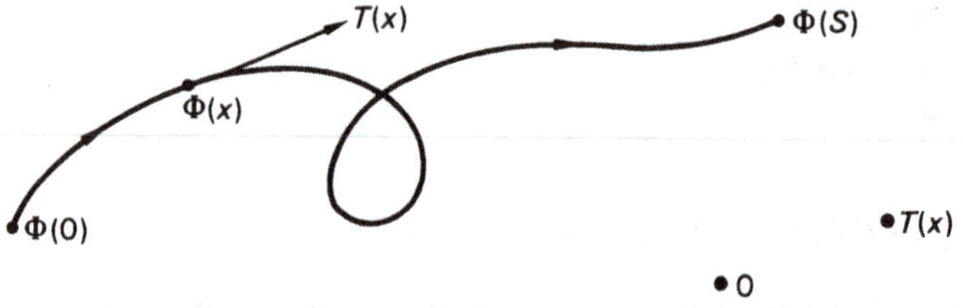

FIG. 2.14

The arrow notation requires further comment. To start with, notice that the set of all arrows in $\mathbf{R}^n$, with initial point the origin, is in 1:1 correspondence with $\mathbf{R}^n$. Indeed the arrow $0 \longrightarrow x$ is naturally associated with the point x (Fig. 2.15).

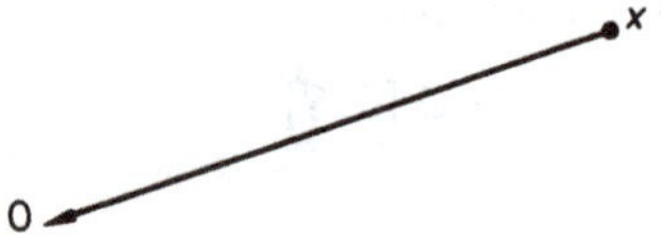

FIG. 2.15

In general, an arrow $x \longrightarrow z$ should be thought of as defining the vector z in the vector space isomorphic to $\mathbf{R}^n$ with origin at x (Fig. 2.16). The isomorphism between this vector space centred at x and $\mathbf{R}^n$ is given explicitly by

$$(x \longrightarrow z) \longmapsto z \qquad \text{for all } z \in \mathbf{R}^n.$$

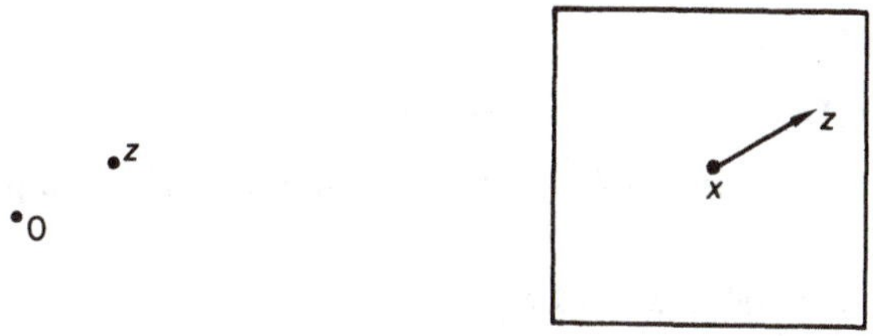

FIG. 2.16

In particular, we define addition of arrows, with the *same initial points*, by the rule (Fig. 2.17)

$$(x \longrightarrow z_1) + (x \longrightarrow x_2) = (x \longrightarrow z_1 + z_2).$$

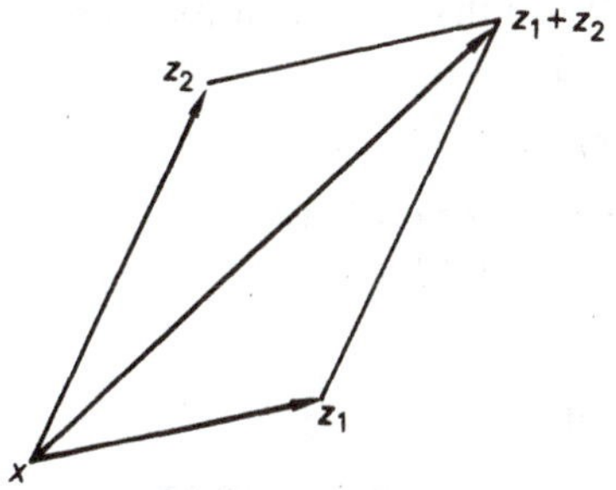

FIG. 2.17

Let us denote the set of all arrows in $\mathbf{R}^n$ by Arr $(\mathbf{R}^n)$. Then Arr $(\mathbf{R}^n)$ may be put in 1:1 correspondence with the set $\mathbf{R}^n \times \mathbf{R}^n$ in a natural way. Indeed

$$\text{Arr}\,(\mathbf{R}^n) = \{x \longrightarrow y : x, y \in \mathbf{R}^n\}$$

and we map the arrow $x \longrightarrow y$ to the point $(x, y) \in \mathbf{R}^n \times \mathbf{R}^n$. If we let $\text{Arr}_x\,(\mathbf{R}^n)$ denote the set of arrows in $\mathbf{R}^n$ with initial point x, then the above map restricts to give a bijection between $\text{Arr}_x\,(\mathbf{R}^n)$ and $\{x\} \times \mathbf{R}^n$ for all $x \in \mathbf{R}^n$. As indicated above, $\text{Arr}_x\,(\mathbf{R}^n)$ has a natural vector space structure for all $x \in \mathbf{R}^n$. For reasons that will become clearer later we shall make no attempt to put a vector-space structure on the *whole* of Arr $(\mathbf{R}^n)$. (The reader should note that the vector-space structure on Arr $(\mathbf{R}^n)$ induced from that on $\mathbf{R}^n \times \mathbf{R}^n$ by the above bijection is *not* compatible with the vector-space structure we have defined on $\text{Arr}_x\,(\mathbf{R}^n)$). We shall regard Arr $(\mathbf{R}^n)$ as a parametrized family of vector spaces isomorphic to $\mathbf{R}^n$. That is

$$\text{Arr}\,(\mathbf{R}^n) = \{\text{Arr}_x\,(\mathbf{R}^n) : x \in \mathbf{R}^n\}.$$

Let $\phi: [a, b] \longrightarrow \mathbf{R}^n$ be a C^2 parametrization of the smooth curve C. It follows from formula A above that the length function $L: [a, b] \longrightarrow [0, S]$ is also C^2. Moreover, by the inverse function theorem, $P: [0, S] \longrightarrow [a, b]$ is C^2 (see Chapter 3) and so the canonical parametrization Φ of C is C^2. Let us consider the constant map $\gamma: [0, S] \longrightarrow \mathbf{R}$ defined by

$$\gamma(x) = \langle T(x), T(x) \rangle, \qquad x \in [0, S].$$

Since Φ is C^2, T is C^1 and so

$$\gamma'(x) = 2\langle T(x), T'(x) \rangle, \qquad x \in [0, S].$$

Since γ is constant, $\gamma' = 0$ and so $T(x) \perp T'(x)$ for all $x \in [0, S]$. Two possibilities can occur:

(a) $T'(x) = 0$ for some values $x \in [0, S]$.
(b) $T'(x) \neq 0$ on $[0, S]$.

Case (a) occurs, for example, if C is a straight line. We shall be most interested in case (b).

Definition 2.13.7

Let $\Phi : [0, S] \longrightarrow \mathbf{R}^n$ be the canonical parametrization of the smooth curve C. Suppose that Φ is of class C^2 and that $T' \neq 0$ on $[0, S]$. We define the continuous function $N: [0, S] \longrightarrow \mathbf{R}^n$ by

$$N(x) = T'(x)/\|T'(x)\|_2.$$

$N(x)$ is called the (unit) *principal normal* to C at x.

We also define $R: [0, S] \longrightarrow \mathbf{R}$ by

$$R(x) = \|T'(x)\|_2.$$

$R(x)$ is called the *curvature* of C at x.

From the remarks preceding this definition we have the following proposition.

Proposition 2.13.8

Let $\Phi: [0, S] \longrightarrow \mathbf{R}^n$ be the canonical parametrization of the smooth curve C. Suppose that Φ is of class C^2 and that $T' \neq 0$ on C. Then

$$N(x) \perp T(x) \qquad \text{for all } x \in [0, S].$$

Examples

1. Consider the circle radius r, centre 0, in $\mathbf{R}^2$, parametrized by

$$\phi(t) = (r \cos t, r \sin t), \qquad t \in [0, 2\pi].$$

The canonical parametrization of this circle is easily seen to be given by

$$\Phi(x) = (r \cos (x/r), r \sin (x/r)), \qquad x \in [0, 2\pi r].$$

Hence the unit tangent and principal normal to the circle at x are given by

$$T(x) = (-\sin (x/r), \cos (x/r)), \qquad N(x) = (-\cos (x/r), -\sin (x/r)).$$

Pictorially we may represent $T(x)$ and $N(x)$ as in Fig. 2.18. Notice that the principal normal points inwards. The curvature of the circle is constant and equal to $1/r$.

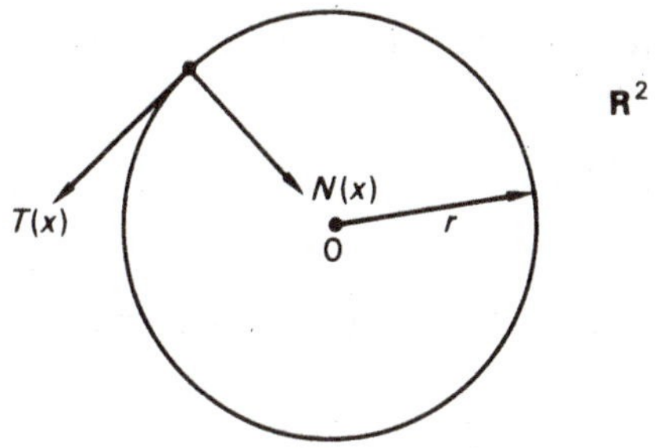

FIG. 2.18

2. Consider the helix of radius r in $\mathbf{R}^3$ parametrized by

$$\phi(t) = (r \cos t, r \sin t, t), \qquad t \in [0, c], \quad c > 0.$$

The reader may verify that the canonical parametrization of the helix is given by

$$\Phi(x) = (r \cos (ax), r \sin (ax), ax), \qquad x \in [0, c/a],$$

where $a = 1/\sqrt{(2r^2 + 1)}$. Now $T'(x) = -ra^2(\cos (ax), \sin (ax), 0)$ for all $x \in [0, c/a]$. Hence $T' \neq 0$ on $[0, c/a]$, and the principal normal and curvature are given by

$$N(x) = -(\cos (ax), \sin (ax), 0), \qquad R(x) = r/(2r^2 + 1).$$

Suppose that $\phi : [a, b] \longrightarrow \mathbf{R}^n$ describes the motion of a particle moving in $\mathbf{R}^n$ along the smooth curve C. The distance travelled in time t is given by $L(t)$, and the 'speed' of the particle by $L'(t) = \| \phi'(t) \|_2$. From Proposition 2.13.6, the velocity of the particle ϕ' and the unit tangent to C are related by the equation

$$\phi' = L'T.$$

That is

$$\phi'(t) = \| \phi'(t) \|_2 T(L(t)).$$

This equation enables us to find the unit tangent to C at $\phi(t)$ without having to find the function L explicitly. Indeed, the unit tangent at $\phi(t)$ is obviously $\phi'(t)/\| \phi'(t) \|_2$. Differentiating the above equation we obtain

$$\phi''(t) = L''(t)T(L(t)) + (L'(t))^2 T'(L(t)).$$

If $T' \neq 0$ on C, we may rewrite this equation in terms of the normal to C as

$$\phi'' = L''T + (L'^2 R)N. \qquad \text{(D)}$$

Hence the 'acceleration' ϕ'' of the particle is made up of a component along the tangent and a component along the principal normal. From eq. (D) we have

$$RN = (1/L'^2)(\phi'' - L''T). \qquad \text{(E)}$$

Since $L' = \| \phi' \|_2$ we also have

$$L''(t) = \langle \phi'(t), \phi''(t) \rangle / \| \phi'(t) \|_2.$$

Hence we can compute the curvature and principal normal, R and N, from eq. (E) in terms of ϕ and its derivatives.

Example. Find the curvature and the principal normal at the point $t = 0$ for the curve $\phi : \mathbf{R} \longrightarrow \mathbf{R}^3$ defined by

$$\phi(t) = (t^2 + 3 \sin t, t^2, 4t).$$

We have

$$\phi'(t) = (2t + 3 \cos t, 2t, 4), \qquad \phi''(t) = (2 - 3 \sin t, 2, 0).$$

Hence

$$\phi'(0) = (3, 0, 4), \qquad L'(0) = \| \phi'(0) \|_2 = 5,$$
$$\phi''(0) = (2, 2, 0), \qquad \langle \phi'(0), \phi''(0) \rangle = 6.$$

Substituting in the equation for $L''(t)$ we find that

$$L''(0) = 6/5.$$

It follows from eq. (E) that

$$RN = (1/25)((2, 2, 0) - (6/25)(3, 0, 4)) = -(2/25)(8, -1, 12).$$

Computing we find that

$$R = (2\sqrt{209})/25, \qquad N = (1/\sqrt{209})(-8, 1, -12).$$

Exercises

1. Consider the helix in $\mathbf{R}^3$ defined by

$$\phi(t) = (r \cos t, r \sin t, kt), \qquad t \in [t_0, t_1].$$

   Find the length of the helix and hence show that it is directly proportional to $t_1 - t_0$.

2. Find the length of the smooth curve in $\mathbf{R}^3$ parametrized by

$$\phi(t) = (r \cosh (t/r), r \sinh (t/r), t), \qquad t \in [0, a].$$

3. Consider the infinite spiral in $\mathbf{R}^2$ parametrized by

$$\phi(t) = (e^{-t} \cos t, e^{-t} \sin t), \qquad t \geqslant 0.$$

   Show that the length of the curve from $t = 0$ to $t = T$ is $\sqrt{2}(1 - e^{-T})$. Hence find an expression for the length of the complete spiral.

4. Find T, N, R at $t = 0$ for the curves

   (a) $\phi(t) = (t, t^2, t^3)$,
   (b) $\phi(t) = (e^t \cos t, e^t \sin t)$,
   (c) $\phi(t) = (t + t^2, t^2, \sin t, 2t)$.

5. Let $\Phi: [0, S] \longrightarrow \mathbf{R}^3$ be the canonical parametrization of the smooth curve C. Suppose that Φ is C^3 and that the principal normal N is everywhere defined. Define the vector B to be equal to $T \times N$ (that is, B is the unit vector orthogonal to T and N such that (T, N, B) forms a right-handed system). B is called the *binormal* to C. Prove the following:

(a) $B'(x) = T'(x) \times N(x)$ for all $x \in [0, S]$.
(b) $B' \perp B$.
(c) $B' = -\tau N$, where $\tau : [0, S] \longrightarrow \mathbf{R}$ (τ may be zero).
(d) $N' = \tau B - RT$.

τ is called the *torsion* of C. Show that if $\tau = 0$ the curve C lies completely inside a two-dimensional affine subspace of $\mathbf{R}^3$.

CHAPTER 3

The inverse- and implicit-function theorems

3.1 Topological linear isomorphisms

If E and F are normed vector spaces and $T:E \longrightarrow F$ is a continuous linear map with a linear inverse $T^{-1}:F \longrightarrow E$ it does not, in general, follow that T^{-1} need be continuous. For example, if $E = F = s$ (the space of finitely non-zero sequences) the map $T:s \longrightarrow s$ defined by

$$T((x_1, \ldots, x_n, 0, 0, \ldots)) = (x_1, x_2/2, \ldots, x_n/n, 0, 0, \ldots)$$

is certainly continuous linear with a linear inverse T^{-1} defined by

$$T^{-1}((x_1, \ldots, x_n, 0, 0, \ldots)) = (x_1, 2x_2, \ldots, nx_n, 0, 0, 0, \ldots).$$

T^{-1} is obviously unbounded on the unit disc in s and is therefore not continuous.

In this section we wish to study continuous linear maps between normed vector spaces which do have a continuous linear inverse. It turns out that to get a satisfactory theory one must assume that the normed vector spaces are complete, that is, Banach spaces. Indeed it follows from the closed-graph theorem that if E and F are Banach spaces and $T \in L(E, F)$ has a linear inverse T^{-1}, then T^{-1} is necessarily continuous. We shall not, however, make any use of this deep result about Banach spaces in this text.

For the rest of this chapter all normed vector spaces will be assumed complete and over either the real or complex field.

Definition 3.1.1

Let E and F be Banach spaces and let $T \in L(E, F)$. We shall say that T is a *topological linear isomorphism* or *linear homeomorphism* if there exists a map $S \in L(F, E)$ such that

$$T.S = I_F, \qquad S.T = I_E.$$

The map S is necessarily unique and is called the inverse of T. In the sequel we denote it by T^{-1}.

We denote the set of all topological linear isomorphisms from E to F by Iso (E, F).

Remarks

1. We shall abbreviate the term 'topological linear isomorphism' to 'toplinear isomorphism' in the sequel.
2. If $T \in \text{Iso}(E, F)$ then $T^{-1} \in \text{Iso}(F, E)$.

Examples

1. If E is any Banach space, then $I_E \in \text{Iso}(E, E)$. Hence $\text{Iso}(E, E)$ is non-empty. If E and F are arbitrary Banach spaces, then $\text{Iso}(E, F)$ is often empty. For example, if E and F have different dimensions.
2. Let $\mathbf{R}^n$ be given any norm. Then

$$\text{Iso}(\mathbf{R}^n, \mathbf{R}^m) = \begin{cases} \emptyset, & n \neq m \\ \{A \in L(\mathbf{R}^n, \mathbf{R}^n): \det_{\mathbf{R}}(A) \neq 0\}, & n = m. \end{cases}$$

If we fix a basis for $\mathbf{R}^n$ and look at the matrix of $A \in L(\mathbf{R}^n, \mathbf{R}^n)$, we see that $\det[A] = \det(A)$ is a polynomial in the components of $[A]$. It follows that $\det: L(\mathbf{R}^n, \mathbf{R}^n) \longrightarrow \mathbf{R}$ is actually a C^∞ function. In particular, det is continuous and so $\text{Iso}(\mathbf{R}^n, \mathbf{R}^n)$ is an *open* subset of $L(\mathbf{R}^n, \mathbf{R}^n)$. We shall soon see that this result generalizes to arbitrary Banach spaces.

Similarly, if $\mathbf{C}^n$ is given any norm

$$\text{Iso}(\mathbf{C}^n, \mathbf{C}^n) = \{A \in L(\mathbf{C}^n, \mathbf{C}^n): \det_{\mathbf{C}}(A) \neq 0\}.$$

and $\text{Iso}(\mathbf{C}^n, \mathbf{C}^n)$ is an open subset of $L(\mathbf{C}^n, \mathbf{C}^n)$.

Remark. $\text{Iso}(\mathbf{R}^n, \mathbf{R}^n)$, whilst certainly not a vector subspace of $L(\mathbf{R}^n, \mathbf{R}^n)$, has the structure of a group if we define composition as composition of linear maps. With this group structure, $\text{Iso}(\mathbf{R}^n, \mathbf{R}^n)$ is usually referred to as the *general linear group of order n* and is denoted by $GL(n, \mathbf{R})$. We have a similar definition of $GL(n, \mathbf{C})$ and indeed of $GL(n, F)$, where F is any field.

In the remainder of this section we will usually drop the subscript E from our notation for the identity map I_E of E.

Proposition 3.1.2

Let E be a Banach space and suppose that $T \in L(E, E)$ has norm less than $1: \|T\| < 1$. Then we have the following properties of the map $I + T$.

1. $I + T \in \text{Iso}(E, E)$.
2. $(I + T)^{-1} = I + \Sigma_{j=1}^\infty (-1)^j T^j$.
3. $\|(I + T)^{-1}\| \leqslant 1/(1 - \|T\|)$.
4. $\|(I + T)^{-1} - (I - T)\| \leqslant \|T\|^2/(1 - \|T\|)$.

Proof. We define the sequence $(\phi_n)_{n=1}^\infty$ in $L(E, E)$ by the formula

$$\phi_n = I + \sum_{j=1}^{n} (-1)^j T^j, \qquad n = 1, 2, \ldots.$$

We claim that (ϕ_n) is a Cauchy sequence. For $m > n$ we have the equality

$$\phi_n - \phi_m = \sum_{j=n+1}^{m} (-1)^j T^j.$$

Taking norms it follows that

$$\|\phi_n - \phi_m\| \leqslant \sum_{j=n+1}^{m} \|T^j\|$$

$$\leqslant \sum_{j=n+1}^{m} \|T\|^j \qquad \text{(Proposition 1.6.4)}$$

$$\leqslant \sum_{j=n+1}^{\infty} \|T\|^j$$

$$= \|T\|^{n+1}/(1 - \|T\|) \longrightarrow 0 \text{ as } n, m \longrightarrow \infty.$$

Hence (ϕ_n) is a Cauchy sequence. By Theorem 1.9.7, ϕ_n converges to some point $\phi \in L(E, E)$ and

$$\phi = I + \sum_{j=1}^{\infty} (-1)^j T^j.$$

We shall now prove that $\phi . (I + T) = I$ and $(I + T) . \phi = I$ and hence that ϕ is the inverse of $I + T$. First notice that

$$\|(I + T) . \phi_n - (I + T) . \phi\| \leqslant \|(I + T)\| \, \|\phi - \phi_n\| \longrightarrow 0 \text{ as } n \longrightarrow \infty.$$

Hence

$$(I + T) . \phi_n \longrightarrow (I + T) . \phi \qquad \text{as } n \longrightarrow \infty.$$

A simple computation yields

$$(I + T) . \phi_n = I \pm T^{n+1}.$$

Since $\|T\| < 1$,

$$\|T^{n+1}\| \leqslant \|T\|^{n+1} \longrightarrow 0 \qquad \text{as } n \longrightarrow \infty.$$

Therefore,

$$(I + T) . \phi_n \longrightarrow I \qquad \text{as } n \longrightarrow \infty.$$

Hence $(I + T) . \phi = I$. Similarly, $\phi . (I + T) = I$ and so $\phi = (I + T)^{-1}$.

Condition (3) of the proposition follows from (2) by taking norms. (4) follows from (2) by noting that

$$(I + T)^{-1} - I + T = \left(I + \sum_{j=1}^{\infty} (-1)^j T^j\right) - I + T$$

$$= T^2(I - T + T^2 - \cdots)$$

$$= T^2 . (I + T)^{-1}$$

and then taking norms and applying (3). ∎

Our first application of Proposition 3.1.2 is to the following generalization of Example 2 above.

Proposition 3.1.3

Let E and F be Banach spaces. Then $\text{Iso}(E, F)$ is an open subset of $L(E, F)$.

Proof. Let $T \in \text{Iso}(E, F)$; then $T^{-1} \in \text{Iso}(F, E)$. Set $r = 1/\| T^{-1} \|$. We shall show that

$$B_r(T) \subset \text{Iso}(E, F).$$

To do this it is sufficient to show that if $S \in B_r(0)$, then $T + S \in \text{Iso}(E, F)$. If $S \in B_r(0)$, $\| S \| < 1/\| T^{-1} \|$ and so $T^{-1} . S \in L(E, E)$ has norm less than 1:

$$\| T^{-1} . S \| < 1.$$

Hence, by Proposition 3.1.2, we have

$$I + T^{-1} . S \in \text{Iso}(E, E).$$

We claim that the map $(I + T^{-1} . S)^{-1} . T^{-1}$ is the inverse of $T + S$. We have

$$(I + T^{-1} . S)^{-1} . T^{-1} . (T + S) = (I + T^{-1} . S)^{-1} . (I + T^{-1} . S) = I$$

and

$$(T + S) . (I + T^{-1} S)^{-1} . T^{-1} = T . (I + T^{-1} . S) . (I + T^{-1} . S)^{-1} . T^{-1} = I.$$

Hence $T + S \in \text{Iso}(E, F)$ for all $S \in B_r(0)$ and so $B_r(T) \subset \text{Iso}(E, F)$. ∎

Notation. Let E and F be Banach spaces. We let $\beta : \text{Iso}(E, F) \longrightarrow \text{Iso}(F, E)$ denote the map defined by taking inverses: $\beta(T) = T^{-1}$, $T \in \text{Iso}(E, F)$.

Proposition 3.1.4

Let E and F be Banach spaces; then the map $\beta : \text{Iso}(E, F) \longrightarrow \text{Iso}(F, E)$ is continuous.

Proof. Let $T \in \text{Iso}(E, F)$. From Proposition 3.1.3, $\exists r > 0$ such that $B_r(T) \subset \text{Iso}(E, F)$. Let $S \in B_r(0)$; then

$$\begin{aligned} (T + S)^{-1} - T^{-1} &= (T + S)^{-1} . (I - (T + S) . T^{-1}) \\ &= -(T + S)^{-1} . S . T^{-1} \\ &= -(I + T^{-1} S)^{-1} . T^{-1} . S . T^{-1}. \end{aligned}$$

If $r < 1/\| T^{-1} \|$, $\| T^{-1} S \| < 1$ and so by (3) of Proposition 3.1.2,

$$\| (I + T^{-1} . S)^{-1} \| \leqslant 1/(1 - \| T^{-1} S \|).$$

It follows that if $r < 1/\| T^{-1} \|$ we have the following inequalities:

$$\begin{aligned} \| (T + S)^{-1} - T^{-1} \| &\leqslant \| T^{-1} \|^2 \| S \| / (1 - \| T^{-1} S \|) \\ &\leqslant \| T^{-1} \|^2 \| S \| / (1 - \| T^{-1} \| \| S \|). \end{aligned}$$

Since this last term tends to zero as $S \longrightarrow 0$, this is sufficient to prove the continuity of β at T. ∎

Remark. If $E = F = \mathbf{R}^n$ in Proposition 3.1.4, then we could have proved the continuity of β by taking a basis for $\mathbf{R}^n$ and using the fact that we have an explicit algebraic formula for $[T^{-1}] = [T]^{-1}$ which clearly depends continuously on the components of $[T]$. A similar remark applies also to the following theorem.

Theorem 3.1.5

Let E and F be Banach spaces; then the map

$$\beta : \text{Iso}\,(E, F) \longrightarrow \text{Iso}\,(F, E)$$

is C^∞ and the derivative $D\beta : \text{Iso}\,(E, F) \longrightarrow L(L(E, F), L(F, E))$ is given by the formula

$$D\beta_T(A) = -T^{-1} . A . T^{-1}, \qquad T \in \text{Iso}\,(E, F), \quad A \in L(E, F).$$

Proof. We shall start by showing that β is differentiable on Iso (E, F). Fix $T_0 \in$ Iso (E, F) and choose $S \in L(E, F)$ such that $\| S \| < 1/\| T_0^{-1} \|$. As in the proof of Proposition 3.1.4 we may show that

$$(T_0 + S)^{-1} - T_0^{-1} = ((I + T_0^{-1} . S)^{-1} - I) . T_0^{-1},$$

and so

$$(T_0 + S)^{-1} - T_0^{-1} - (-T_0^{-1} . S . T_0^{-1}) = [(I + T_0^{-1} . S)^{-1} - I + T_0^{-1} . S] . T_0^{-1}.$$

Applying (4) of Proposition 3.1.2 with $T = T_0^{-1} . S$, we obtain the estimate

$$\| (T_0 + S)^{-1} - T_0^{-1} + T_0^{-1} . S . T_0^{-1} \| \leqslant \| T_0^{-1} \|^3 \, \| S \|^2 / (1 - \| T_0^{-1} \| \, \| S \|).$$

Clearly the term on the right-hand side of this inequality is $o(S)$, and so β is differentiable at T_0 with derivative the map defined by

$$D\beta_{T_0}(A) = -T_0^{-1} . A . T_0^{-1}, \qquad A \in L(E, F).$$

We now prove that β is C^∞. For simplicity of notation let us set $L(E, F) = X$ and $L(F, E) = Y$. Then we have a bilinear map

$$\zeta : Y \times Y \longrightarrow L(X, Y)$$

defined by

$$\zeta(U, V)(A) = -U . A . V \qquad \text{for all } U, V \in Y \text{ and } A \in X.$$

That is, if $U, V \in L(F, E)$ and $A \in L(E, F)$, the map $\zeta(U, V)(A) \in L(F, E)$ is the vertical dotted arrow formed by the composition of maps in the following diagram:

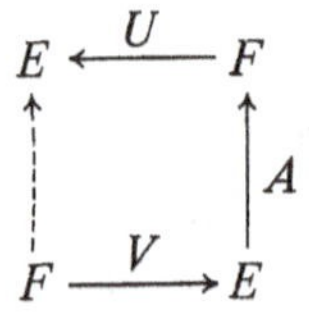

ζ is clearly continuous, since $\| \zeta(U, V) \| \leqslant \| U \| \, \| V \|$. Hence ζ is C^∞. $D\beta$ may be written in terms of ζ by the following formula:

$$D\beta = \zeta(\gamma, \gamma). \tag{A}$$

That is,

$$D\beta_T = \zeta(T^{-1}, T^{-1}) = \zeta(\beta(T), \beta(T)).$$

We have already shown that β is continuous, and so it follows from formula (A) that β is C^1. Proceeding inductively, suppose that we have shown that β is p times differentiable on Iso (E, F). By Theorem 2.10.3 applied to the function $\zeta(\beta, \beta)$, it follows that β is $p + 1$ times differentiable. Hence β is C^∞. ∎

Example. Let $T \in L(E, E)$, where E is a complex Banach space. We define the subset $\Omega \subset \mathbf{C}$ as follows

$$\Omega = \{\lambda \in \mathbf{C} : T - \lambda I \in \mathrm{Iso}(E, E)\}.$$

Since $T - \lambda I = -\lambda(I - T/\lambda)$, it follows from Proposition 3.1.2 that $\lambda \in \Omega$, for $|\lambda| > \|T\|$. From Proposition 3.1.3 it therefore follows that Ω is a non-empty open subset of $\mathbf{C}$. We let $J : \Omega \longrightarrow \mathrm{Iso}(E, E)$ denote the map defined by

$$J(\lambda) = (T - \lambda I)^{-1}.$$

We claim that J is C^∞. This follows since J can be expressed as the composite βP, where $P : \Omega \longrightarrow \mathrm{Iso}(E, E)$ is defined by $P(\lambda) = T - \lambda I$. Since P is affine linear and we have already shown that β is C^∞, it follows that J is C^∞. The derivative of J is given by the composite-mapping formula as

$$\begin{aligned} J'(\lambda) &= D_{\beta P(\lambda)} \,.\, DP_\lambda(1) \\ &= -(T - \lambda I)^{-1} \,.\, (-I) \,.\, (T - \lambda I)^{-1} = (T - \lambda I)^{-2} = J(\lambda)^2. \end{aligned}$$

In the exercises we indicate the main steps in the proof of the important result that $\Omega \neq \mathbf{C}$. That is, we can always find a complex number λ such that $T - \lambda I \notin \mathrm{Iso}(E, E)$.

Exercises

1. Let E be a complex Banach space and let $T \in L(E, E)$. In Chapter 1 we defined an eigenvalue and eigenvector of T. Show that if E is infinite dimensional, T need not have any eigenvalues. (*Hint*: Consider the endomorphism of l^2 defined by mapping $(x_1, x_2, \ldots)$ to $(0, x_1, x_2, \ldots)$.
2. Let E be a complex Banach space and let $T \in L(E, E)$. We say that $\lambda \in \mathbf{C}$ is a *proper* or *spectral value* of T if $T - \lambda I \notin \mathrm{Iso}(E, E)$. We call the set of proper values of T the *spectrum of* T, spec (T). Show that if $E(T)$ denotes the set of eigenvalues of T, then $E(T) \subset$ spec (T).
3. Continuing with the notation and assumptions of Exercise 2, set $\Omega = \mathbf{C} \setminus \mathrm{spec}\,(T)$. In the example preceding these exercises we showed that Ω was a non-empty open subset of $\mathbf{C}$ and that the map $J(\lambda) = (T - \lambda I)^{-1}$ was C^∞ on Ω. Prove that

   (a) $(\mu T - I) \in \mathrm{Iso}(E, E)$, for all $\mu \in \mathbf{C}$ satisfying $|\mu| < 1/\|T\|$. Deduce that

   $$\Omega \supset \{\lambda \in \mathbf{C} : |\lambda| > \|T\|\}.$$

   (b) If $\Omega = \mathbf{C}$, then $\|J(\lambda)\|$ is bounded above. That is, $\exists M \geqslant 0$ such that

   $$\|J(\lambda)\| \leqslant M \qquad \text{for all } \lambda \in \mathbf{C}.$$

   (c) If $f \in (L(E, E))^*$, show that $fJ : \Omega \longrightarrow \mathbf{C}$ is analytic (that is, fJ is complex differentiable on Ω).

   Liouville's theorem states that an analytic function defined on the complex plane which is bounded in modulus is a constant. Show, using Liouville's theorem together with the Hahn-Banach theorem and (b) and (c), that spec $(T) \neq \emptyset$.
4. Continuing with the assumptions of Exercise 2, prove that spec (T) is a compact subset of $\mathbf{C}$. (*Hint*: Use (a) of Exercise 3.)

5. Let $T \in L(E,E)$, where E is a real or complex Banach space. We define a map $\exp(T): E \longrightarrow E$ by the formula

$$\exp(T) = \sum_{j=0}^{\infty} T^j/j!$$

Prove that $\exp(T)$ is defined for all $T \in L(E,E)$ and that $\exp: L(E,E) \longrightarrow L(E,E)$ is C^∞.

Show also that if $T, S \in L(E,E)$ commute ($T.S = S.T$), then $\exp(T+S) = \exp(T).\exp(S)$.

6. With the assumptions of Exercise 5, show that $\exists r > 0$ such that the series

$$\log(T) = (T-I) - (T-I)^2/2 + \cdots$$

converges to a point in $L(E,E)$ for all $T \in B_r(I)$. Prove that the map $\log: B_r(I) \longrightarrow L(E,E)$ is C^∞.

Show also that if T and S commute, $\log(TS) = \log(T) + \log(S)$.

7. Continuing with the assumptions and notation of Exercises 5 and 6, show that $\exists t > 0$ such that, if $A \in B_t(0)$,

$$\log(\exp(A)) = A,$$
$$\exp(\log(I+A)) = I + A.$$

8. Let H be a Hilbert space and suppose that $A \in L(H,H)$ is symmetric ($A = A^*$). Prove that $\exp(A)$ is positive definite. That is, show that $\exists c > 0$ such that

$$\langle \exp(A)x, x\rangle \geqslant c\,\|x\|^2 \qquad \text{for all } x \in H.$$

9. Let E be a real or complex Banach space. With the definitions of the miscellaneous exercises at the end of Chapter 1, let

$$GL_F(E) = \{A \in \operatorname{Iso}(E,E): A = I + B, \text{ where } \operatorname{Im}(B) \text{ is finite dimensional}\},$$
$$GL_c(E) = \{A \in \operatorname{Iso}(E,E): A = I + B, \text{ where } B \text{ is a compact operator}\},$$
$$GL_N(E) = \{A \in \operatorname{Iso}(E,E): A = I + B, \text{ where } B \text{ is a nuclear operator}\}.$$

Prove that $GL_F(E) \subset GL_N(E) \subset GL_c(E)$ and that the three spaces are subgroups of $\operatorname{Iso}(E,E)$. Show that $GL_F(E)$ is not a closed subgroup of $\operatorname{Iso}(E,E)$ if $E = l^2$, and that $GL_c(E)$ is always a closed subgroup of $\operatorname{Iso}(E,E)$. (By 'closed' subgroup, we mean closed as a topological subspace of $\operatorname{Iso}(E,E)$).

3.2 The inverse-function theorem

Suppose that E and F are Banach spaces, that U is an open subset of E and that $f: U \longrightarrow F$ is differentiable. Let us set $f(U) = V$. In this section we shall be interested in finding when f has a differentiable inverse. That is, when there exists a differentiable map $g: V \longrightarrow U$ such that

$$gf = I_E, \qquad fg = I_F.$$

If g exists it is necessarily unique. One condition that follows from the existence of g is that f is a homeomorphism onto its image, V, and we shall always assume that V is an open subset of F. This allows us to talk about the differentiability of g.

Suppose then that g is the differentiable inverse of f. By the composite-mapping formula we must have

$$Df_{g(x)} \cdot Dg_x = I_F \qquad \text{for all } x \in V,$$

$$Dg_{f(x)} \cdot Df_x = I_E \qquad \text{for all } x \in U.$$

It follows that $Df_x \in \text{Iso}(E, F)$ for all $x \in U$. Hence a *necessary* condition for f to have a differentiable inverse is that

$$Df: U \longrightarrow \text{Iso}(E, F).$$

Notice that this implies that $E \cong F$.

If U is connected and $E = F = \mathbf{R}$, it is in fact true that this is also a sufficient condition for f to have an inverse. It is, however, far from true generally. For example, let $U = E = F = \mathbf{R}^2$ and define f by

$$f(x, y) = (e^x \cos y, e^x \sin y), \qquad (x, y) \in \mathbf{R}^2.$$

The reader may verify that the Jacobian of f at (x, y) is equal to $e^{2x} \neq 0$ and hence that $Df: \mathbf{R}^2 \longrightarrow \text{Iso}(\mathbf{R}^2, \mathbf{R}^2)$. But f is certainly not injective and so f cannot have an inverse.

On the other hand, if $f: U \longrightarrow F$ is differentiable and $Df_x \in \text{Iso}(E, E)$ for some point $x \in U$, we might expect that f is invertible 'near' x since the behaviour of f near x is dominated by its linear approximation which is given by Df_x. More precisely, we might hope to find an open subset W of x in U such that $f \mid W$ has a differentiable inverse. It is the purpose of this section to make this statement of local invertibility more precise and prove a basic theorem which gives sufficient conditions for local invertibility. First we introduce some new notation.

Notation. Let U and V be open subsets of the Banach spaces E and F respectively. For $1 \leqslant p \leqslant \infty$, we set

$$C^p(U, V) = \{f: U \longrightarrow F: f \text{ is } C^p \text{ and } f(U) \subset V\},$$

$$\text{Diff}^p(U, V) = \{f \in C^p(U, V): f(U) = V$$

$$\text{and } \exists g \in C^p(V, U) \text{ which is the inverse of } f\}.$$

In general, $\text{Diff}^p(U, V)$ may be empty.

Definition 3.2.1

Let U and V be open subsets of the Banach spaces E and F respectively. An element $f \in \text{Diff}^p(U, V)$ is called a (C^p-) *diffeomorphism.*

We may now state the main theorem of this section.

Theorem 3.2.2 (Inverse-function theorem)

Let E and F be Banach spaces and let U be an open subset of E. Suppose that $f \in C^p(U, F)$, $p \geqslant 1$, and that there exists $x_0 \in U$ such that $Df_{x_0} \in \text{Iso}(E, F)$. Then we may find an open neighbourhood W of x_0 in U such that

1. $f(W)$ is an open subset of F.
2. $f \mid W \in \text{Diff}^p(W, f(W))$.
3. If we denote the inverse of $f \mid W$ by f^{-1}, then $D(f^{-1})_y = (Df_x)^{-1}$, where $x = f^{-1}(y)$ and $y \in f(W)$.

The proof of this theorem, like that of most theorems which assert the existence of an inverse function, makes essential use of a fixed-point theorem. In this case the relevant result is the contraction lemma.

Lemma (Contraction-mapping lemma)

Let (M, d) be a complete metric space and let $f: M \longrightarrow M$ satisfy the following condition:

$$\exists K, 0 < K < 1, \text{ such that for all } x, y \in M$$

$$d(f(x), f(y)) \leqslant K d(x, y).$$

Then there exists a unique fixed point $x_0 \in M$ for f and

$$\lim_{n \to \infty} f^n(x) = x_0 \qquad \text{for all } x \in M.$$

Proof. We shall only sketch the proof of this result, which should already be known, in some form, by the reader. Let x be any point of M. We define a sequence $(x_n)_{n=1}^{\infty}$ in M by the condition $x_n = f^n(x)$, $n = 1, 2, \ldots$. It is easily shown that this sequence is Cauchy and that the limit of the sequence is a fixed point for f. The uniqueness follows from the condition $d(f(x), f(y)) < d(x, y)$. ■

Proof of Theorem 3.2.2. In our proof of the inverse-function theorem, we shall make essential use of the completeness of E and F in the construction of an inverse for f. Once we have an inverse for f, the continuity and differentiability properties of the inverse will all follow from the mean-value theorem.

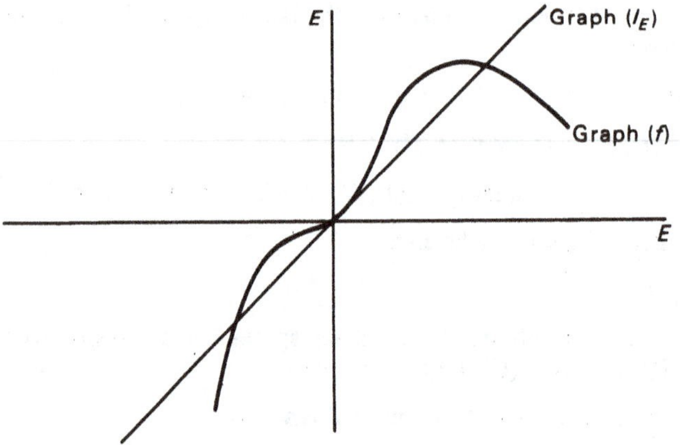

FIG. 3.1

Let us start by proving a special case of the theorem: We shall assume that $E = F$, $x_0 = f(x_0) = 0$ and $Df_{x_0} = I_E$. Thus we have the situation illustrated in Fig. 3.1. We shall prove the special case of the theorem in a number of steps.

Step 1. We define $\phi: U \longrightarrow E$ by $\phi(x) = x - f(x)$. Since $\phi = I_E - f$, it follows that

$$D\phi_0 = 0.$$

The continuity of Df implies that $\exists r > 0$ such that

$$\| D\phi_x \| \leqslant \tfrac{1}{2} \qquad \text{for all } x \in \bar{B}_r(0). \tag{A}$$

We shall apply the mean-value theorem to ϕ. From (A) it follows that

$$\| \phi(x_1) - \phi(x_2) \| \leqslant \tfrac{1}{2} \| x_1 - x_2 \| \qquad \text{for all } x_1, x_2 \in \bar{B}_r(0). \tag{B}$$

In particular, since $\phi(0) = 0$, we have

$$\| \phi(x) \| \leqslant \tfrac{1}{2} \| x \| \qquad \text{for all } x \in \bar{B}_r(0). \tag{C}$$

Hence

$$\phi(\bar{B}_r(0)) \subset \bar{B}_{r/2}(0).$$

Step 2. Construction of the inverse. We claim that, given $y \in \bar{B}_{r/2}(0)$, there exists a unique $x \in \bar{B}_r(0)$ such that $f(x) = y$. Notice that $f(x) = y$ if and only if $(x - f(x)) + y = x$. That is, $f(x) = y$, if and only if x is a fixed point of the map $\Psi_y : \bar{B}_r(0) \longrightarrow E$ defined by

$$\Psi_y(x) = y + x - f(x) = \phi(x) + y.$$

If $y \in \bar{B}_{r/2}(0)$ and $x \in \bar{B}_r(0)$, then

$$\begin{aligned} \| \Psi_y(x) \| &= \| \phi(x) + y \| \\ &\leqslant \| \phi(x) \| + \| y \| \\ &\leqslant r, \qquad \text{by (C) of step 1.} \end{aligned}$$

Hence

$$\Psi_y : \bar{B}_r(0) \longrightarrow \bar{B}_r(0).$$

Since $\bar{B}_r(0)$ is a complete metric space, we are now in a position to apply the contraction-mapping lemma once we have shown that Ψ_y is a contraction. Suppose that $x_1, x_2 \in \bar{B}_r(0)$; then

$$\begin{aligned} \| \Psi_y(x_1) - \Psi_y(x_2) \| &= \| \phi(x_1) - \phi(x_2) \| \\ &\leqslant \tfrac{1}{2} \| x_1 - x_2 \|, \qquad \text{by (B) of step 1.} \end{aligned}$$

Hence we may apply the contraction-mapping lemma to Ψ_y with $K = 1/2$, and there exists a unique point $x \in \bar{B}_r(0)$ such that $\Psi_y(x) = x$. That is, $f(x) = y$.

Let us set $x = f^{-1}(y)$, $y \in \bar{B}_{r/2}(0)$.

Step 3. Continuity of the inverse. From step 2 we have a local inverse

$$f^{-1} : \bar{B}_{r/2}(0) \longrightarrow \bar{B}_r(0)$$

for f. We shall now show that f^{-1} is continuous. Let $y_1, y_2 \in \bar{B}_{r/2}(0)$. Then $f^{-1}(y_1)$, $f^{-1}(y_2) \in \bar{B}_r(0)$ and

$$\begin{aligned} \| f^{-1}(y_1) - f^{-1}(y_2) \| &= \| f(f^{-1}(y_1)) - f(f^{-1}(y_2)) + \phi(f^{-1}(y_1)) - \phi(f^{-1}(y_2)) \| \\ &\leqslant \| y_1 - y_2 \| + \| \phi(f^{-1}(y_1)) - \phi(f^{-1}(y_2)) \| \\ &\leqslant \| y_1 - y_2 \| + \tfrac{1}{2} \| f^{-1}(y_1) - f^{-1}(y_2) \|, \qquad \text{by (B) of step 1.} \end{aligned}$$

That is

$$\|f^{-1}(y_1) - f^{-1}(y_2)\| \leqslant 2\|y_1 - y_2\|, \qquad y_1, y_2 \in \bar{B}_{r/2}(0).$$

Therefore, f^{-1} is continuous.

Step 4. Differentiability of the inverse. We shall show that $f^{-1}: B_{r/2}(0) \longrightarrow B_r(0)$ is differentiable and that $Df_y^{-1} = (Df_{f^{-1}(y)})^{-1}, y \in B_{r/2}(0)$. Let $y_1, y_2 \in B_{r/2}(0)$ and set $f^{-1}(y_1) = x_1, f^{-1}(y_2) = x_2$. Then

$$\begin{aligned}
&\|f^{-1}(y_1) - f^{-1}(y_2) - (Df_{f^{-1}(y)})^{-1}(y_1 - y_2)\| \\
&= \|(Df_{x_1})^{-1}(Df_{x_1}(x_1 - x_2) - y_1 + y_2)\| \\
&\leqslant \|(Df_{x_1})^{-1}\| \|Df_{x_1}(x_1 - x_2) - f(x_1) + f(x_2)\|.
\end{aligned}$$

Since f is differentiable,

$$Df_{x_1}(x_1 - x_2) - f(x_1) + f(x_2) = o(x_1 - x_2).$$

Since f^{-1} is continuous (step 3),

$$o(x_1 - x_2) = o(f^{-1}(y_1) - f^{-1}(y_2)) = o(y_1 - y_2).$$

Therefore the right-hand side of the above inequality is less than or equal to

$$\|(Df_{x_1})^{-1}\| o(y_1 - y_2) = o(y_1 - y_2).$$

Hence f^{-1} is differentiable at y with derivative

$$(Df_{f^{-1}(y)})^{-1}.$$

Step 5. f^{-1} is C^p. Recall from Section 3.1 that $\beta: \operatorname{Iso}(E, E) \longrightarrow \operatorname{Iso}(E, E)$ was defined by $\beta(T) = T^{-1}$ and that β was C^∞. Now

$$D(f^{-1}) = \beta((Df)f^{-1}).$$

Since we have shown that f^{-1} is continuous and, by assumption, Df is continuous, it follows that Df^{-1} is continuous and that f^{-1} is C^1. Suppose that we have shown that f^{-1} is $C^r, r < p$. Then Df is C^{p-1} and so, applying the composite-mapping theorem to $\beta((Df)f^{-1}) = D(f^{-1})$, it follows that $D(f^{-1})$ is C^r. Hence f^{-1} is C^{r+1}. It follows inductively that f^{-1} is C^p.

Step 6. Let us set $W = f^{-1}(B_{r/2}(0))$. Then, since f^{-1} is continuous, W is open. Since $f(W) = B_{r/2}(0)$ is open it follows that

$$f \in \operatorname{Diff}^p(W, f(W)).$$

Step 7. General case. We may reduce the general case to the special one treated above as follows. Since $Df_{x_0} \in \operatorname{Iso}(E, F)$, we may form the map

$$\tilde{f} = (Df_{x_0})^{-1} f.$$

Then $\tilde{f}: U \longrightarrow E$ is C^p and $D\tilde{f}_{x_0} = I_E$. Set $U' = U - x_0$. We define $\hat{f}: U' \longrightarrow E$ by

$$\hat{f}(x) = \tilde{f}(x + x_0) - \tilde{f}(x_0).$$

Then $\hat{f}$ satisfies the conditions of the special case. Hence there exists an open neighbourhood W' of 0 in U' such that

$$\hat{f} \in \operatorname{Diff}^p(W', \hat{f}(W')).$$

We leave it to the reader to verify that if we set $W = W' + x_0$, then

$$f \in \text{Diff}^p(W, f(W)). \quad \blacksquare$$

Corollary 3.2.2.1

Let U and V be open subsets of the Banach spaces E and F respectively and suppose that $f \in C^p(U, V)$ and that $f \in \text{Diff}^1(U, V)$. Then $f \in \text{Diff}^p(U, V)$.

Proof. Apply the proof of step 5 of Theorem 3.2.2. ■

Examples

1. Let (a, b) be an open subinterval of $\mathbf{R}$ and suppose that $f \in C^p((a, b), \mathbf{R})$ has a non-vanishing derivative. Then we claim that there exists an open interval $(c, d) \subset \mathbf{R}$ such that $f \in \text{Diff}^p((a, b), (c, d))$. Indeed, since $f' \neq 0$ on (a, b), it follows that f is either strictly increasing or strictly decreasing on (a, b). In either case f must be injective. Since f is continuous, the image of f is connected. Let Image $(f) = V$. Now for each point $x \in (a, b)$, there exists, by Theorem 3.2.2, an open neighbourhood W_x of x in (a, b) such that $f \in \text{Diff}^p(W_x, f(W_x))$ and $f(W_x)$ is open. Clearly,

$$V = \bigcup_{x \in (a, b)} f(W_x)$$

and so V is open. Since V is connected, $V = (c, d)$ (we include the possibility of c or $d = \pm\infty$). Now f defines a bijection between (a, b) and (c, d). By the inverse-function theorem f^{-1} is locally differentiable and hence differentiable on (c, d).
2. The map $\tan: (-\pi/2, \pi/2) \longrightarrow \mathbf{R}$ is well known to be C^∞, with derivative $\sec^2 t$, $t \in (-\pi/2, \pi/2)$. Hence, the derivative of tan is nowhere zero and by the previous example there exists a C^∞ map $\tan^{-1}: \mathbf{R} \longrightarrow (-\pi/2, \pi/2)$ which is the inverse to tan. The derivative of $\tan^{-1}$ at $t \in \mathbf{R}$ is easily computed to be $1/(1 + t^2)$.
3. Let $P: \mathbf{R}^2 \longrightarrow \mathbf{R}^2$ be defined by

$$P(u, v) = (u \cos v, u \sin v).$$

P is familiarly known as defining 'polar coordinates' on $\mathbf{R}^2$. In Section 3.4 we shall make a more general study of coordinate transformations on $\mathbf{R}^n$. In this example we merely emphasize some of the problems that can occur in studying coordinate transformations. We shall first find a connected open subset of $\mathbf{R}^2$ on which P is invertible. Relative to the standard basis of $\mathbf{R}^2$, the derivative of P is given in matrix form by

$$[DP_{(u,v)}] = \begin{bmatrix} \cos v & -u \sin v \\ \sin v & u \cos v \end{bmatrix}.$$

Hence the Jacobian of P at (u, v) is equal to u. Thus, $DP_{(u,v)} \in \text{Iso}(\mathbf{R}^2, \mathbf{R}^2)$ for $u \neq 0$. Since P is clearly periodic in v, period 2π, it follows that a natural candidate for a region of $\mathbf{R}^2$ on which P is invertible is given by the half-infinite strip

$$D = (0, \infty) \times (0, 2\pi).$$

For fixed $v_0 \in (0, 2\pi)$ the P-image of the line $(0, \infty) \times \{v_0\}$ is the ray in $\mathbf{R}^2$ making an angle v_0 with the x-axis. For fixed $u_0 \in (0, \infty)$, the image of the interval $\{u_0\} \times (0, 2\pi)$ is the circle, centre 0, radius u_0 in $\mathbf{R}^2$, omitting the point $(u_0, 0)$. It follows that $P|D$ is injective. Now the image of $P|D$ is $\mathbf{R}^2 \setminus \{(u, 0) : u \geqslant$ Let P^{-1} denote the inverse to $P|D$. Since $DP: D \longrightarrow \text{Iso}(\mathbf{R}^2, \mathbf{R}^2)$, it follows from the inverse-function theorem that P^{-1} is C^∞.

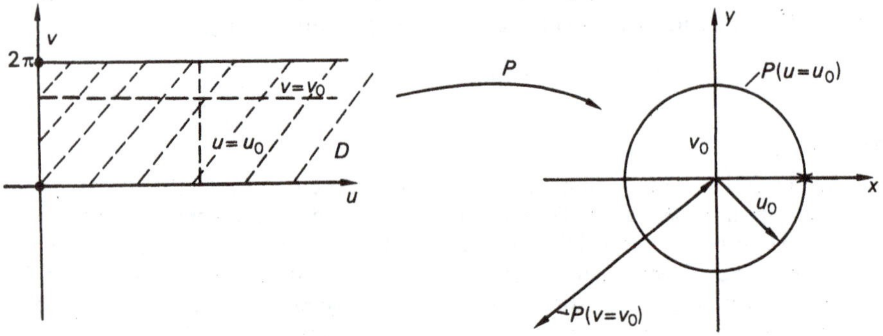

FIG. 3.2

Let

$$P^{-1}(x, y) = (u(x, y), v(x, y)).$$

Then the reader may verify that

$$u(x, y) = \sqrt{(x^2 + y^2)}$$

$$v(x, y) = \begin{cases} \tan^{-1}(y/x), & y, x > 0 \\ \pi/2, & x = 0, y > 0 \\ \pi + \tan^{-1}(y/x), & x < 0 \\ 3\pi/2, & x = 0, y < 0 \\ 2\pi + \tan^{-1}(y/x), & x > 0, y < 0 \end{cases}$$

and that the Jacobian of P^{-1} at (x, y) is equal to $1/\sqrt{(x^2 + y^2)}$.

In the above examples we made use of the following result which is an immediate consequence of the inverse-function theorem.

Corollary 3.2.2.2

Let E and F be Banach spaces and let U be an open subset of E. Suppose that $f: U \longrightarrow F$ is a C^p map which is injective and whose derivative satisfies the condition

$$Df: U \longrightarrow \text{Iso}(E, F).$$

Then

1. $f(U)$ is an open subset of F.
2. $f \in \text{Diff}^p(U, f(U))$.

Exercises

1. Show, by example, that a map which is a homeomorphism and differentiable need not have a *differentiable* inverse.

2. Assuming the fact that the exponential function is the unique C^∞ function defined on $\mathbf{R}$ which satisfies the equation $y' = y$ and the condition $y(0) = 1$, investigate the differentiability and existence properties of the inverse of the exponential function, 'log', using the inverse-function theorem.
3. Consider the map $f: \mathbf{R}^2 \longrightarrow \mathbf{R}^2$ defined by $f(x, y) = (e^x \cos y, e^x \sin y)$. Find a domain $D \subset \mathbf{R}^2$ such that $f|D$ is injective with image $\mathbf{R}^2 \setminus \{(x, 0): x \geqslant 0\}$. Describe f and f^{-1} geometrically in the spirit of Example 3 preceding these exercises.
4. Suppose that $f: \mathbf{R} \longrightarrow \mathbf{R}$ is a polynomial of degree n. Investigate the existence of an inverse for f (assume that f has at most n roots).

3.3 The implicit-function theorem

In this section we shall study a number of important corollaries to the inverse-function theorem. These corollaries may either be thought of as asserting the existence of a solution to a functional equation or as 'straightening out' theorems. Where possible, we shall emphasize the geometrical nature of the theorems with a view to applications in Chapter 4.

Theorem 3.3.1 (Implicit-function theorem)

Let E and F be Banach spaces and let U be an open subset of $E \times F$. Suppose that $f \in C^p(U, F)$ and that

$$D_2 f_{(a,b)} \in \operatorname{Iso}(E, F)$$

for some point $(a, b) \in U$.

Then there exists an open neighbourhood $V_1 \times V_2$ of $(a, f(a, b)) \in E \times F$, an open neighbourhood W of (a, b) in U and an element $h \in \operatorname{Diff}^p(V_1 \times V_2, W)$ such that the composite

$$V_1 \times V_2 \xrightarrow{h} U \xrightarrow{f} F$$

is the restriction of the projection map $p_2: E \times F \longrightarrow F$ to $V_1 \times V_2$. That is, $fh = p_2$. In addition, there exists a map $h_2 \in C^p(V_1 \times V_2, F)$ such that

$$h(x, y) = (x, h_2(x, y)) \qquad \text{for all } (x, y) \in V_1 \times V_2$$

('h preserves the first coordinate'). The derivative of h_2 is given in terms of the partial derivatives of f by the formula

$$Dh_2 = -[(D_2 f)h]^{-1} . [(D_1 f)h] . p_1 + [(D_2 f)h]^{-1} . p_2.$$

Before starting the proof we make some remarks about the theorem. Just as in the case of the proof of the inverse-function theorem, it is no loss of generality to suppose that $(a, b) = 0$ and $f(a, b) = 0$. This being so we have the situation illustrated in Fig. 3.3.

The 'wavy' lines in W represent the family $\{f^{-1}(z): z \in V_2\}$. The geometric content of the theorem is that we can find a diffeomorphism h^{-1} of some neighbourhood W of the origin in $E \times F$ which 'straightens out' the lines $f^{-1}(z)$ into the open subsets $V_1 \times \{z\}$ of the affine spaces $E \times \{z\}$. Furthermore, the condition on h_2 given in the statement of the theorem implies that this straightening-out process is achieved by stretching and contracting W parallel to F. That is, $h(\{x\} \times V_2) \subset \{x\} \times F$, for all $x \in V_1$.

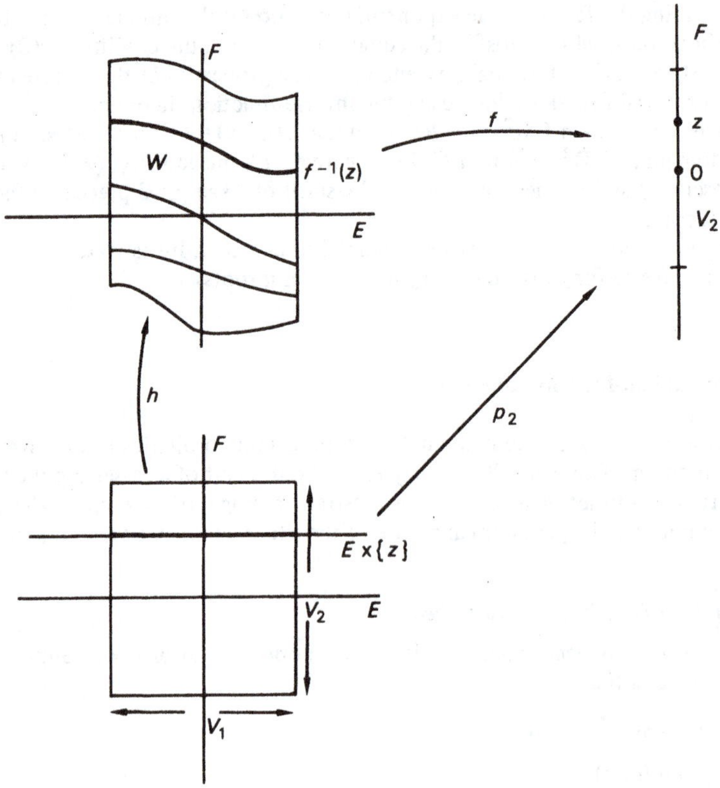

FIG. 3.3

Proof of Theorem 3.3.1. We define a map $\phi: U \longrightarrow E \times F$ by

$$\phi(x, y) = (x, f(x, y)).$$

ϕ is certainly C^p and the derivative of ϕ at (0, 0) may be represented by the 'matrix'

$$D\phi_{(0,0)} = \begin{pmatrix} I_E & 0 \\ D_1 f_{(0,0)} & D_2 f_{(0,0)} \end{pmatrix} \in \operatorname{Iso}(E \times F, E \times F).$$

Therefore, by the inverse-function theorem, ϕ has a C^p inverse h defined on some open neighbourhood $V_1 \times V_2$ of (0, 0) in $E \times F$. We set $W = h(V_1 \times V_2)$. By choice of ϕ,

$$p_2 \phi = f.$$

Hence, composing on the right by $\phi^{-1} = h$, we have

$$p_2 = f\phi^{-1} = fh.$$

h certainly preserves first coordinates since ϕ does. Set $h = (p_1, h_2)$. The formula for Dh_2 follows by differentiating the identity

$$f(p_1, h_2) = p_2$$

and using the fact that $D_2 f$ is invertible. ■

As a special case of Theorem 3.3.1 we have

Corollary 3.3.1.1

Let E and F be Banach spaces and let U be an open subset of $E \times F$. Suppose that $f \in C^p(U, F)$ and that

1. $D_2 f_{(a,b)} \in \mathrm{Iso}(F, F)$, for some $(a, b) \in U$.
2. $f(a, b) = 0$.

Then there exist open neighbourhoods V of $a \in E$ and W of $(a, b) \in E \times F$ together with a function $u \in C^p(V, F)$ such that

1. $f(x, u(x)) = 0$ for all $x \in V$.
2. The only solutions of the equation $f(x, y) = 0$ in W are those given by (1).

The derivative of u is given by the formula

$$Du_x = -(D_2 f_{(x,u(x))})^{-1} . D_1 f_{(x,u(x))}, \qquad x \in V.$$

Proof. From Theorem 3.3.1 we may find open neighbourhoods $V_1 \times V_2$ of $(a, 0)$ and W of (a, b) together with a C^p diffeomorphism $h: V_1 \times V_2 \longrightarrow W$ such that $fh = p_2$ and $h(x, y) = (x, h_2(x, y))$. We define $u: V_1 \longrightarrow F$ by

$$u(x) = h_2(x, 0) \qquad \text{for all } x \in V_1$$

and set $V = V_1$. Then if $x \in V$,

$$f(x, u(x)) = f(x, h(x, 0)) = p_2(x, 0) = 0.$$

Clearly u gives the only solutions of $f(x, y) = 0$ in W, since $f|W = p_2 h^{-1}$. Finally the formula for Du follows easily from the formula for Dh_2 given in Theorem 3.3.1. ■

Figure 3.4 describes the conclusion of the corollary in the case when $(a, b) = (0, 0)$.

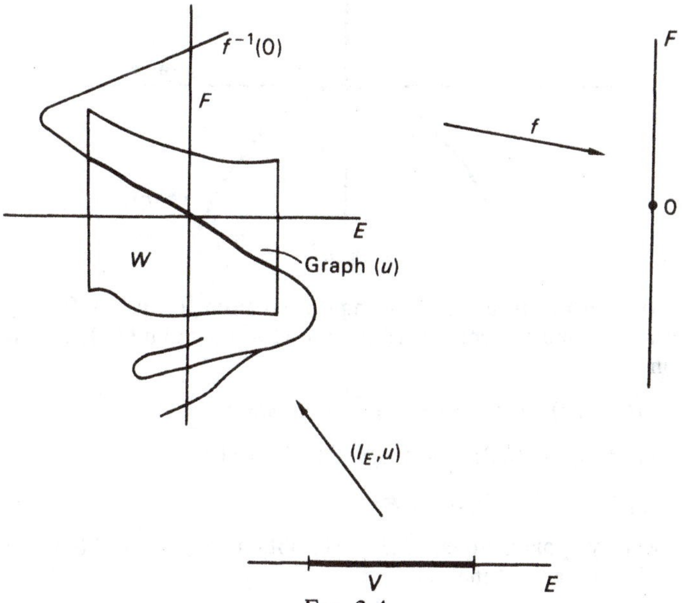

FIG. 3.4

Notice that the corollary says nothing about the solutions of the equation $f(x, y) = 0$ outside the set W. Indeed, as indicated in the above diagram, the equation $f(x, y) = 0$ may have many solutions for each fixed value of x.

Examples

1. Consider the map $f: \mathbf{R}^2 \longrightarrow \mathbf{R}$ defined by

$$f(x, y) = x^2 \cos y + \sin y$$

$f_{;2}(0, 0) = 1$. Hence we may apply Corollary 3.3.1.1 and there exists an open interval (a, b), containing 0, and a C^∞ map $u:(a, b) \longrightarrow \mathbf{R}$ such that

$$f(x, u(x)) = x^2 \cos (u(x)) + \sin (u(x)) = 0, \qquad x \in (a, b).$$

Furthermore, u gives the only solutions to the equation $f(x, y) = 0$ in some open neighbourhood W of $(0, 0) \in \mathbf{R}^2$.

Notice that $f_{;1}(0, 0) = 0$ and so we cannot apply Corollary 3.3.1.1 to find uniqu solutions of the equation for each fixed y close enough to 0. In fact, if $x^2 \cos y + \sin y = 0$, $x = \pm\sqrt{(-\tan y)}$. Hence there are *no* solutions of this equation if $y > 0$. If $y < 0$, we have *two* solutions and no uniqueness. Hence we cannot solve this equation uniquely for x in terms of y in any neighbourhood of $y = 0$. Close to the origin of $\mathbf{R}^2$, $x^2 \cos y + \sin y \triangleq x^2 + y$ and we have the situation shown in Fig. 3.5.

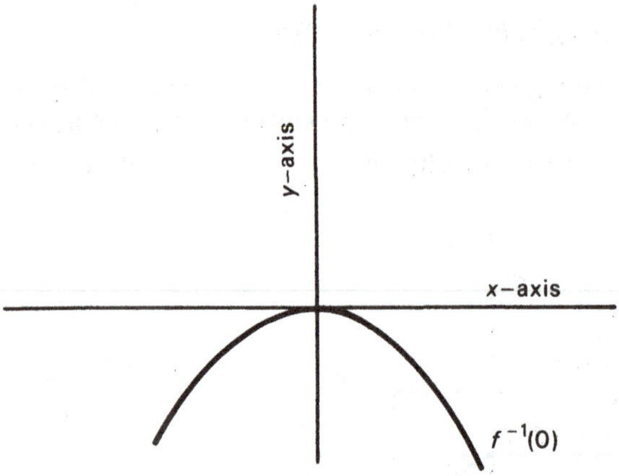

FIG. 3.5

2. Let us consider the problem of solving the equation $f(x, y, z) = 0$ uniquely for x in terms of y and z in some neighbourhood of the point $(1, 1, 1)$ when f is the function

$$f(x, y, z) = (x^2 + y^2 + z^2)^{1/2} - \sin \pi(x + y + z) - \sqrt{3}.$$

f is C^∞ on $\mathbf{R}^3 \setminus \{(0, 0, 0)\}$ and $f(1, 1, 1) = 0$. We have

$$f_{;1}(1, 1, 1) = 1/\sqrt{3} + \pi \neq 0.$$

Hence we may apply Corollary 3.3.1.1 to obtain an open neighbourhood V of $(1, 1)$ in $\mathbf{R}^2$ and a C^∞ function

$$u: V \longrightarrow \mathbf{R}$$

and an open neighbourhood W of $(1, 1, 1)$ such that the solutions of $f(x, y, z) = 0$ in W are given by

$$\{(u(y, z), y, z) : (y, z) \in V\}.$$

That is,

$$f(u(y, z), y, z) = 0 \qquad \text{for all } (y, z) \in V.$$

3. Let $F : \mathbf{R}^3 \longrightarrow \mathbf{R}$ be C^1 and suppose that $F_{;3}(0, 0, 0) \neq 0$. Then the map $u : V \subset \mathbf{R}^2 \longrightarrow \mathbf{R}$ given by Corollary 3.3.1.1 has derivative

$$Du_x = -(D_2F_{(x,u(x))})^{-1} \cdot D_1F_{(x,u(x))}, \qquad x \in V.$$

Composing on the left by D_2F, it follows that Du satisfies the equation

$$D_2F_{(x,u(x))} \cdot Du_x + D_1F_{(x,u(x))} = 0.$$

We therefore have the following relation between the partial derivatives of u and F

$$F_{;3}(x, u(x))(u_{;1}(x), u_{;2}(x)) + (F_{;1}(x, u(x)), F_{;2}(x, u(x)) = 0.$$

That is, the partial derivatives $u_{;1}$ and $u_{;2}$ of u satisfy the equations

$$F_{;1}(x, u(x)) + F_{;3}(x, u(x))u_{;1}(x) = 0$$

$$F_{;2}(x, u(x)) + F_{;3}(x, u(x))u_{;2}(x) = 0.$$

Solving, we obtain the following formulae for $u_{;1}$ and $u_{;2}$:

$$u_{;1}(x) = -F_{;1}(x, u(x))/F_{;3}(x, u(x)), \qquad x \in V,$$

$$u_{;2}(x) = -F_{;2}(x, u(x))/F_{;3}(x, u(x)), \qquad x \in V.$$

4. Consider the equation

$$z^3 + (x^4 + y^4)z + 1 = 0.$$

We claim that we can find a unique solution $z = f(x, y)$ of this equation defined on the whole of $\mathbf{R}^2$ and that f is C^∞. First notice that, for fixed x and y, the equation has at least one real root since the sign of the expression $z^3 + (x^4 + y^4)z + 1$ is that of z, for $|z|$ large. We define the map $F : \mathbf{R}^3 \longrightarrow \mathbf{R}$ by

$$F(x, y, z) = z^3 + (x^4 + y^4)z + 1.$$

F is C^∞ and

$$F_{;3}(x, y, z) = 3z^2 + (x^4 + y^4)$$
$$= 0 \text{ if and only if } (x, y, z) = 0.$$

From the positivity of $F_{;3}$ it follows easily that there can be at most one real root of the equation $F(x, y, z) = 0$ for fixed x and y. Hence there is exactly one real root of this equation for each $(x, y) \in \mathbf{R}^2$. We let $f : \mathbf{R}^2 \longrightarrow \mathbf{R}$ be the function giving this root. Thus

$$F(x, y, f(x, y)) = 0 \qquad \text{for all } x, y \in \mathbf{R}.$$

It follows from Corollary 3.3.1.1 that f is C^∞ except possibly where $x = y = 0$.

Computing the partial derivatives of f, using Example 3, we find that for $(x, y) \neq (0, 0)$

$$f_{;1}(x, y) = -4x^3 z/(3z^2 + x^4),$$

$$f_{;2}(x, y) = -4y^3 z/(3z^2 + y^4).$$

Now $f(0, 0) = -1$ and substituting $x = y = 0, z = -1$ in the above expressions for $f_{;1}$ and $f_{;2}$ it follows that if we *define* $f_{;1}(0, 0) = f_{;2}(0, 0) = 0$, then $f_{;1}$ and $f_{;2}$ are continuous functions on $\mathbf{R}^2$. To complete the proof it is therefore sufficient, by Theorem 2.8.3, to show that f is partially differentiable at (0, 0) with both partial derivatives zero. It will then follow inductively from the above formulae for $f_{;1}$ and $f_{;2}$ that f is C^∞.

We shall show that $f(x, 0) = -1 + o(x), f(0, y) = -1 + o(y)$. This will clearly imply that f is partially differentiable at (0, 0). To see that $f(x, 0) = -1 + o(x)$, we make the substitution $p = (z + 1)/x$ for z in $z^3 + x^4 z + 1$ to obtain a cubic polynomial in p with coefficients depending continuously on x. We leave it for the reader to verify that $p = 0$ is a root of this polynomial, when $x = 0$. Since the roots of a polynomial depend continuously on the coefficients, it follows that

$$f(x, 0) + 1\ (= z + 1 = xp) = o(x).$$

The special case of Corollary 3.3.1.1 when $E = \mathbf{R}^n$ and $F = \mathbf{R}$ is of particular interest and, in this case, we have

Corollary 3.3.1.2

Let U be an open subset of $\mathbf{R}^n \times \mathbf{R}$ and $f \in C^p(U, \mathbf{R})$. Suppose that there exists a point $(a_1, \ldots, a_n, b) \in U$ such that

$$f_{;n+1}(a_1, \ldots, a_n, b) \neq 0.$$

Then there exists an open neighbourhood V of $(a_1, \ldots, a_n)$ in $\mathbf{R}^n$, a function $u \in C^p(V, \mathbf{R})$ and an open neighbourhood W of $(a_1 \ldots, a_n, b)$ such that

$$f(x_1, \ldots, x_n, u(x_1, \ldots, x_n)) = 0 \qquad \text{for all } (x_1, \ldots, x_n) \in V,$$

and these are the only solutions of $f(x_1, \ldots, x_n, y) = 0$ in W.

The partial derivatives of u are given by the formulae

$$u_{;j}(x) = -f_{;j}(x, u(x))/f_{;n+1}(x, u(x)), \qquad x \in V, \quad 1 \leqslant j \leqslant n.$$

Proof. Immediate from Corollary 3.3.1.1. ∎

Example. Consider the simultaneous equations in the unknown functions f and g

$$[f(x, y)]^3 + x[g(x, y)]^2 + y = 0,$$

$$[g(x, y)]^3 + yg(x, y) + [f(x, y)]^2 - x = 0.$$

We may regard the solution of these equations as being equivalent to the study of the zero set of the function $F: \mathbf{R}^4 \longrightarrow \mathbf{R}^2$ defined by

$$F(x, y, u, v) = (u^3 + xv^2 + y, v^3 + yv + u^2 - x).$$

That is, to find f and g it is sufficient to solve the equation

$$F(x, y, u, v) = 0$$

for u and v in terms of x and y.

Let us try to find such a solution subject to the conditions that $f(1, 1) = -1$, $g(1, 1) = 0$. That is, we shall investigate the function F in a neighbourhood of the point $(1, 1, -1, 0) \in \mathbf{R}^4$. Observe that $F(1, 1, -1, 0) = (0, 0)$. Writing $\mathbf{R}^4 = \mathbf{R}^2 \times \mathbf{R}^2$, we find that

$$[D_2f_{(x,y,u,v)}] = \begin{bmatrix} 3u^2 & 2vx \\ 2u & 3v^2 + y \end{bmatrix}$$

and hence

$$\det [D_2f_{(1,1,-1,0)}] = \det \begin{bmatrix} 3 & 0 \\ -2 & 1 \end{bmatrix} \neq 0.$$

Hence we may apply Corollary 3.3.1.1 and there exists a neighbourhood V of $(1, 1) \in \mathbf{R}^2$ and C^∞ functions $f, g: V \longrightarrow \mathbf{R}$ satisfying the condition

$$F(x, y, f(x, y), g(x, y)) = 0.$$

Furthermore, f and g give unique solutions to the equation $F = 0$ in some neighbourhood of $(1, 1, -1, 0)$.

We may evaluate the derivatives of g and f by making use of the formula of Corollary 3.3.1.1. Indeed, if we set $u = (f, g)$, we have

$$Du = (D_2F(1, u))^{-1} . D_1F(1, u).$$

For example, if $(x, y) = (1, 1)$ we find that

$$[D_2F] = \begin{bmatrix} 3 & 0 \\ -2 & 1 \end{bmatrix}, \qquad [D_1F] = \begin{bmatrix} 0 & 1 \\ -1 & 0 \end{bmatrix},$$

and so

$$\begin{bmatrix} \partial f/\partial x & \partial f/\partial y \\ \partial g/\partial x & \partial g/\partial y \end{bmatrix} = \begin{bmatrix} 3 & 0 \\ -2 & 1 \end{bmatrix}^{-1} \begin{bmatrix} 0 & 1 \\ -1 & 0 \end{bmatrix} = \begin{bmatrix} 0 & 1/3 \\ -1 & 5/3 \end{bmatrix}.$$

We remark for future reference the following important special case of Theorem 3.3.1.

Theorem 3.3.2

Let U be an open subset of $\mathbf{R}^m$ and let $f \in C^p(U, \mathbf{R}^n)$, $m \geqslant n$. Suppose that for some point $x_0 \in U$, Df_{x_0} is surjective and $f(x_0) = 0$.

Then we may find open neighbourhoods W of x_0 in U, $V_1 \times V_2$ of $(0, 0) \in \mathbf{R}^{m-n} \times \mathbf{R}^n$ and a map $h \in \mathrm{Diff}^p(V_1 \times V_2, W)$ such that

$$fh(x, y) = y \qquad \text{for all } (x, y) \in V_1 \times V_2.$$

In particular, h maps $V_1 \times \{0\}$ homeomorphically onto $f^{-1}(0) \cap W$.

Proof. Since Df_{x_0} is surjective, $\mathrm{Ker}\,(Df_{x_0})$ is an $(m - n)$-dimensional linear subspace of $\mathbf{R}^m$. Let us start by assuming that $x_0 = 0$ and

$$\mathrm{Ker}\,(Df_{x_0}) = \mathbf{R}^{m-n} \times \{0\} \subset \mathbf{R}^m.$$

Then $\mathrm{Ker}\,(Df_{x_0})$ has the complementary subspace $\{0\} \times \mathbf{R}^n$ and the result follows by applying Theorem 3.3.1 with $E = \mathbf{R}^{m-n}$ and $F = \mathbf{R}^n$. If $x_0 \neq 0$ and

$$\mathrm{Ker}\,(Df_{x_0}) \neq \mathbf{R}^{m-n} \times \{0\},$$

we can always find an affine linear isomorphism

$$A:\mathbf{R}^m \longrightarrow \mathbf{R}^{m-n} \times \mathbf{R}^n$$

which maps x_0 to $(0, 0)$ and $\operatorname{Ker}(Df_{x_0})$ to $\mathbf{R}^{m-n} \times \{0\}$. By the special case of the theorem proved above, we can find open neighbourhoods $V_1 \times V_2$ and $\tilde{W}$ of $(0, 0) \in \mathbf{R}^{m-n} \times \mathbf{R}^n$ and a map $\tilde{h} \in \operatorname{Diff}^p(V_1 \times V_2, \tilde{W})$ such that

$$(fA^{-1})\tilde{h} = p_2.$$

To complete the proof we set $W = A^{-1}(\tilde{W})$ and $h = A^{-1}\tilde{h}$. ∎

The following theorem may be regarded as a 'dual' version of Theorem 3.3.1.

Theorem 3.3.3

Let E and F be Banach spaces and let U be an open subset of E. Suppose that $f \in C^p(U, E \times F)$ and that

$$Df_{x_0} \in \operatorname{Iso}(E, E \times \{0\}), \qquad \text{for some } x_0 \in U.$$

Then there exist open neighbourhoods $V_1 \times V_2$ of $(x_0, 0)$ and W of $f(x_0)$ in $E \times F$ and a map $g \in \operatorname{Diff}^p(W, V_1 \times V_2)$ such that the composite $gf: V_1 \longrightarrow E \times F$ is equal to I_1, where I_1 is the inclusion of E into $E \times F$. That is,

$$gf(x) = (x, 0) \qquad \text{for all } x \in V_1.$$

The conclusion of this theorem is illustrated pictorially in Fig. 3.6 for the case $x_0 = 0, f(x_0) = 0$.

The map g 'straightens out' the image of $f \mid V_1$. The reader should observe that the conclusion of the theorem implies that $f \mid V_1$ is injective.

Proof of Theorem 3.3.3. Without loss of generality, take $x_0 = 0, f(x_0) = 0$. We define a map $\phi: U \times F \longrightarrow E \times F$ by

$$\phi(x, y) = f(x) + (0, y).$$

That is, $\phi = fp_1 + p_2$, where p_1 and p_2 denote the appropriate projection maps. ϕ is obviously C^p and

$$D\phi_0 = Df_0 \cdot p_1 + p_2 = \begin{pmatrix} Df_0 & 0 \\ 0 & I_F \end{pmatrix}.$$

By assumption, $Df_0 \in \operatorname{Iso}(E, E)$ and so $D\phi_0 \in \operatorname{Iso}(E \times F, E \times F)$. Hence, by the inverse-function theorem, we may find an open neighbourhood $V_1 \times V_2$ of $0 \in E \times F$ such that

$$\phi \in \operatorname{Diff}^p(V_1 \times V_2, \phi(V_1 \times V_2)).$$

Set $\phi(V_1 \times V_2) = W$ and $\phi^{-1} = g$. We claim that g satisfies the conditions of the theorem. Indeed, on V_1, we clearly have

$$\phi I_1 = f.$$

Composing on the left by $g = \phi^{-1}$, the result follows. ∎

Remark. Notice that if f preserves first coordinates (that is, $f(x) = (x, f_2(x))$), then so does g. This just follows from the definition of ϕ.

The following special case of Theorem 3.3.3 will be useful in Chapter 4.

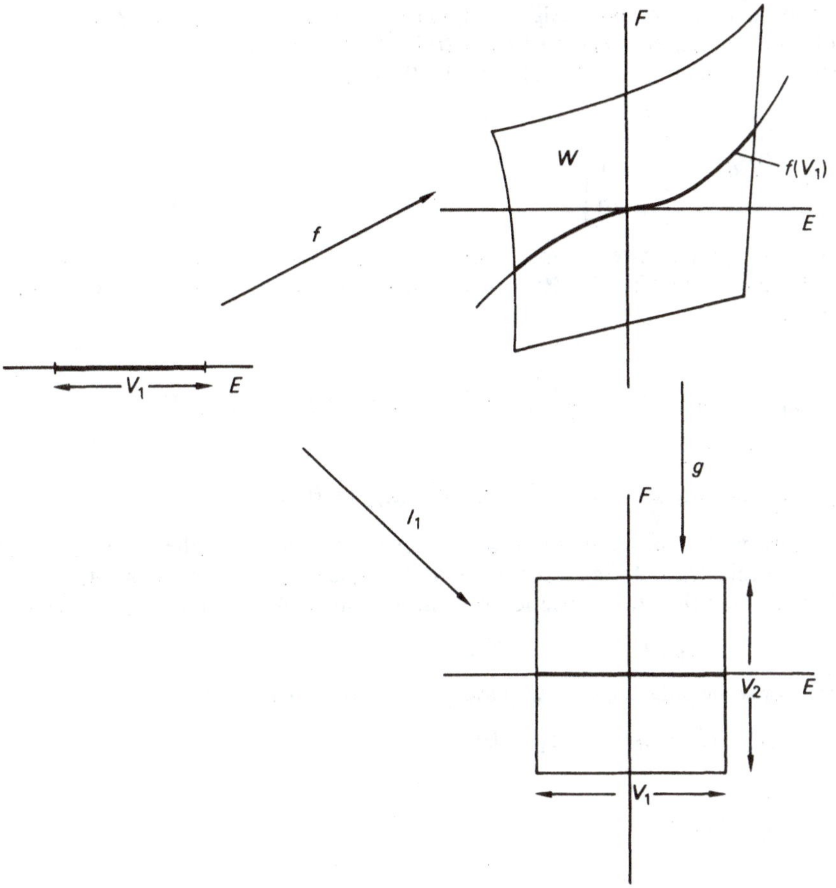

FIG. 3.6

Theorem 3.3.4

Let U be an open subset of $\mathbf{R}^n$ and let $f \in C^p(U, \mathbf{R}^m)$, $m \geqslant n$. Suppose that Df_{x_0} is injective at some point $x_0 \in U$.

Then there exist open neighbourhoods $V_1 \times V_2$ of $(x_0, 0) \in \mathbf{R}^n \times \mathbf{R}^{m-n}$ and W of $f(x_0) \in \mathbf{R}^m$ and $g \in \mathrm{Diff}^p(W, V_1 \times V_2)$ such that

$$gf = I_1,$$

where $I_1 : \mathbf{R}^n \longrightarrow \mathbf{R}^n \times \mathbf{R}^{m-n}$ is the inclusion map on the first factor.

Proof. Since Df_{x_0} is injective, $\mathrm{Im}\,(Df_{x_0})$ is an n-dimensional linear subspace of $\mathbf{R}^m$. Let us start by assuming that $f(x_0) = 0$ and $\mathrm{Im}\,(Df_{x_0}) = \mathbf{R}^n \times \{0\} \subset \mathbf{R}^m$. Setting $E = \mathbf{R}^n$ and $F = \mathbf{R}^{m-n}$, the result now follows immediately from Theorem 3.3.3. The general case now follows in much the same way as in the proof of Theorem 3.3.2 and we leave details to the reader. ∎

Example. Let $f: \mathbf{R}^2 \longrightarrow \mathbf{R}^3$ be defined by

$$f(x, y) = (x \sin (xy), ye^x \cos y, x + y^2).$$

We claim that we can find a neighbourhood V of $(0, 0)$ in $\mathbf{R}^2$ such that $f|V$ is injective and the only solution of $f(x, y) = (0, 0, 0)$ in V is $(x, y) = (0, 0)$.

$(0, 0)$ is certainly a solution of $f(x, y) = 0$. Now

$$Df_{(0,0)} = \begin{bmatrix} 0 & 0 \\ 0 & 1 \\ 1 & 0 \end{bmatrix}.$$

Hence $Df_{(0,0)}$ is injective. We may apply Theorem 3.3.4 to find open neighbourhoods $V_1 \times V_2$ and W of $(0, 0, 0) \in \mathbf{R}^2 \times \mathbf{R}$, together with a map $g \in \mathrm{Diff}^\infty(V_1 \times V_2, W)$ such that

$$gf = I_1,$$

where I_1 is the inclusion of $\mathbf{R}^2$ into $\mathbf{R}^2 \times \mathbf{R}$. Set $V = V_1$. Then on V

$$gf(x, y) = (x, y, 0)$$

and hence the only solution of $f = 0$ on V is $(x, y) = (0, 0)$.

We now wish to prove an important theorem which includes Theorems 3.3.1 and 3.3.3 as special cases. In these theorems we assumed that Df_{x_0} was either injective or surjective, and this is the restriction we wish to remove. We shall be studying maps

$$f: U \subset E_1 \times E_2 \longrightarrow E_1 \times F_2,$$

where, for some point $x_0 \in U$, we have $E_2 = \mathrm{Ker}\,(Df_{x_0})$ and also

$$Df_{x_0}|E_1 \in \mathrm{Iso}\,(E_1, E_1 \times \{0\}).$$

That is,

$$Df_{x_0} = \begin{pmatrix} A & 0 \\ B & 0 \end{pmatrix},$$

where $A \in \mathrm{Iso}\,(E_1, E_1)$ and $B \in L(E_1, F_2)$. If we are to hope for any theorem which straightens out the image of f and the inverse images $f^{-1}(z)$, it is not hard to see that we must have some form of stability for either $\mathrm{Ker}\,(Df_x)$ or $\mathrm{Im}\,(Df_x)$ for x near x_0. This happens automatically in Theorems 3.3.1 and 3.3.3, since injectivity and surjectivity are 'open' conditions. Here we have to assume, a priori, an openness condition on the derivative of f. The problem is related to the fact that if $A \in L(\mathbf{R}^m, \mathbf{R}^n)$ is a linear map of rank less than m, then we can approximate A arbitrarily closely by linear isomorphisms. Our a priori condition will essentially be that the 'rank' of Df_x remains constant in a neighbourhood of x_0.

First, however, we have some preliminary technicalities.

Definition 3.3.5

Let F be a closed subspace of the Banach space E. A subspace G of E is said to be a *topological complement* for F if

1. G is closed.
2. $G \cap F = \{0\}$.
3. $G + F = E$.

If G is a topological complement of F, we shall write $G \oplus F$ to denote the vector sum of G and F, and call $G \oplus F$ the *direct sum* of G and F.

For the next few paragraphs we shall be considering the products $E_1 \times E_2$ and $E_1 \times F_2$, where E_1, E_2 and F_2 are Banach spaces. We let p_1, p_2, I_1, I_2 denote the projection and inclusion maps associated with the product $E_1 \times E_2$ and p'_1, p'_2, I'_1, I'_2 the corresponding maps for $E_1 \times F_2$. Any map $A \in L(E_1 \times E_2, E_1 \times F_2)$ can be written uniquely in 'matrix' form as

$$\begin{pmatrix} A_{11} & A_{12} \\ A_{21} & A_{22} \end{pmatrix},$$

where $A_{ij} = p'_i . A . I_j$, $1 \leqslant i, j \leqslant 2$. For example, $A_{21} \in L(E_1, F_2)$ (see also Section 1.6).

Lemma 3.3.6

Let E_1 and F_2 be Banach spaces and let

$$\mathrm{Inj}_1\,(E_1, E_1 \times F_2) = \{A \in L(E_1, E_1 \times F_2): p'_1 . A \in \mathrm{Iso}\,(E_1, E_1)\}.$$

Then $\mathrm{Inj}_1\,(E_1, E_1 \times F_2)$ is an open subset of $L(E_1, E_1 \times F_2)$ and there exists a C^∞ map

$$\theta : \mathrm{Inj}_1\,(E_1, E_1 \times F_2) \longrightarrow L(E_1, F_2)$$

such that

1. $\mathrm{graph}\,(\theta(T)) = \mathrm{Im}\,(T)$ for all $T \in \mathrm{Inj}_1\,(E_1, E_1 \times F_2)$,
2. $(I_{E_1}, \theta(T)) . p'_1 . T = T$ for all $T \in \mathrm{Inj}_1\,(E_1, E_1 \times F_2)$.
3. $\mathrm{Im}\,(T)$ is a topological complement to $E_1 \times \{0\}$ in $E_1 \times F_2$ for all

$$T \in \mathrm{Inj}_1\,(E_1, E_1 \times F_2).$$

Proof. Let $T \in \mathrm{Inj}_1\,(E_1, E_1 \times F_2)$. The situation is as shown in Fig. 3.7. The main point of the lemma is to show that $\mathrm{Im}\,(T) \subset E_1 \times F_2$ may always be represented as the graph of an element $\theta(T) \in L(E_1, F_2)$ and in such a way that θ is C^∞ and condition (2) of the lemma holds.

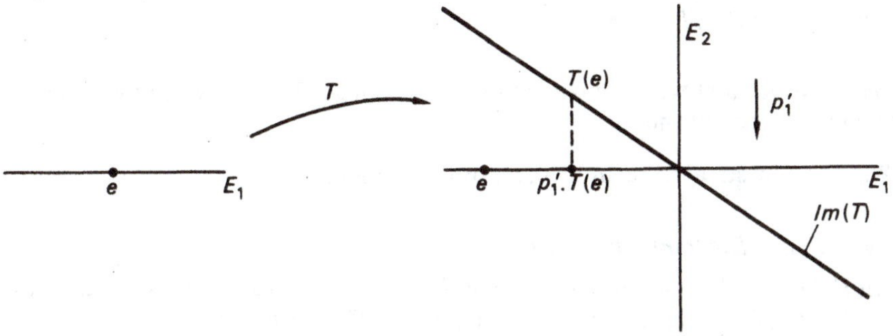

FIG. 3.7

Since $p'_1 . T \in T \in \mathrm{Iso}\,(E_1, E_1)$, condition (2) of the lemma implies that

$$\mathrm{Im}\,(T) = \mathrm{Im}\,(I_{E_1}, \theta(T)) = \mathrm{graph}\,(\theta(T))$$

and so (2) implies (1). On the other hand the graph of a map in $L(E_1, F_2)$ obviously defines a topological complement to $E_1 \times \{0\}$ in $E_1 \times F_2$ and so (1) implies (3). $\mathrm{Inj}_1\,(E_1, E_1 \times F_2)$ is clearly open since $\mathrm{Iso}\,(E_1, E_1)$ is open in $L(E_1, E_1)$ and the map $T \longmapsto p'_1 . T$ is continuous. It therefore remains to construct the map θ and show that it is C^∞ and satisfies condition (2).

In condition (2) we require that

$$(I_{E_1}, \theta(T)) . p'_1 . T = T$$

and, since $p'_1 . T \in \text{Iso}(E_1, E_1)$, this condition is clearly satisfied if and only if

$$\theta(T) = p'_2 . T . (p'_1 . T)^{-1}.$$

Let this equation define θ. Since the map $T \longmapsto (p'_1 . T)^{-1}$ is C^∞ it follows from the composite-mapping theorem that θ is C^∞. ∎

Lemma 3.3.7

Let E_1, E_2, F_2 be Banach spaces and let $A \in L(E_1 \times E_2, E_1 \times F_2)$. Representing A in the 'matrix' form

$$\begin{pmatrix} A_{11} & A_{21} \\ A_{12} & A_{22} \end{pmatrix}.$$

Let us suppose that $A_{11} \in \text{Iso}(E_1, E_1)$. Then $\text{Ker}(A)$ is a topological complement to E_1 in $E_1 \times E_2$ if and only if $\text{Im}(A)$ is a topological complement to F_2 in $E_1 \times F_2$.

Proof. Suppose first that $\text{Im}(A)$ is a topological complement of F_2. The condition on $A_{11} = p'_1 . A . I_1$ implies that $A . I_1 \in \text{Inj}_1(E_1, E_1 \times F_2)$ and therefore, by Lemma 3.3.6, $\text{Im}(A . I_1)$ is also a topological complement of F_2. Since $\text{Im}(A . I_1) \subset \text{Im}(A)$, it follows immediately that $\text{Im}(A) = \text{Im}(A . I_1)$.

Hence, $A . I_1 : E_1 \longrightarrow \text{Im}(A)$ is a *bijection.* If y is any point of $E_1 \times E_2$ there must therefore exist a (unique) point $x \in E_1$ such that $A(x) = A(y)$. Hence $y - x \in \text{Ker}(A)$ and so $\text{Ker}(A)$ is a complement to E_1. Since A is continuous, $\text{Ker}(A)$ is closed and is a topological complement.

Now assume that $\text{Ker}(A)$ is a topological complement of E_1. As above, $\text{Im}(A . I_1)$ is a topological complement of F_2. But if $\text{Ker}(A)$ is a complement of E_1, $\text{Im}(A)$ obviously equals $\text{Im}(A . I_1)$. ∎

Remark. Notice that we give no conditions in Lemma 3.3.7 as to when $\text{Ker}(A)$ or $\text{Im}(A$ are topological complements.

We may now state the main theorem of this section.

Theorem 3.3.8 (The 'rank' theorem)

Let E_1, E_2, F_2 be Banach spaces and let U be an open subset of $E_1 \times E_2$. Suppose that $f \in C^p(U, E_1 \times F_2)$ and there exists a point $x_0 \in U$ such that

1. $\text{Ker}(Df_{x_0}) = \{0\} \times E_2 \subset E_1 \times E_2$ and $\text{Im}(Df_{x_0}) = E_1 \times \{0\} \subset E_1 \times F_2$.
2. One or other of the following conditions holds:

   (a) $\text{Ker}(Df_x)$ is a topological complement to $E_1 \times \{0\}$ for all $x \in U$;
   (b) $\text{Im}(Df_x)$ is a topological complement to $\{0\} \times F_2$ for all $x \in U$.

Then there exist open neighbourhoods W_1 of x_0, W_2 of $f(x_0)$, open sets $V_1 \subset E_1$, $V_2 \subset E_2$ and $U_2 \subset F_2$ and maps $h \in \text{Diff}^p(V_1 \times V_2, W_1)$, $g \in \text{Diff}^p(W_2, V_1 \times U_2)$ such that

$$gfh(x, y) = (x, 0) \qquad \text{for all } x \in V_1 \text{ and } y \in V_2.$$

If, for $y \in V_2$, we let $u_y : V_1 \longrightarrow E_1 \times F_2$ be the map defined by

$$u_y(x) = fh(x, y),$$

then u_y is independent of y and maps V_1 bijectively onto the image $f(W_1)$ of $f|W_1$.

Finally, if f preserves first coordinates (that is, there exists $f_2 \in C^p(U, F_2)$ such that $f(x, y) = (x, f_2(x, y))$ for all $(x, y) \in U$), then so do the maps h and g.

In spite of the technical nature of the statement of the rank theorem, its geometrical meaning is easily seen from Fig. 3.8, where we have taken $x_0 = 0$ and $f(x_0) = 0$ and let $u : V_1 \longrightarrow E_1 \times F_2$ denote the map $u_y, y \in V_2$.

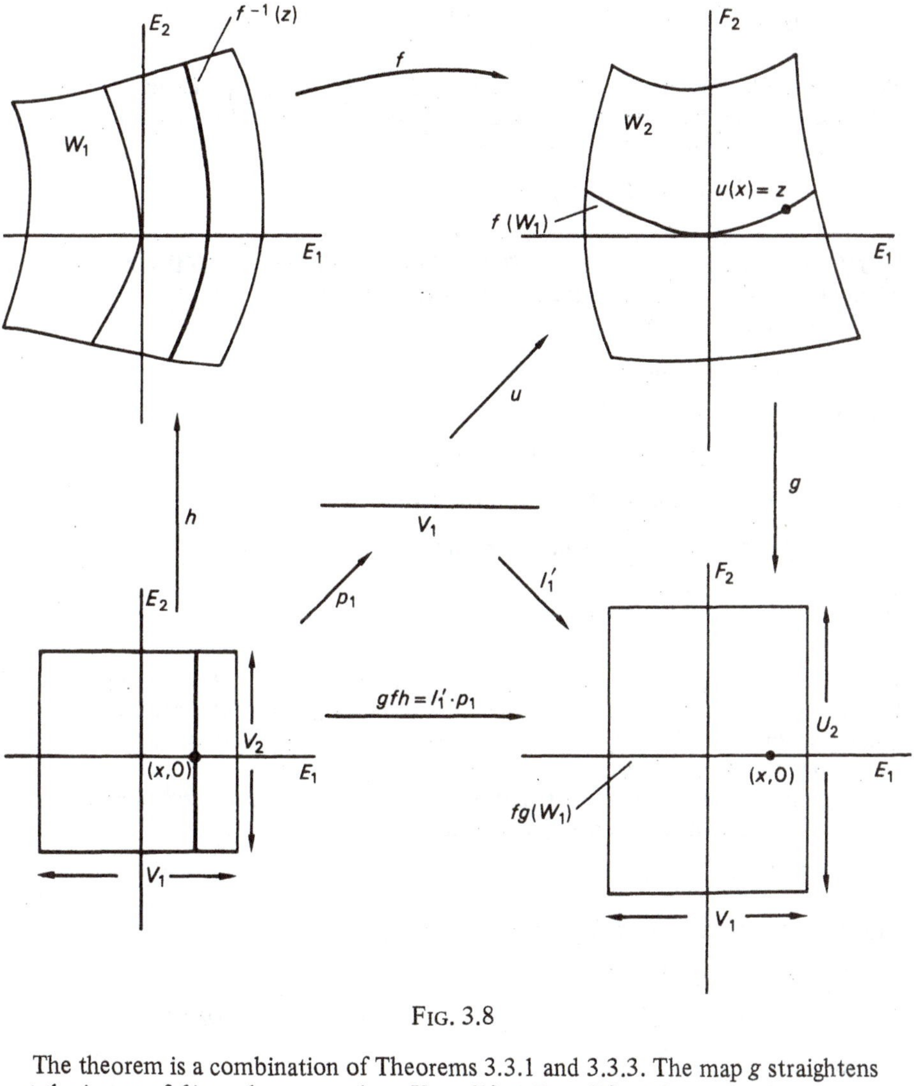

FIG. 3.8

The theorem is a combination of Theorems 3.3.1 and 3.3.3. The map g straightens out the image of f into the open subset $V_1 \times \{0\} \subset E_1 \times \{0\}$ whilst h straightens out the set of inverse images $\{f^{-1}(z) : z \in f(W_1)\}$. In particular, h maps $f^{-1}(u(x))$ bijectively onto $\{x\} \times V_2$ for all $x \in V_1$.

Proof of Theorem 3.3.8. The proof is a combination of that of Theorems 3.3.1 and 3.3.3. Without loss of generality we shall take $x_0 = 0$ and $f(x_0) = 0$. Shrinking U if necessary, Lemma 3.3.6 implies that we may assume

$$Df_x . I_1 \in \mathrm{Inj}_1 (E_1, E_1 \times F_2) \qquad \text{for all } x \in U.$$

We define the map $\phi : U \longrightarrow E_1 \times E_2$ by

$$\phi(x) = p'_1 f(x) + p_2(x), \qquad x \in U.$$

Then $D\phi_0 = p'_1 Df_0 + p_2$ and, in matrix form,

$$D\phi_0 = \begin{pmatrix} p'_1 . Df_0 . I_1 & p'_1 . Df_0 . I_2 \\ 0 & I_{E_2} \end{pmatrix}.$$

Hence $D\phi_0 \in \mathrm{Iso}(E_1 \times E_2, E_1 \times E_2)$ since $p'_1 . D_1 f_0 \in \mathrm{Iso}(E_1, E_1)$. By the inverse-function theorem, there exist open neighbourhoods $\tilde{V}_1 \times \tilde{V}_2$ and $\tilde{W}$ of $0 \in E_1 \times E_2$ such that $\phi \in \mathrm{Diff}^p(\tilde{W}_1, \tilde{V}_1 \times \tilde{V}_2)$. We set

$$h = \phi^{-1} \in \mathrm{Diff}^p(\tilde{V}_1 \times \tilde{V}_2, \tilde{W}_1).$$

So far we have only used assumptions on Df_x at $x = 0$. Since we are assuming that $p'_1 . Df_x . I_1 \in \mathrm{Iso}(E_1, E_2)$ for all $x \in U$, it follows from Lemma 3.3.7 that conditions (a) and (b) in the statement of the rank theorem are equivalent.

We claim that there exists a C^∞ map

$$\sigma_f : \tilde{W}_1 \longrightarrow \mathrm{Inj}_1 (E_1, E_1 \times F_2)$$

such that

$$\sigma_f(x) . p'_1 . Df_x = Df_x \qquad \text{for all } x \in \tilde{W}_1. \tag{X}$$

With the notation of Lemma 3.3.6, we shall define

$$\sigma_f(x) = (I_{E_1}, \theta(Df_x . I_1)), \qquad x \in \tilde{W}_1.$$

We certainly then have

$$\sigma_f(x) . p'_1 . Df_x . I_1 = Df_x . I_1.$$

Since $\mathrm{Ker}(Df_x)$ is a topological complement for $E_1 \times \{0\}$ in $E_1 \times E_2$, it follows that $Df_x . I_1 . p_x = Df_x$ where $p_x : E_1 \times E_2 \longrightarrow E_1$ is the projection associated with the decomposition $E_1 \oplus \mathrm{Ker}(Df_x)$ of $E_1 \times E_2$. Composing on the right by p_x in the above equation yields formula (X).

Let us set $F = fh$. We shall show that $F : \tilde{V}_1 \times \tilde{V}_2 \longrightarrow E_1 \times F_2$ is *independent* of the $\tilde{V}_2$ variable. We shall do this by proving that

$$D_2 F = 0 \qquad \text{on } \tilde{V}_1 \times \tilde{V}_2.$$

Now $F\phi = f$. Differentiating this identity we obtain

$$\begin{aligned} Df &= (DF)\phi . D\phi \\ &= (D_1 F)\phi . p'_1 . Df + (D_2 F)\phi . p_2, \qquad \text{since } D\phi = p'_1 . Df + p_2. \end{aligned}$$

Hence

$$\begin{aligned} (D_2 F)\phi . p_2 &= Df - (D_1 F)\phi . p'_1 . Df \\ &= (\sigma_f - (D_1 F)\phi) . p'_1 . Df. \end{aligned}$$

It follows immediately from this relation that

$$(\sigma_f - (D_1F)\phi) . p_1' . Df)(x) \mid E_1 = 0 \qquad \text{for all } x \in \tilde{W}_1.$$

But $\operatorname{Im}(p_1' . Df_x) = E_1$ for all $x \in \tilde{W}_1$. Therefore,

$$\sigma_f - (D_1F)\phi \equiv 0.$$

Hence, $(D_2F)\phi . p_2 = 0$. But this obviously implies that $(D_2F)\phi = 0$.

Let us define $u_1 : \tilde{V}_1 \longrightarrow E_1$ and $u_2 : \tilde{V}_2 \longrightarrow F_2$ by

$$F(x, y) = (u_1(x), u_2(x)) \qquad \text{for all } (x, y) \in \tilde{V}_1 \times \tilde{V}_2.$$

Then

$$u = (u_1, u_2) : \tilde{V}_1 \longrightarrow E_1 \times F_2$$

and $u(\tilde{V}_1) = f(\tilde{W}_1)$.

Now $Du_0 \in \operatorname{Iso}(E_1, E_1 \times \{0\})$, since $fh \mid \tilde{V}_1 \times \{0\} = u$. Hence we may apply Theorem 3.3.3 to the map $u : \tilde{V}_1 \longrightarrow E_1 \times F_2$ to obtain open sets $V_1 \subset \tilde{V}_1$, $W_2 \subset E_1 \times F_2$, $U_2 \subset F_2$ and a map $g \in \operatorname{Diff}^p(W_2, V_1 \times U_2)$ such that

$$gu(x) = (x, 0) \qquad \text{for all } x \in V_1.$$

Setting $V_2 = \tilde{V}_2$ and $W_1 = h(V_1 \times V_2)$, we obtain the main conclusion of the theorem.

Notice that the condition $gu(x) = (x, 0)$ implies that u is injective. The remaining assertion follows from the definition of ϕ and the remark following Theorem 3.3.3. ■

Theorem 3.3.8 is best known when E_1, E_2 and F_2 are finite dimensional. Recall that the *rank* of a linear map between finite-dimensional vector spaces is the dimension of its image. We have the following special case of the rank theorem.

Theorem 3.3.9

Let U be an open neighbourhood of $x_0 \in \mathbf{R}^m$ and let $f \in C^p(U, \mathbf{R}^n)$. Suppose that Df_x is of constant rank p on U.

Then there exist open neighbourhoods W_1 of x_0, W_2 of $f(x_0)$, open subsets $V_1 \subset \mathbf{R}^p$, $V_2 \subset \mathbf{R}^{m-p}$, $U_2 \subset \mathbf{R}^{n-p}$ and maps $h \in \operatorname{Diff}^p(V_1 \times V_2, W_1)$, and $g \in \operatorname{Diff}^p(W_2, V_1 \times U_2)$ such that

$$gfh(x, y) = (x, 0) \qquad \text{for all } x \in V_1 \text{ and } y \in V_2.$$

In particular, $f^{-1}(f(x_0)) \cap W_1$ is mapped by h^{-1} homeomorphically onto the open set $\{h^{-1}(x_0)\} \times V_2 \subset \{h^{-1}(x_0)\} \times \mathbf{R}^{m-p}$.

Proof. Follows from Theorem 3.3.8 in the same way that Theorems 3.3.2 and 3.3.4 followed from Theorems 3.3.1 and 3.3.3. We leave details to the reader. ■

Exercises

1. Show that the equation $xe^y - 2 \sin y - \cos y = 0$ has a unique solution of the form $y = f(x)$ in some neighbourhood of $x = 1, y = 0$.
2. Show that there exist C^∞ functions f and g, defined on some neighbourhood of $(0, 0) \in \mathbf{R}^2$, that satisfy the equations

$$(8 + x^2)f(x, y) - (y + 1)g(x, y)^3 + y^2 = 0,$$
$$g(x, y)^2 - (y + 1)f(x, y)g(x, y) - 2 = 0,$$

subject to the condition that $f(0, 0) = 1$ and $g(0, 0) = 2$.

3. Show that one can prove Corollary 3.3.1.1 directly (without using the inverse-function theorem) in the special case $f(a, b) = 0, D_2f_{(a,b)} = I_F, (a, b) = (0, 0)$, by showing that the map $x \longmapsto y - f(x, y)$ is a contraction map for x and y sufficiently small. Hence deduce Corollary 3.3.1.1. Can you generalize this method to give a proof of Theorem 3.3.1?
4. Show, by means of examples, that the conditions on Ker (Df_x) in the statement of the rank theorem cannot be weakened.
5. Using the rank theorem, show that a C^1 map $f:\mathbf{R}^m \longrightarrow \mathbf{R}^n$ can only be injective if $m \leqslant n$.
6. Let U be an open subset of the Banach space E and suppose that $f: U \longrightarrow L(\mathbf{R}^m, \mathbf{R}^n)$ is C^p and that $f(x)$ is of rank q for all $x \in U$. Show that if x_0 is a fixed point of U, then we may find C^p maps

$$\alpha: U_0 \longrightarrow GL(m; \mathbf{R}), \qquad \beta: U_0 \longrightarrow GL(n; \mathbf{R}),$$

defined on an open neighbourhood U_0 of x_0, such that

$$\beta . f . \alpha: U_0 \longrightarrow L(\mathbf{R}^m, \mathbf{R}^n)$$

is the constant map π_q defined by $\pi_q(x_1, \ldots, x_m) = (x_1, \ldots, x_q, 0, \ldots, 0)$.

(*Hint:* Suppose that $x_0 = 0$ and $f(0) = \pi_q$. Using the rank theorem study the map

$$F: U \times (L(\mathbf{R}^m, \mathbf{R}^m) \times L(\mathbf{R}^n, \mathbf{R}^n)) \longrightarrow U \times L(\mathbf{R}^m, \mathbf{R}^n)$$

defined by $F(x, A, B) = (x, B . f(x) . A)$. Alternatively, study the map $\tilde{f}: U \times \mathbf{R}^m \longrightarrow U \times \mathbf{R}^n$ defined by $\tilde{f}(x, t) = (x, f(x)(t))$.)

3.4 Change of variables

One of the examples in Section 3.2 concerned the map $P: \mathbf{R}^2 \longrightarrow \mathbf{R}^2$ defined by

$$P(r, \theta) = (r \cos \theta, r \sin \theta), \qquad (r, \theta) \in \mathbf{R}^2.$$

P is an example of a *coordinate transformation* on $\mathbf{R}^2$ and is usually referred to as 'defining polar coordinates on $\mathbf{R}^2$'. Let us list some of the characteristic properties of the polar-coordinate transformation P.

1. P is a C^∞ map from $\mathbf{R}^2$ to $\mathbf{R}^2$ which is onto but not injective.
2. We may find a closed subset $E \subset \mathbf{R}^2$ and an open region $D \subset \mathbf{R}^2$ such that $P|D \in \text{Diff}^\infty (D, \mathbf{R}^2 \setminus E)$. For example, we may set

$$E = \{(x, 0) : x \geqslant 0\}, \qquad D = \{(r, \theta) : r > 0, \theta \in (0, 2\pi)\}.$$

An explicit inverse for $P \mid D$ is given in Section 3.2 (Example 3). In general, there exist many possible choices for D and E. Notice, though, that $\mathbf{R}^2 \setminus E$ is open and dense in $\mathbf{R}^2$.

We now wish to give a general definition of what is meant by a coordinate transformation of $\mathbf{R}^n$. Before doing this, however, we remark that our choice of $\mathbf{R}^n$, as opposed to an open subset of $\mathbf{R}^n$, is unnecessary and the definition may easily be generalized to include coordinate transformations of arbitrary open subsets of $\mathbf{R}^n$.

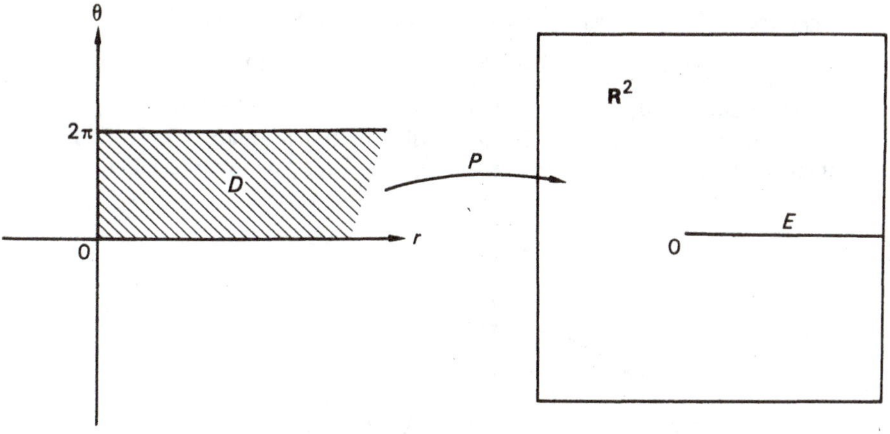

FIG. 3.9

Definiton 3.4.1

A (differentiable) coordinate transformation of $\mathbf{R}^n$ is a triple (ϕ, D, E) where

1. $\phi : \mathbf{R}^n \longrightarrow \mathbf{R}^n$ is a C^∞ surjective map.
2. E is a closed subset of $\mathbf{R}^n$ (possibly empty) such that $\mathbf{R}^n \setminus E$ is open and dense in $\mathbf{R}^n$; D is an open subset of $\mathbf{R}^n$ such that $\phi(\bar{D}) = \mathbf{R}^n$.
3. $\phi \mid D \in \text{Diff}^\infty(D, \mathbf{R}^n \setminus E)$.

Remarks

1. Any element of $\text{Diff}^\infty(\mathbf{R}^n, \mathbf{R}^n)$ defines a coordinate transformation of $\mathbf{R}^n$ with $E = \phi$ and $D = \mathbf{R}^n$.
2. D and E will not, in general, be uniquely defined by ϕ.

Examples

1. *Polar coordinates.* We have already discussed polar coordinates on $\mathbf{R}^2$. The map $P : \mathbf{R}^2 \longrightarrow \mathbf{R}^2$, defined by $P(r, \theta) = (r \cos \theta, r \sin \theta)$ is said to define polar coordinates on $\mathbf{R}^2$. We shall always take $E = \{(x, 0) : x \geqslant 0\}$ and $D = (0, \infty) \times (0, 2\pi)$.
2. *Cylindrical coordinates.* We define the map $Q : \mathbf{R}^3 \longrightarrow \mathbf{R}^3$ by

$$Q(r, \theta, z) = (r \cos \theta, r \sin \theta, z), \qquad (r, \theta, z) \in \mathbf{R}^3.$$

Q is said to define cylindrical coordinates on $\mathbf{R}^3$. Since

$$Q = P \times I : \mathbf{R}^2 \times \mathbf{R} \longrightarrow \mathbf{R}^2 \times \mathbf{R},$$

it follows easily from our results on the polar-coordinate transformation that Q is a coordinate transformation with

$$D = (0, \infty) \times (0, 2\pi) \times \mathbf{R}, \qquad E = \{(x, 0, z) : x \geqslant 0, z \in \mathbf{R}\}.$$

We use the term 'cylindrical' to describe Q since, for fixed $r > 0$, the set

$$Q_r = Q(\{r\} \times \mathbf{R}^2) = \{(r \cos \theta, r \sin \theta, z) : \theta \in \mathbf{R}, z \in \mathbf{R}\}$$

is a cylinder in $\mathbf{R}^3$, radius r and axis the z-axis.

3. *Spherical polar coordinates.* We define the map $S:\mathbf{R}^3 \longrightarrow \mathbf{R}^3$ by

$$S(r, \theta, \psi) = (r \cos \theta \sin \psi, r \sin \theta \sin \psi, r \cos \psi), \qquad (r, \theta, \psi) \in \mathbf{R}^3.$$

The point $S(r, \theta, \psi)$ clearly lies on the sphere in $\mathbf{R}^3$ of radius r and centre the origin. We may utilize Fig. 3.10 to describe the transformation S. To find the image of (r, θ, ψ) by S we proceed as follows. Let S_r^2 denote the sphere of

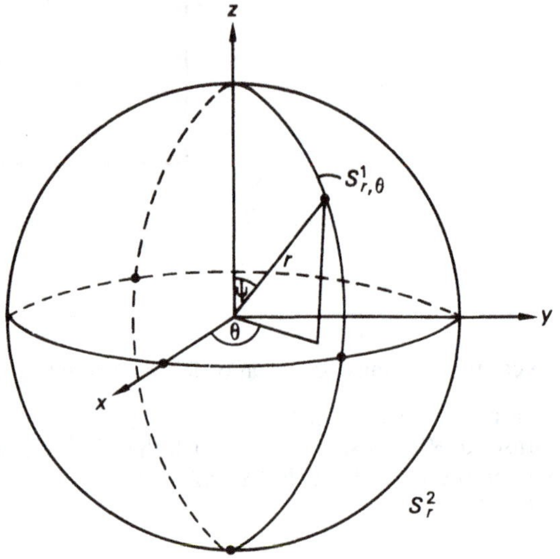

FIG. 3.10

radius r and centre 0 in $\mathbf{R}^3$. Then $S(r, \theta, \psi) \in S_r^2$. Suppose that $0 \leqslant \theta < 2\pi$ and $0 \leqslant \psi < \pi$. If we rotate the xz-plane through θ about the z-axis and denote the resulting plane by P_θ, then $S(r, \theta, \psi) \in S_r^2 \cap P_\theta$. That is, $S(r, \theta, \psi)$ lies on the circle of radius r and centre 0 in P_θ, $S^1_{r,\theta}$. Finally, $S(r, \theta, \psi)$ is the unique point on $S^1_{r,\theta}$ determined by rotating the z-axis through an angle ψ in an anticlockwise direction (viewed from the y-axis).

The above description shows that

$$S([0, \infty) \times [0, 2\pi) \times [0, \pi)) = \mathbf{R}^3$$

and it follows easily that if we set

$$E = \{(x, 0, z) : x \geqslant 0, z \in \mathbf{R}\}, \qquad D = (0, \infty) \times (0, 2\pi) \times (0, \pi),$$

S maps D bijectively onto $\mathbf{R}^3 \setminus E$. Since $DS_{(r,\theta,\psi)} \in GL(3, \mathbf{R})$ for all $(r, \theta, \psi) \in D$ it follows from the inverse-function theorem that $S \in \mathrm{Diff}^\infty(D, \mathbf{R}^3 \setminus E)$ and hence that S is a coordinate transformation on $\mathbf{R}^3$.

Definition 3.4.2

Suppose that (ϕ, D, E) is a coordinate transformation on $\mathbf{R}^n$ and that $f: U \longrightarrow \mathbf{R}$ is a continuous function defined on an open subset of $\mathbf{R}^n$. We shall say that the map $f^\phi = f\phi : \phi^{-1}(U) \longrightarrow \mathbf{R}$ is the ϕ-transform of f or, alternatively, 'f in ϕ-coordinates'.

Examples

1. Let $f: \mathbf{R}^2 \longrightarrow \mathbf{R}$ be the function

$$f(x, y) = (x^2 + y^2)^{3/2} - (x^2 + y^2)^{1/2}x.$$

Then f in polar coordinates is the map $f^P : \mathbf{R}^2 \longrightarrow \mathbf{R}$ defined by

$$f^P(r, \theta) = r^3 - r^2 \cos \theta.$$

2. Let $f: \mathbf{R}^3 \longrightarrow \mathbf{R}$ be the function

$$f(x, y, z) = (x^2 + y^2)^3 z^2 - (x^2 + y^2).$$

Then f in cylindrical coordinates is the map $f^Q : \mathbf{R}^3 \longrightarrow \mathbf{R}$ defined by

$$f^Q(r, \theta, z) = r^6 z^2 - r^2.$$

3. Let $f: \mathbf{R}^3 \longrightarrow \mathbf{R}$ be the function

$$f(x, y, z) = (x^2 + y^2 + z^2)^{1/2}(x^2 + y^2)^{1/2}.$$

Then f in spherical polar coordinates is the map $f^S : \mathbf{R}^3 \longrightarrow \mathbf{R}$ defined by

$$f^S(r, \theta, \psi) = r^2 \sin \psi.$$

Remark. The above examples show that by a suitable choice of coordinate transformation a function can be considerably simplified. In effect what we are doing in the above examples is exploiting the fact that all the functions presented have a degree of symmetry and we therefore choose a coordinate transformation which exhibits the same type of symmetry. For example, the function $(x^2 + y^2 + z^2)^{1/2}$ is constant on spheres, and so the natural choice of coordinate transformation would be spherical polar coordinates.

For most of the remainder of this section we wish to show how derivatives behave under coordinate transformations, and give one or two applications to the study of partial differential equations. For examples, we shall continue to use the three best-known coordinate transformations: polar coordinates, cylindrical coordinates and spherical polar coordinates. First, however, let us suppose that (ϕ, D, E) is an arbitrary coordinate transformation of $\mathbf{R}^n$ and consider the set $\tilde{D} \subset \mathbf{R}^n$ defined by

$$\tilde{D} = \{x \in \mathbf{R}^n : D\phi_x \in GL(n; \mathbf{R})\}.$$

Since $\tilde{D} = \{x \in \mathbf{R}^n : \det(D\phi_x) \neq 0\}$, it follows that $\tilde{D}$ is an open subset of $\mathbf{R}^n$. Clearly $\tilde{D} \supset D$ and for all interesting coordinate transformations of $\mathbf{R}^n$, $\tilde{D}$ will actually be dense in $\mathbf{R}^n$.

Examples

1. For the polar coordinate map P on $\mathbf{R}^2$, we have

$$[DP_{(r,\theta)}] = \begin{bmatrix} P_{1;1}(r, \theta) & P_{1;2}(r, \theta) \\ P_{2;1}(r, \theta) & P_{2;2}(r, \theta) \end{bmatrix} = \begin{bmatrix} \cos \theta & -r \sin \theta \\ \sin \theta & r \cos \theta \end{bmatrix},$$

since $P_1(r, \theta) = r \cos \theta$ and $P_2(r, \theta) = r \sin \theta$. It follows that

$$\det(DP_{(r,\theta)}) = r \qquad \text{for all } (r, \theta) \in \mathbf{R}^2$$

and hence

$$\tilde{D} = \{(r, \theta) : r \neq 0\}.$$

2. For the cylindrical coordinate map Q on $\mathbf{R}^3$ we find that

$$DQ_{(r,\theta,z)} = \begin{bmatrix} \cos\theta & -r\sin\theta & 0 \\ \sin\theta & r\cos\theta & 0 \\ 0 & 0 & 1 \end{bmatrix}.$$

Hence $\det(DQ_{(r,\theta,z)}) = r$ for all $(r, \theta, z) \in \mathbf{R}^3$ and so

$$\tilde{D} = \{(r, \theta, z) : r \neq 0\}.$$

3. For the spherical polar coordinate map S on $\mathbf{R}^3$ we find that

$$[DS_{(r,\theta,\psi)}] = \begin{bmatrix} \cos\theta\sin\psi & -r\sin\theta\sin\psi & r\cos\theta\cos\psi \\ \sin\theta\sin\psi & r\cos\theta\sin\psi & r\sin\theta\cos\psi \\ \cos\psi & 0 & -r\sin\psi \end{bmatrix}.$$

Hence

$$\det(DS_{(r,\theta,\psi)}) = -r^2 \sin\psi \qquad \text{for all } (r, \theta, \psi) \in \mathbf{R}^3.$$

It follows that

$$\tilde{D} = \{(r, \theta, \psi) : r \neq 0, \psi \neq n\pi, n \in \mathbf{N}\}.$$

Notation. Suppose that (ϕ, D, E) is a coordinate transformation on $\mathbf{R}^n$ and suppose that $x \in \tilde{D}$. We shall let

$$\phi_{i;j}^{-1}(\phi(x))$$

denote the ijth component of the matrix $[D\phi_x]^{-1}$. To justify this notation, we remark that if we restrict ϕ to $D \subset \mathbf{R}^n$ and let ϕ^{-1} denote the inverse of $\phi | D : D \longrightarrow \mathbf{R}^n \setminus E$, then

$$(\phi^{-1})_{i;j}(\phi(x)) = \phi_{i;j}^{-1}(\phi(x)) \qquad \text{for all } x \in D.$$

We may now state the main result of this section which shows how partial derivatives behave under coordinate transformations.

Proposition 3.4.3

Let (ϕ, D, E) be a coordinate transformation of $\mathbf{R}^n$ and suppose that $f : U \longrightarrow \mathbf{R}$ is a differentiable map on the open subset U of $\mathbf{R}^n$. Then, for $x \in \tilde{D} \cap \phi^{-1}(U)$, we have the relation

$$(f_{;i})^{\phi}(x) = \sum_{j=1}^{n} f_{;j}^{\phi}(x)\phi_{j;i}^{-1}(\phi(x)), \qquad 1 \leqslant i \leqslant n.$$

Proof. $f^{\phi} = f\phi$ and so, by the composite-mapping formula, f^{ϕ} is differentiable on $\phi^{-1}(U$ with derivative given by

$$D(f^{\phi})_x = Df_{\phi(x)} . D\phi_x, \qquad x \in \phi^{-1}(U).$$

It follows that, if $x \in \phi^{-1}(U) \cap \tilde{D}$, we have the equation

$$D(f^\phi)_x \cdot (D\phi_x)^{-1} = Df_{\phi(x)}. \tag{A}$$

By definition,

$$(f_{;i})^\phi(x) = (f_{;i}\phi)(x) = Df_{\phi(x)}(e_i),$$

where e_i is the ith element of the standard basis of $\mathbf{R}^n$. Evaluating eq. (A) at the point e_i yields

$$(f_{;i})^\phi(x) = D(f^\phi)_x \cdot (D\phi_x)^{-1}(e_i) = Df_x^\phi\left(\sum_{j=1}^n \phi^{-1}_{j;i}(\phi(x))e_j\right) = \sum_{j=1}^n f^\phi_{;j}(x)\phi^{-1}_{j;i}(\phi(x)). \quad \blacksquare$$

Examples

1. Let $f: \mathbf{R}^2 \longrightarrow \mathbf{R}$ be differentiable. Let us find $f_{;1}$ and $f_{;2}$ in terms of $f^P_{;1}$ and $f^P_{;2}$. In the examples preceding Proposition 3.4.3 we showed that

$$[DP_{(r,\theta)}] = \begin{bmatrix} \cos\theta & -r\sin\theta \\ \sin\theta & r\cos\theta \end{bmatrix}.$$

It follows easily that for $r \neq 0$

$$[DP_{(r,\theta)}]^{-1} = \begin{bmatrix} \cos\theta & \sin\theta \\ -(\sin\theta)/r & (\cos\theta)/r \end{bmatrix}.$$

Let $(r, \theta) \in \mathbf{R}^2$, $r \neq 0$ and set

$$(x, y) = P(r, \theta) = (r\cos\theta, r\sin\theta).$$

It follows from Proposition 3.4.3 that

$$f_{;1}(x, y) = \cos\theta f^P_{;1}(r, \theta) - \frac{\sin\theta}{r} f^P_{;2}(r, \theta),$$

$$f_{;2}(x, y) = \sin\theta f^P_{;1}(r, \theta) + \frac{\cos\theta}{r} f^P_{;2}(r, \theta).$$

In classical notation we write

$$f_{;1}(x, y) = \partial f/\partial x \qquad f_{;2}(x, y) = \partial f/\partial y,$$
$$f^P_{;1}(r, \theta) = \partial f/\partial r \qquad f^P_{;2}(r, \theta) = \partial f/\partial\theta,$$

and the above relations take the form

$$\frac{\partial f}{\partial x} = \cos\theta\frac{\partial f}{\partial r} - \frac{\sin\theta}{r}\frac{\partial f}{\partial\theta}, \qquad \frac{\partial f}{\partial y} = \sin\theta\frac{\partial f}{\partial r} - \frac{\cos\theta}{r}\frac{\partial f}{\partial\theta}.$$

It is useful to interpret the above in terms of differential operators. Thus we write

$$\frac{\partial}{\partial x} \equiv \cos\theta\frac{\partial}{\partial r} - \frac{\sin\theta}{r}\frac{\partial}{\partial\theta}, \qquad \frac{\partial}{\partial y} \equiv \sin\theta\frac{\partial}{\partial r} + \frac{\cos\theta}{r}\frac{\partial}{\partial\theta},$$

where it is understood that the operators on the right-hand side act on

P-transformed functions in the coordinates (r, θ). For example, since

$$\frac{\partial^2 f}{\partial x^2} = \frac{\partial}{\partial x}\frac{\partial f}{\partial x},$$

we have

$$\frac{\partial^2 f}{\partial x^2} = \left(\cos\theta\,\frac{\partial}{\partial r} - \frac{\sin\theta}{r}\,\frac{\partial}{\partial\theta}\right)\left(\cos\theta\,\frac{\partial f}{\partial r} - \frac{\sin\theta}{r}\,\frac{\partial f}{\partial\theta}\right)$$

$$= \cos^2\theta\,\frac{\partial^2 f}{\partial r^2} + \frac{2\cos\theta\sin\theta}{2}\,\frac{\partial f}{\partial\theta} - \frac{2\cos\theta\sin\theta}{r}\,\frac{\partial^2 f}{\partial r}\,\partial\theta$$

$$+ \frac{\sin^2\theta}{r}\,\frac{\partial f}{\partial r} + \frac{\sin^2\theta}{r^2}\,\frac{\partial^2 f}{\partial\theta^2}.$$

We similarly find that

$$\frac{\partial^2 f}{\partial y^2} = \sin^2\theta\,\frac{\partial^2 f}{\partial r^2} - \frac{2\cos\theta\sin\theta}{2}\,\frac{\partial f}{\partial\theta} + \frac{2\cos\theta\sin\theta}{r}\,\frac{\partial^2 f}{\partial r\,\partial\theta}$$

$$+ \frac{\cos^2\theta}{r}\,\frac{\partial f}{\partial r} + \frac{\cos^2\theta}{r^2}\,\frac{\partial^2 f}{\partial\theta^2}.$$

In particular, if we add these expressions we obtain the following important identity:

$$\frac{\partial^2 f}{\partial x^2} + \frac{\partial^2 f}{\partial y^2} = \frac{\partial^2 f}{\partial r^2} + \frac{1}{r}\,\frac{\partial f}{\partial r} + \frac{1}{r^2}\,\frac{\partial^2 f}{\partial\theta^2}.$$

The differential operator $\partial^2/\partial x^2 + \partial^2/\partial y^2$ is known as *Laplace's operator* (in two variables). The partial differential equation

$$\partial^2 f/\partial x^2 + \partial^2 f/\partial y^2 = 0$$

associated with Laplace's operator is known as Laplace's equation and solutions of this equation are called *harmonic functions.* For example, the real or imaginary parts of an (entire) analytic function are harmonic.

The above computations show that a function on $\mathbf{R}^2$ is harmonic only if when we transform it into polar coordinates it satisfies the equation

$$\frac{\partial^2 f}{\partial r^2} + \frac{1}{r}\,\frac{\partial f}{\partial r} + \frac{1}{r^2}\,\frac{\partial^2 f}{\partial\theta^2} = 0.$$

Let us look at a characteristic application of the above ideas. We shall find a harmonic function on $\mathbf{R}^2$ which depends only on the distance from the origin. That is, we shall find a map $f\colon \mathbf{R}^2 \longrightarrow \mathbf{R}$ such that $\partial^2 f/\partial x^2 + \partial^2 f/\partial y^2 = 0$ and $f = f((x^2 + y^2)^{1/2})$. Let $\tilde{f}$ denote f in polar coordinates. $\tilde{f}$ must satisfy the equation

$$\frac{\partial^2 \tilde{f}}{\partial r^2} + \frac{1}{r}\,\frac{\partial \tilde{f}}{\partial r} + \frac{1}{r^2}\,\frac{\partial^2 \tilde{f}}{\partial\theta^2} = 0.$$

By assumption, $\tilde{f}$ is independent of θ and depends only on r. Hence $\partial^2\tilde{f}/\partial\theta^2 = 0$ and $\tilde{f}$ satisfies the ordinary differential equation

$$\frac{d^2\tilde{f}}{dr^2} + \frac{1}{r}\frac{d\tilde{f}}{dr} = 0.$$

The general solution of this equation is given by

$$\tilde{f}(r) = A \operatorname{Log} r + Br^2$$

Since Log r is not defined for $r = 0$, it follows that the function $Br^2 = B(x^2 + y^2)$ defines a harmonic function on $\mathbf{R}^2$ satisfying our requirements. It is not hard to show that the constant multiples of $(x^2 + y^2)$ are the only harmonic functions on $\mathbf{R}^2$ which are bounded at the origin and depend only on the distance from the origin.

2. The Laplace operator on $\mathbf{R}^3$,

$$\frac{\partial^2}{\partial x^2} + \frac{\partial^2}{\partial y^2} + \frac{\partial^2}{\partial z^2},$$

transforms under cylindrical coordinates to the differential operator

$$\frac{\partial^2}{\partial r^2} + \frac{1}{r}\frac{\partial}{\partial r} + \frac{1}{r^2}\frac{\partial^2}{\partial r^2} + \frac{\partial^2}{\partial z^2}.$$

The details are similar to those of the first example and we leave them to the exercises.

3. The Laplace operator on $\mathbf{R}^3$ transforms under spherical polar coordinates to the differential operator

$$\frac{1}{r^2 \sin\theta}\left[\frac{\partial}{\partial r}\left(r^2 \sin\theta \frac{\partial}{\partial r}\right) + \frac{1}{\sin\theta}\frac{\partial^2}{\partial\psi^2} + \frac{\partial}{\partial\theta}\left(\sin\theta\frac{\partial}{\partial\theta}\right)\right].$$

We leave details to the exercises.

4. Associated with Laplace's operator on $\mathbf{R}^3$ we may define Laplace's equation

$$\frac{\partial^2 f}{\partial x^2} + \frac{\partial^2 f}{\partial y^2} + \frac{\partial^2 f}{\partial z^2} = 0.$$

Let us find a solution of Laplace's equation which is defined on $\mathbf{R}^3 \setminus \{0\}$ and depends only on the distance from the origin of $\mathbf{R}^3$. If f satisfies Laplace's equation it must satisfy it in spherical polar coordinates. Hence f must satisfy the equation

$$\frac{1}{r^2 \sin\theta}\left[\frac{\partial}{\partial r}\left(r^2 \sin\theta \frac{\partial f}{\partial r}\right) + \frac{1}{\sin\theta}\frac{\partial^2 f}{\partial\psi^2} + \frac{\partial}{\partial\theta}\left(\sin\theta\frac{\partial f}{\partial\theta}\right)\right] = 0.$$

Now, by assumption, f depends only on $r = (x^2 + y^2 + z^2)^{1/2}$ and so $\partial^2 f/\partial\psi^2 = \partial f/\partial\theta = 0$. Hence f satisfies the ordinary differential equation

$$\frac{d}{dr}\left(r^2\frac{df}{dr}\right) = 0.$$

The general solution of this equation is easily seen to be

$$f(r) = -c/r + k,$$

where c and k are arbitrary constants. The required solution of Laplace's equation is therefore given by

$$f(x, y, z) = -c(x^2 + y^2 + z^2)^{-1/2} + k.$$

With $k = 0$, f is the familiar Newtonian potential function.

In this section we have only given a very brief outline of the theory and applications of coordinate transformations. The reader should notice that the material we develop in Chapter 4 is closely related to that of this section. For more details about coordinate transformations of $\mathbf{R}^n$, and their applications to the solution of partial differential equations, we refer the reader to [20].

Exercises

1. Let $a \geqslant b > 0$. Show that the map $T: \mathbf{R}^2 \longrightarrow \mathbf{R}^2$, defined by

$$T(t, \theta) = (ta \cos\theta, tb \sin\theta), \qquad (t, \theta) \in \mathbf{R}^2,$$

is a coordinate transformation of $\mathbf{R}^2$. What is the Laplacian in the coordinates defined by T?

2. Find a solution of Laplace's equation on $\mathbf{R}^3$ which depends only on (1) Distance from the xy-plane; (2) Distance from the z-axis (axial symmetry).
3. Try to find the most general solution of Laplace's equation on $\mathbf{R}^2$ of the form $u(r)\, v(\theta)$ (u and v are differentiable real-valued functions defined on $\mathbf{R}$).
4. Let $A \subset \mathbf{R}^3$ denote the region $[-1, +1] \times \mathbf{R}^2$. Consider the map $\phi: A \longrightarrow \mathbf{R}^3$ defined by

$$\phi(r, \theta, t) = ((1 - t^2)^{1/2}(1 + r^2)^{1/2} \cos\theta, (1 - t^2)^{1/2}(1 + r^2)^{1/2} \sin\theta, rt),$$
$$(r, \theta, t) \in A.$$

ϕ is said to define 'oblate-spheroidal coordinates' on $\mathbf{R}^3$. Find an open subset $D \subset A$ and closed subset $E \subset \mathbf{R}^3$ such that

$$\phi \mid D \in \mathrm{Diff}^\infty(D, \mathbf{R}^3 \setminus E).$$

CHAPTER 4

Differential manifolds

4.1 Introduction

Above all the theory of differentiation is valuable for its applications to other branches of mathematics and the physical sciences. For example, it is basic to the study of differential equations and the theory of optimization and extremal values of functions. Yet the theory we have developed so far is noteworthy for its general inapplicability to most practical problems. Realistic problems in mechanics, electromagnetics, economics, etc. are but rarely naturally formulated in terms of functions defined on open subsets of a vector space. Instead, most such problems are naturally formulated on a far larger class of topological spaces including, for example, spheres, tori and products of these compact spaces with Euclidean spaces. By trying to formulate such problems on open domains of Euclidean space we often throw away the geometrical and topological constraints of the natural domain of definition of the problem and consequently lose much insight into its solution.

For example, let us consider the natural domain of definition of the following mechanical system: Suppose we have a pendulum with rigid rod and fixed support with bob allowed to move in $\mathbf{R}^3$ (that is, a 'spherical pendulum'). Then if the support is at the origin of $\mathbf{R}^3$ and the rod is of length 1, the bob naturally moves on the unit sphere S^2 of $\mathbf{R}^3$. That is, the position of the bob is uniquely defined by a point in the compact topological space S^2. If the bob is at the point $x \in S^2$, its velocity is uniquely determined by a point in the tangent plane to S^2 at x. If we denote the tangent plane to S^2 at x by T_xS^2 and form the disjoint union

$$TS^2 = \bigcup_{x \in S^2} T_xS^2$$

then TS^2 is the space which naturally parametrizes position and velocity of the bob. In a natural way TS^2 has the structure of a topological space and it can be shown that TS^2 is *not* homeomorphic to the product $S^2 \times \mathbf{R}^2$ and that there is no way to embed TS^2 as an open subset of $\mathbf{R}^n$, $n \geqslant 1$. In many classical studies of the spherical pendulum all that is attempted is a study of its motion, under perturbation, near an equilibrium position.

Many phenomena in nature are periodic or cyclic and are naturally associated with functions defined on the circle S^1 or on products of circles.

Continual exposure to the analysis of functions defined on open subsets of $\mathbf{R}^n$ tends to build up the conviction that they are the most 'natural' type of domain to work with. This is far from the truth. For example, our natural domain of experience is the surface of a two-dimensional sphere, S^2. To treat the surface of the earth as though it were an open subset of $\mathbf{R}^2$ (a 'flat earth') and so ignore its intrinsic geometry leads to absurd results, and not just in navigation! Thus, all of the following results are true for S^2 but not, in general, true for open subsets of $\mathbf{R}^2$: If $x \in S^2$ there exists a unique point $\hat{x} \in S^2$ such that there are infinitely many *distinct* paths of shortest length between x and $\hat{x}$; every continuous function on S^2 is bounded; no curve on S^2 ever meets a boundary; it is not possible to assign in a continuous way a *non-zero* tangent vector to every point of S^2. (One cannot comb a hairy ball flat.)

The fact that an open domain of $\mathbf{R}^n$ is never compact means that it is always hard to prove that a function actually attains (or even has) a maximum or minimum value on such a domain. Thus, even though we may be able to establish the existence of local extrema on such a domain it is no easy matter to prove that they are absolute extrema.

For these and many other reasons it becomes essential to extend our theory of differentiation to a larger class of topological spaces. This will be the purpose of much of this chapter.

In all of this chapter topological spaces will always be *Hausdorff* and vector spaces will always be real and finite dimensional. Most of what we say, however, generalizes easily to real or complex Banach spaces.

4.2 Charts and atlases

The title of this section suggests 'cartography' and indeed a differential manifold is essentially a topological space provided with a set of 'maps' or 'charts' which cover every point of the space (that is, an 'atlas') and which fit together in a differentiable way. For this reason we shall motivate our discussion by taking as an example of a topological space the unit sphere, S^2, of $\mathbf{R}^3$.

Suppose that $f: S^2 \longrightarrow \mathbf{R}$ is a continuous function and that $x_0 \in S^2$. What can we mean by the statement: 'f is differentiable at x_0'? To answer this question, we first remark that S^2 *locally* 'looks like' an open subset of $\mathbf{R}^2$ (this is a fact of our own experience). More precisely, given $x_0 \in S^2$, we may find an open disc U, centre x_0, in S^2, together with a homeomorphism ϕ which maps U onto an open subset of $\mathbf{R}^2$. For example, we may define ϕ by restricting to U the orthogonal projection of $\mathbf{R}^3$ onto the tangent plane to S^2 at x_0 as illustrated in Fig. 4.1. Then the map ϕ^{-1} induces a *local system of coordinates* on U (Fig. 4.2).

Notice that the above is just a mathematical description of one of the standard ways in which a map is drawn to represent part of the earth's surface. Notice also, that it is not possible to obtain *one* map of the whole of the earth's surface without introducing discontinuities in the chart map ϕ.

We call the pair (U, ϕ), as constructed above, a *chart.* We make the following formal definition.

Definition 4.2.1

Let M be a Hausdorff topological space and E be a finite-dimensional real normed

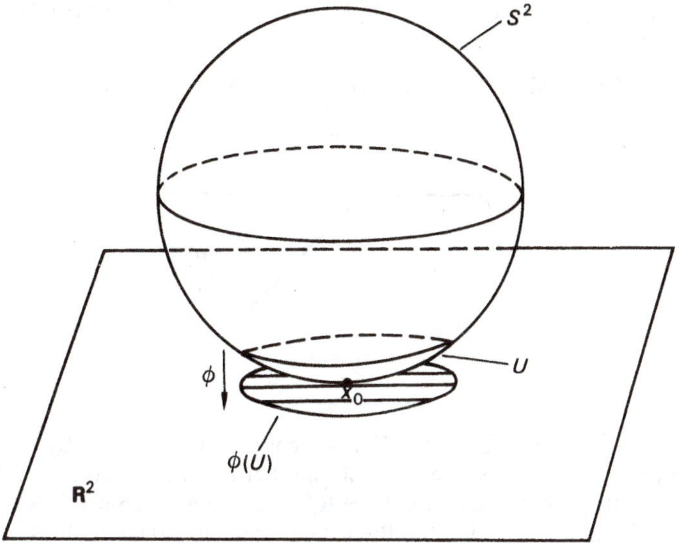

FIG. 4.1

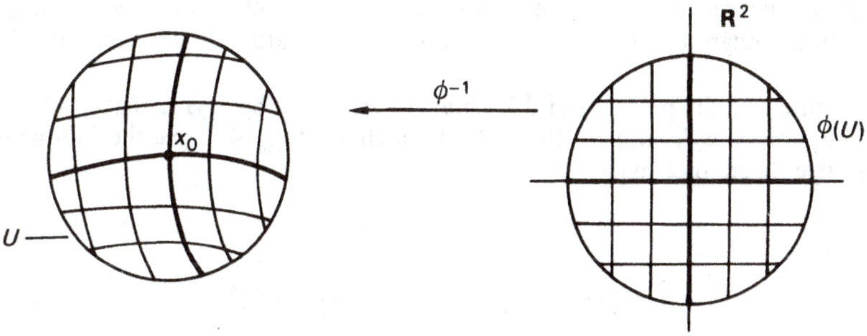

FIG. 4.2

vector space. If U is an open subset of M and $\phi : U \longrightarrow E$ is a homeomorphism onto the open subset $\phi(U)$ of E, then the pair (U, ϕ) is called a *chart* on M.

Remarks

1. We recall the basic assumptions of this chapter: All topological spaces are assumed to be *Hausdorff* and all normed vector spaces are *real* and *finite* dimensional. In the sequel, we shall not usually write these assumptions into the statement of every definition, proposition and theorem.
2. Notice that the very existence of a chart (U, ϕ) for M implies facts about the topology of M. For example, U is a locally compact subset of M.

We have already indicated that every point $x_0 \in S^2$ has a neighbourhood U and associated map $\phi : U \longrightarrow \mathbf{R}^2$ such that (U, ϕ) is a chart. This is certainly not true for arbitrary topological spaces. For example consider the subset of the plane (Fig. 4.3) defined by

$$X = \{(t, 0): -1 < t < 1\} \cup \{(t, -t): 0 < t < 1\}.$$

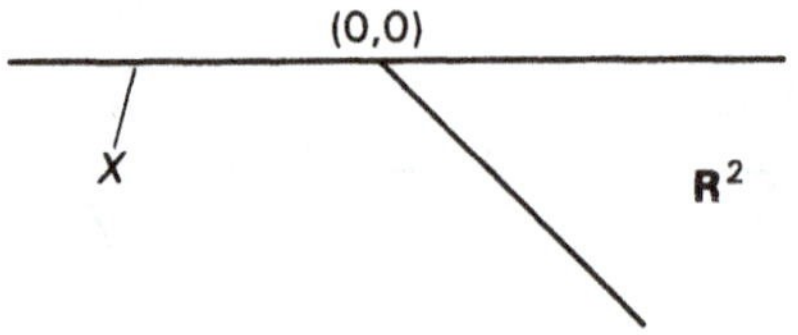

FIG. 4.3

We certainly cannot find a chart for X which contains the point (0, 0). To see that this is so, suppose that (U, ϕ) is a chart for X and that U is a connected open neighbourhood of (0, 0). Suppose that $\phi : U \longrightarrow \mathbf{R}^n$ for some $n \geqslant 1$. Since U is connected, $\phi(U)$ is connected. But $X \setminus \{(0, 0)\}$ has three connected components and so, therefore, does $U \setminus \{(0, 0)\}$. But $\phi(U \setminus \{(0, 0)\})$ can clearly have at most two connected components, since $\phi(U) \subset \mathbf{R}^n$ is open and connected. Contradiction.

Another example of a space which does not have charts containing every point is the closed unit interval of $\mathbf{R}$: We cannot find charts of [0, 1] which contain the end points. It is, however, possible to modify our definition of chart to allow topological spaces with boundary like the unit interval. We shall not attempt this generalization here.

Let us return to the problem of deciding what is meant by saying that $f : S^2 \longrightarrow \mathbf{R}$ is 'differentiable at x_0'. Suppose that (U, ϕ) is a chart at x_0. We have the following commutative diagram of maps:

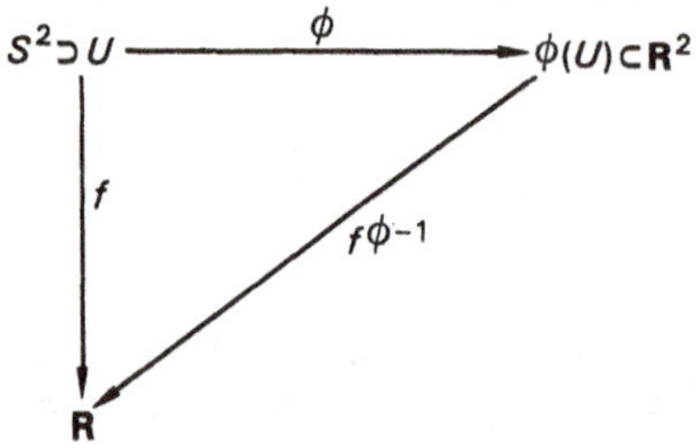

A first definition of the differentiability of f at x_0 might be to require that $f\phi^{-1}$ is differentiable at $\phi(x_0) \in \phi(U) \subset \mathbf{R}^2$. That is, to require that f be differentiable with respect to the local coordinate system induced on U by ϕ. However, this definition of differentiability of f clearly depends on the choice of the chart (U, ϕ). Suppose that (V, ψ) is another chart of S^2 which contains the point x_0. Figure 4.4 illustrates the situation.

If f is differentiable at x_0, with respect to the chart (U, ϕ), we would like to find a relation between the charts (U, ϕ) and (V, ψ) which would imply the differentiability

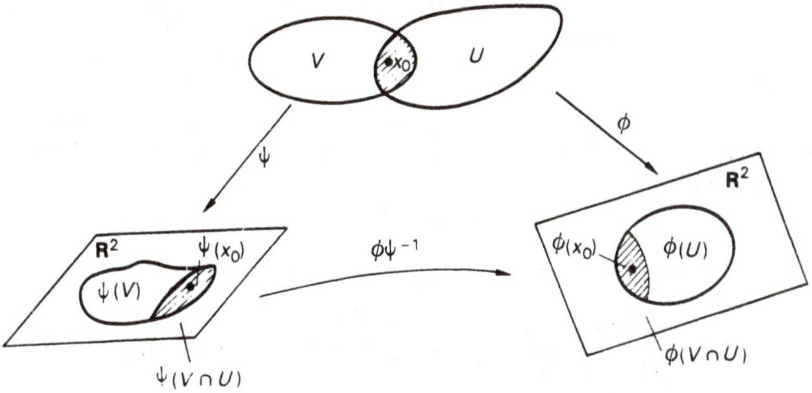

FIG. 4.4

of f at x_0 with respect to the chart (V, ψ). That is, we wish to find conditions which imply that

$$f\psi^{-1}: \psi(V) \longrightarrow \mathbf{R}$$

is differentiable at $\psi(x_0)$, given that $f\phi^{-1}$ is differentiable at $\phi(x_0)$.

We have the identity

$$f\psi^{-1} = (f\phi^{-1})(\phi\psi^{-1}) \text{ on } \psi(V \cap U).$$

By assumption, $f\phi^{-1}$ is differentiable at $\phi(x_0)$. In general, $f\psi^{-1}$ is not differentiable at $\psi(x_0)$ unless the map

$$\phi\psi^{-1}: \psi(V \cap U) \longrightarrow \phi(V \cap U)$$

is differentiable at $\psi(x_0)$. Consequently, if the map $\phi\psi^{-1}$ is differentiable on the overlap $\psi(V \cap U)$, it follows that if f is differentiable on $V \cap U$, with respect to the chart (U, ϕ), then f is differentiable on $V \cap U$ with respect to the chart (V, ψ). Similarly, if the map $\psi\phi^{-1}$ is differentiable on $\phi(V \cap U)$, then differentiability with respect to (V, ψ) implies differentiability with respect to (U, ϕ). More generally we have

Lemma 4.2.2

Let $f: M \longrightarrow \mathbf{R}$ be a continuous function on the topological space M and suppose that (U, ϕ) and (V, ψ) are charts on M such that $V \cap U \neq \emptyset$. Suppose that

$$\phi\psi^{-1} \in \text{Diff}^{\infty}(\psi(V \cap U), \phi(V \cap U)).$$

Then we have

1. $f\phi^{-1}$ is differentiable at $\phi(x_0)$ if and only if $f\psi^{-1}$ is differentiable at $\psi(x_0)$.
2. $f\phi^{-1}$ is differentiable on $\phi(U \cap V)$ if and only if $f\psi^{-1}$ is differentiable on $\psi(V \cap U)$.
3. $f\phi^{-1}$ is C^r on $\phi(V \cap U)$ if and only if $f\psi^{-1}$ is C^r on $\psi(V \cap U)$, $r = 1, 2, \ldots$.

Proof. (1), (2) and (3) follow immediately from the composite-mapping theorem applied to the identities

$$f\phi^{-1} = (f\psi^{-1})(\psi\phi^{-1}), \qquad f\psi^{-1} = (f\phi^{-1})(\phi\psi^{-1}). \quad ■$$

We now give the main definition of this section, which will allow us to generalize the above ideas to the whole of M.

Definition 4.2.3

Let M be a topological space and suppose that we are given a collection of charts $\mathscr{A} = \{(U_\alpha, \phi_\alpha)\}_{\alpha \in \Lambda}$ on M, where $\phi_\alpha: U_\alpha \longrightarrow E_\alpha$, $\alpha \in \Lambda$. Suppose that $\mathscr{A}$ satisfies the following conditions:

1. $\{U_\alpha\}_{\alpha \in \Lambda}$ covers M: $\bigcup_{\alpha \in \Lambda} U_\alpha = M$.
2. $\phi_\alpha \phi_\beta^{-1} \in \text{Diff}^\infty(\phi_\beta(U_\beta \cap U_\alpha), \phi_\alpha(U_\alpha \cap U_\beta))$ for all $\alpha, \beta \in \Lambda$ such that $U_\alpha \cap U_\beta \neq \emptyset$

Then the collection $\mathscr{A}$ is called a (differentiable) *atlas* on M.

M, together with the atlas $\mathscr{A}$, is called a *differential manifold.* Alternatively, $\mathscr{A}$ is said to give M the structure of a differential manifold.

Let us first look at some consequences of Definition 4.2.3. Suppose (U_α, ϕ_α), $(U_\beta, \phi_\beta) \in \mathscr{A}$ and $U_\alpha \cap U_\beta \neq \emptyset$. Then

$$D(\phi_\alpha \phi_\beta^{-1})_x \in \text{Iso}\,(E_\beta, E_\alpha) \qquad \text{for all } x \in \phi_\beta(U_\alpha \cap U_\beta).$$

Hence $E_\alpha \cong E_\beta$. It follows easily that if M is connected, all the vector spaces E_α are isomorphic.

In the sequel we shall *always* assume that M is *connected* and that the spaces E_α are all equal to some fixed normed vector space E. Thus the chart maps ϕ_α will map from U_α to E for all $\alpha \in \Lambda$: $\phi_\alpha \longrightarrow E$. Since $E \cong \mathbf{R}^n$, $n = \dim(E)$, it is no loss of generality to suppose that the chart maps ϕ_α all map to $\mathbf{R}^n$ and, on occasions, we shall do this in the sequel.

We shall call the vector space E the *model* space for M ('M locally looks like an open subset of E'). The *dimension of* M is defined to be the dimension of its model space, E. We write the dimension of M in abbreviated form as dim (M).

Examples

1. We may define an atlas for the vector space $\mathbf{R}^n$ to consist of the single chart $(\mathbf{R}^n, I)$. A far bigger atlas for $\mathbf{R}^n$, which contains this particular chart, is defined to be the set

   $$\mathscr{A}_n = \{(U, \phi): U \text{ and } \phi(U) \text{ are open subsets of } \mathbf{R}^n \text{ and } \phi \in \text{Diff}^\infty(U, \phi(U))\}.$$

   Notice that the model space for $\mathbf{R}^n$ is, as might be expected, just $\mathbf{R}^n$. Observe also that if M is a differential manifold modelled on $\mathbf{R}^n$ with atlas $\mathscr{A} = \{(U_\alpha, \phi_\alpha)\}_{\alpha \in \Lambda}$, then

   $$\phi_\alpha \phi_\beta^{-1} \in \mathscr{A}_n \qquad \text{for all } \alpha, \beta \in \Lambda,$$

   where we have allowed the 'empty chart' to belong to $\mathscr{A}_n$.
2. In this example we shall construct two atlases for the unit circle S^1.

   First atlas for S^1

   We consider S^1 as a closed subset of $\mathbf{R}^2$.

   We define the open subsets U_1, U_2 (Fig. 4.5) of S^1 by

   $$U_1 = \{(\cos\theta, \sin\theta): \theta \in (-\epsilon, \pi + \epsilon)\},$$
   $$U_2 = \{(\cos\theta, \sin\theta): \theta \in (\pi - \epsilon, 2\pi + \epsilon)\},$$

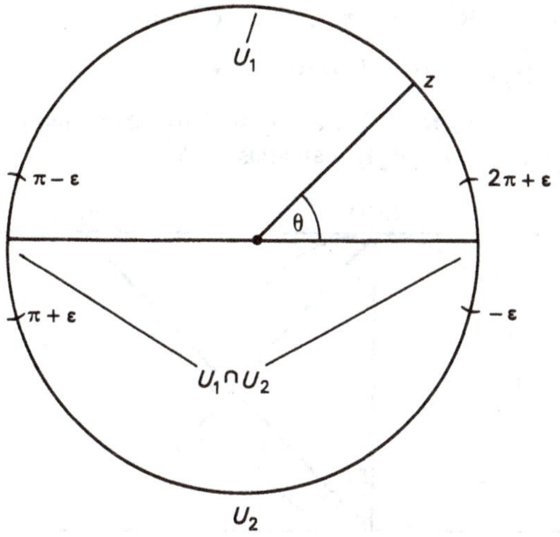

FIG. 4.5

where $0 < \epsilon < \pi$. We define the chart maps ϕ_1 and ϕ_2 by

$$\phi_1 : U_1 \longrightarrow \mathbf{R}; z \longmapsto \theta, \qquad \theta \in (-\epsilon, \pi + \epsilon)$$
$$\phi_2 : U_2 \longrightarrow \mathbf{R}; z \longmapsto \theta, \qquad \theta \in (\pi - \epsilon, 2\pi + \epsilon).$$

Let us set $U_1 \cap U_2 = V \cup W$, where V is the connected component of $U_1 \cap U_2$ containing the point $(1, 0)$ and W is the connected component containing the point $(-1, 0)$. It is then easy to see that

$$\phi_1\phi_2^{-1} \mid V = I - 2\pi; \phi_1\phi_2^{-1} \mid W = I. \qquad \text{(Here } I \text{ denotes the identity map.)}$$

It follows that

$$\phi_1\phi_2^{-1} \in \text{Diff}^\infty((2\pi - \epsilon, 2\pi + \epsilon) \cup (\pi - \epsilon, \pi + \epsilon), (-\epsilon, +\epsilon) \cup (\pi - \epsilon, \pi + \epsilon))$$

and so $\{(U_1, \phi_1), (U_2, \phi_2)\}$ is an atlas for S^1.

Second atlas for S^1

We regard S^1 as the unit circle in the plane and take the tangent lines to S_1 at $(0, 1)$ and $(0, -1)$ as shown in Fig. 4.6. We set

$$V_1 = S^1 \setminus \{(0, 1)\} \qquad \text{and} \qquad V_2 = S^1 \setminus \{(0, -1)\}.$$

We define $h_1 : V_1 \longrightarrow \mathbf{R}$ by projecting from the point $(0, 1)$ onto the tangent line of S^1 at $(0, -1)$. Similarly, $h_2 : V_2 \longrightarrow \mathbf{R}$ is defined by projection from $(0, -1)$. Both h_1 and h_2 are easily seen to be homeomorphisms onto $\mathbf{R}$. We shall show that

$$h_2 h_1^{-1} \in \text{Diff}(\mathbf{R} \setminus \{0\}, \mathbf{R} \setminus \{0\}).$$

With the notation of the diagram we have

$$h_1(z) = 2 \tan \theta; h_2(z) = 2 \cot \theta, \qquad \theta \in (-\pi/2, +\pi/2).$$

Here we have made the obvious identification of the tangent lines to S^1 with $\mathbf{R}$ and used the fact that the diameter of the circle is 2. By Example 2 of Section

3.2 the map $\tan^{-1}: \mathbf{R} \longrightarrow (-\pi/2, \pi/2)$ is C^∞. But

$$h_2 h_1^{-1}(t) = \cot(\tan^{-1}(t/2)), \qquad t \neq 0,$$

and so $h_2 h_1^{-1}$ is C^∞ on $\mathbf{R} \setminus \{0\}$. We may similarly show that $h_1 h_2^{-1}$ is C^∞ and hence that $\{(V_1, h_1), (V_2, h_2)\}$ is an atlas for S^1.

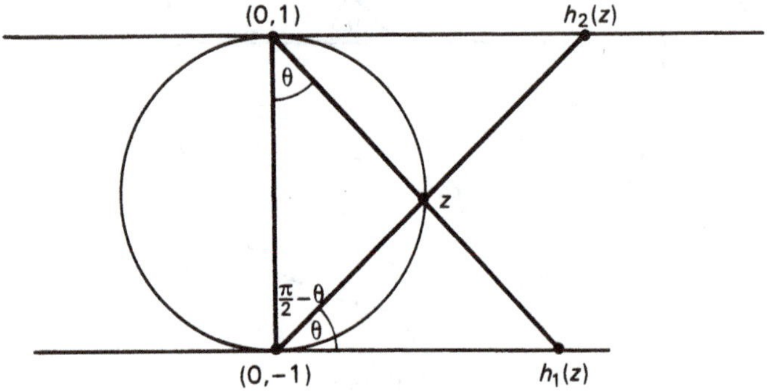

FIG. 4.6

In a sense that will become clear later, the two atlases we have defined on S^1 are equivalent.

3. In this example we shall construct an atlas for the unit two-sphere S^2. Our method generalizes that used to construct the second atlas on S^1. We project S^2 stereographically from the points $(0, 0, 1)$ and $(0, 0, -1)$ onto the xy-plane $\mathbf{R}^2 \times \{0\} \subset \mathbf{R}^3$. Thus we define

$$h_1: S^2 \setminus \{(0, 0, 1)\} \longrightarrow \mathbf{R}^2, \qquad h_2: S^2 \setminus \{(0, 0, -1)\} \longrightarrow \mathbf{R}^2$$

as indicated in Fig. 4.7. Clearly h_1 and h_2 are homeomorphisms onto $\mathbf{R}^2$.

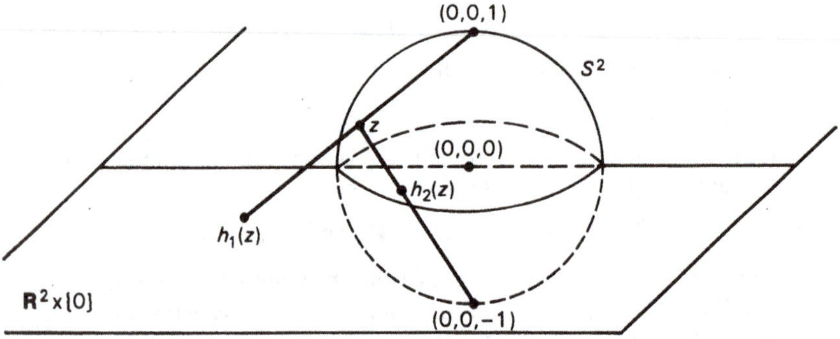

FIG. 4.7

We claim that

$$\{(S^2 \setminus \{(0, 0, 1)\}, h_1), (S^2 \setminus \{(0, 0, -1)\}, h_2)\}$$

is an atlas for S^2. To do this we must show that

$$h_2 h_1^{-1}: \mathbf{R}^2 \setminus \{(0, 0)\} \longrightarrow \mathbf{R}^2 \setminus \{(0, 0)\}$$

is a diffeomorphism.

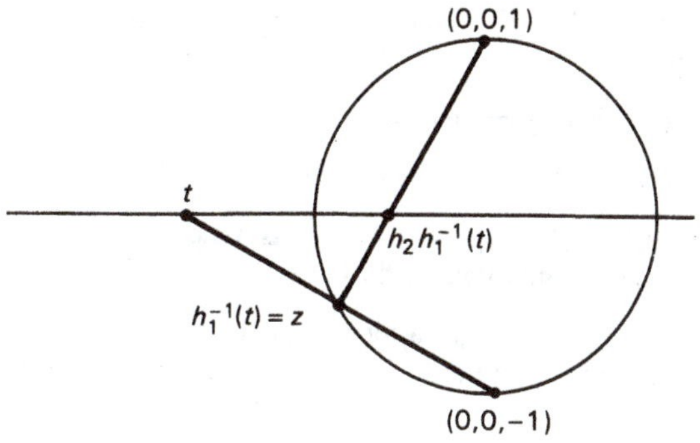

FIG. 4.8

Observe that the plane in $\mathbf{R}^3$ through the points (0, 0, 1), (0, 0, −1) and z contains the points $h_1(z)$ and $h_2(z)$. Let us set $h_1(z) = t$. In this plane we have the picture as in Fig. 4.8. From elementary plane geometry it follows that t and $h_2h_1^{-1}(t)$ are related by

$$h_2h_1^{-1}(t) = t/\|t\|_2^2, \qquad t \neq 0.$$

Hence $h_2h_1^{-1}$ is C^∞ on $\mathbf{R}^2 \setminus \{(0, 0)\}$. Similarly, $h_1h_2^{-1}$ is C^∞, proving our assertion.

The reason that we defined the concept of atlas was so that we could define differentiable functions on a larger class of topological spaces. The following definition makes this precise.

Definition 4.2.4

Let M be a differential manifold with atlas $\mathscr{A} = \{(U_\alpha, \phi_\alpha)\}_{\alpha \in \Lambda}$. A continuous function $f : M \longrightarrow \mathbf{R}$ is said to be differentiable on M, with respect to the atlas $\mathscr{A}$, if

$$f\phi_\alpha^{-1} : \phi_\alpha(U_\alpha) \longrightarrow \mathbf{R}$$

is differentiable for all $\alpha \in \Lambda$. f is said to be C^r, $r \geqslant 1$, if the maps $f\phi_\alpha^{-1}$ are C^r for all $\alpha \in \Lambda$.

Examples

1. In the preceding examples we constructed an atlas $\mathscr{A}_n$ on $\mathbf{R}^n$. We claim that if $f : \mathbf{R}^n \longrightarrow \mathbf{R}$ is differentiable in the usual sense, then f is differentiable in the sense of Definition 4.2.4 and conversely. Suppose that $(U, \phi) \in \mathscr{A}_n$ and that $f : \mathbf{R}^n \longrightarrow \mathbf{R}$ is differentiable in the usual sense. We must show that $f\phi^{-1}$: $\phi(U) \longrightarrow \mathbf{R}$ is differentiable. But this follows immediately from the composite-mapping formula. Conversely, suppose that $f\phi^{-1}$ is differentiable on $\phi(U)$ for all $(U, \phi) \in \mathscr{A}_n$. Since $(\mathbf{R}^n, I) \in \mathscr{A}_n$ it follows immediately that f is differentiable on $\mathbf{R}^n$.

   We have the obvious generalization of the above to C^r functions on $\mathbf{R}^n$, $r \geqslant 1$.
2. Let $f : S^1 \longrightarrow \mathbf{R}$ be the function defined by

$$f((\cos\theta, \sin\theta)) = 2\cos 2\theta, \qquad (\cos\theta, \sin\theta) \in S^1.$$

Let us take the first atlas $\{(U_1, \phi_1), (U_2, \phi_2)\}$ defined on S^1 in the preceding examples. Then

$$f\phi_1^{-1}: (-\epsilon, \pi + \epsilon) \longrightarrow \mathbf{R}; \theta \longmapsto 2 \cos 2\theta$$

$$f\phi_2^{-1}: (\pi - \epsilon, 2\pi + \epsilon) \longrightarrow \mathbf{R}; \theta \longmapsto 2 \cos 2\theta.$$

Hence f is C^∞ on S^1 with respect to this atlas. One may also show that f is C^∞ with respect to the second atlas we defined on S^1.

From our definitions so far it might appear that the 'bigger' the atlas on M the fewer differentiable functions there are on M. This is in fact not the case as the next few paragraphs show.

Definition 4.2.5

If $\mathscr{A}_1$ and $\mathscr{A}_2$ are two atlases on M such that $\mathscr{A}_1 \subset \mathscr{A}_2$ (that is, every chart of $\mathscr{A}_1$ is a chart of $\mathscr{A}_2$) we shall say that $\mathscr{A}_1$ is a *subatlas* of $\mathscr{A}_2$.

Proposition 4.2.6

Suppose that $\mathscr{A}_1$ is a subatlas of $\mathscr{A}_2$ on M and that $f: M \longrightarrow \mathbf{R}$ is differentiable with respect to $\mathscr{A}_1$. Then f is also differentiable with respect to $\mathscr{A}_2$.

Proof. To say that f is differentiable with respect to $\mathscr{A}_1$ means that $f\phi_\alpha^{-1}: \phi_\alpha(U) \longrightarrow \mathbf{R}$ is differentiable for all $(U_\alpha, \phi_\alpha) \in \mathscr{A}_1$. We must show that if (V, ψ) is an arbitrary chart of $\mathscr{A}_2$, then $f\psi^{-1}: \psi(V) \longrightarrow \mathbf{R}$ is differentiable. Let x be an arbitrary point of V. Then, since $\mathscr{A}_1$ is an atlas, we may find a chart $(U_\alpha, \phi_\alpha) \in \mathscr{A}_1$ such that $x \in U$. From Lemma 4.2.2 it follows that $f\phi^{-1}$ is differentiable at $\phi(x)$ if and only if $f\psi^{-1}$ is differentiable at $\psi(x)$. Since x was arbitrary, the result follows. ∎

Remark. The argument of Proposition 4.2.6 proves also that f is C^r with respect to $\mathscr{A}_1$ if and only if f is C^r with respect to $\mathscr{A}_2, r \geqslant 1$.

Proposition 4.2.7

Let M be a differential manifold with atlas $\mathscr{A}$. We define $\hat{\mathscr{A}}$ to be the set of all charts (U, ϕ) on M such that

$$(U, \phi) \in \hat{\mathscr{A}} \text{ if and only if } \mathscr{A} \cup \{(U, \phi)\} \text{ is an atlas on } M.$$

We have the following properties of $\hat{\mathscr{A}}$:

1. $\hat{\mathscr{A}}$ is an atlas.
2. $\mathscr{A}$ is a subatlas of $\hat{\mathscr{A}}$.
3. If $\mathscr{A}'$ is an atlas on M containing $\mathscr{A}$ as a subatlas, then $\mathscr{A}' \subseteq \hat{\mathscr{A}}$.

Proof. (2) is trivial, given that (1) is true. We shall prove (1) and leave the proof of (3) to the exercises.

Let $(U_i, \phi_i) \in \hat{\mathscr{A}}, i = 1, 2$. It is clearly sufficient to show that $\phi_2\phi_1^{-1}$ is C^∞ on $\phi_1(U_1 \cap U_2)$. Let $x \in U_1 \cap U_2$. Then we may find a chart $(V, \psi) \in \mathscr{A}$ such that $x \in V$. On $\phi_1(U_1 \cap U_2 \cap V)$ we have

$$\phi_2\phi_1^{-1} = (\phi_2\psi^{-1})(\psi\phi_1^{-1}).$$

Therefore, by the composite-mapping theorem, $\phi_2\phi_1^{-1}$ is C^∞ on $\phi_1(U_1 \cap U_2 \cap V)$. Since x was chosen arbitrarily, it follows that $\phi_2\phi_1^{-1}$ is C^∞ on $\phi_1(U_1 \cap U_2)$. ∎

Definition 4.2.8

With the notation of Proposition 4.2.7, we shall call $\hat{\mathscr{A}}$ the *completion* of the atlas $\mathscr{A}$. If $\mathscr{A} = \hat{\mathscr{A}}$ we shall call $\mathscr{A}$ a *complete atlas.*

Example. The atlas $\mathscr{A}_n$ defined on $\mathbf{R}^n$ is complete. The completion of the atlas $\{(\mathbf{R}^n, I)\}$ on $\mathbf{R}^n$ is $\mathscr{A}_n$.

Remark. In the sequel we shall generally think of a differential manifold as a topological space together with a complete atlas.

From Proposition 4.2.6 it follows immediately that we have

Proposition 4.2.9

Let $\mathscr{A}$ be an atlas on the topological space M. Then a map $f: M \longrightarrow \mathbf{R}$ is differentiable with respect to $\mathscr{A}$ if and only if f is differentiable with respect to the completion of $\mathscr{A}$.

In the sequel, if we are given a differential manifold with atlas $\mathscr{A}$, we shall often omit reference to the atlas on M and just say that a function f is 'differentiable on M'.

There is no reason why we should restrict attention to real-valued functions on a differential manifold. We have the obvious definition of differentiability for vector-space-valued functions.

Definition 4.2.10

Let M be a differential manifold with atlas $\mathscr{A}$ and let $f: M \longrightarrow E$, where E is a finite-dimensional normed vector space. Then we shall say that f is differentiable on M (with respect to $\mathscr{A}$) if

$$f\phi_\alpha^{-1}: \phi(U_\alpha) \longrightarrow E$$

is differentiable for all $(U_\alpha, \phi_\alpha) \in \mathscr{A}$.

We have a similar definition for C^r functions on M, $r \geqslant 1$.

Proposition 4.2.11

Let M have atlas $\mathscr{A}$ and suppose that $f: M \longrightarrow E$, where E is a finite-dimensional normed vector space. Then f is differentiable with respect to $\mathscr{A}$ if and only if f is differentiable with respect to the completion of $\mathscr{A}$.

Proof. Either directly, as in Proposition 4.2.6, or using Proposition 2.5.1. We leave details to the exercises. ∎

In this section we have indicated that an atlas on a topological space M provides a framework for the definition of differentiable functions on M. In later sections we shall prove theorems that will enable us to find many examples of differential manifolds without having to resort to the ad hoc methods of constructing atlases used in the examples in this section. We remark, however, that a topological space need not have a differential structure even if it is locally homeomorphic to open subsets of $\mathbf{R}^n$. It is also customary to assume that a differential manifold is *paracompact* (though we shall not need this assumption in this volume). This condition may be shown to imply that every differential manifold is metrizable. At a much deeper level, a topological space may have more than one differential structure.

We introduced this subject via the problem of defining differentiable functions on a topological space. It is, therefore, natural that we should show that the atlas of a differential manifold is uniquely determined by the differentiable functions on the manifold. This we shall do in the next section. However, the material in the next section, though important, is not essential for the understanding of the rest of the chapter and the reader may omit it if he wishes.

Exercises

1. Complete the proof of Proposition 4.2.7.
2. Suppose that M has atlas $\mathscr{A}$ and that M' is *homeomorphic* to M. Show that M' may be given the structure of a differential manifold. (*Hint*: Consider charts on M' of the form $(h(U_\alpha), \phi_\alpha h^{-1})$, where $(U_\alpha, \phi_\alpha) \in \mathscr{A}$ and $h: M \longrightarrow M'$ is a homeomorphism.)

   Hence, or otherwise, show that the boundary of the unit square in $\mathbf{R}^2$, $\{x: \|x\|_\infty = 1\}$ may be given the structure of a differential manifold.
3. Show that the completion of the first atlas we defined on S^1 is equal to the completion of the second atlas we defined on S^1.
4. Suppose that $F: \mathbf{R} \longrightarrow \mathbf{R}$ is differentiable and of period 2π: $F(x + 2\pi) = F(x)$ for all $x \in \mathbf{R}$. Show that F induces a differentiable function $\tilde{F}$ on S^1 such that

$$\tilde{F}(\cos x, \sin x) = F(x) \qquad \text{for all } x \in \mathbf{R}.$$

5. Complete the proof of Proposition 4.2.11.
6. Show that the function which measures the square of the distance of a point on the unit sphere $S^2 \subset \mathbf{R}^3$ from the xy-plane is differentiable if S^2 is given the atlas described in this section.

4.3 Equivalent atlases

In this section we wish to show that the structure of a differential manifold is uniquely determined by its differentiable functions. We start with a definition.

Definition 4.3.1

Let $\mathscr{A}_1$ and $\mathscr{A}_2$ be two atlases on the topological space M. We shall say that $\mathscr{A}_1$ and $\mathscr{A}_2$ are *equivalent atlases* if every function $f: M \longrightarrow \mathbf{R}$ is differentiable with respect to $\mathscr{A}_1$ if and only if it is differentiable with respect to $\mathscr{A}_2$. We denote equivalence of the atlases $\mathscr{A}_1$ and $\mathscr{A}_2$ by $\mathscr{A}_1 \sim \mathscr{A}_2$.

$\sim$ is obviously an equivalence relation.

Proposition 4.3.2

Let $\mathscr{A}_1$ and $\mathscr{A}_2$ be two atlases on M. Then

1. $\mathscr{A}_1 \sim \hat{\mathscr{A}}_1$.
2. $\mathscr{A}_1 \sim \mathscr{A}_2$ if and only if $\hat{\mathscr{A}}_1 \sim \hat{\mathscr{A}}_2$.

Proof. (1) follows from Proposition 4.2.9. (2) follows from (1) since $\hat{\mathscr{A}}_2 \sim \mathscr{A}_2 \sim \mathscr{A}_1 \sim \hat{\mathscr{A}}_1$. ■

Example. Let $\mathscr{A} = \{(U_\alpha, \phi_\alpha)\}_{\alpha \in \Lambda}$ be a complete atlas on M and suppose that M has model E. Suppose that F is any normed vector space isomorphic to E and let $\gamma: E \longrightarrow F$ denote a fixed linear isomorphism from E to F. Set $\mathscr{A}' = \{(U_\alpha, \gamma . \phi_\alpha)\}_{\alpha \in \Lambda}$. Then it is easy to verify that $\mathscr{A}'$ is a complete atlas on M and that M has model F relative to this atlas. Obviously, $\mathscr{A}' \sim \mathscr{A}$.

It follows from the above example that we can always replace the original model of M by an isomorphic vector space without changing the differentiable functions on M. Hence we may take the model of M to be $\mathbf{R}^m$, where m = dimension of M, and for the remainder of this section we shall assume that the model of M is fixed and equal to $(\mathbf{R}^m, \| \|_2)$.

Our aim in this section is to prove that $\mathscr{A}_1 \sim \mathscr{A}_2$ if and only if $\hat{\mathscr{A}}_1 = \hat{\mathscr{A}}_2$. That is, the differentiable functions on a manifold uniquely determine the atlas. Before we can do this, however, we need to prove a couple of lemmas which will enable us to construct plenty of differentiable functions on a manifold.

Lemma 4.3.3

Suppose that $\mathbf{R}^n$ is given the Euclidean norm $\| \|_2$. Then, if $0 < a < b$, there exists a C^∞ function $\Psi: \mathbf{R}^n \longrightarrow \mathbf{R}$ such that

1. $\Psi \equiv 1$ on $\bar{B}_a(0)$.
2. $\Psi \equiv 0$ on $\mathbf{R}^n \setminus B_b(0)$.
3. $\Psi(x) \in (0, 1)$ for all $x \in B_b(0) \setminus \bar{B}_a(0)$.

Proof. Define $\gamma: \mathbf{R} \longrightarrow \mathbf{R}$ by

$$\gamma(x) = \begin{cases} e^{-1/x} & x > 0 \\ 0, & x \leqslant 0. \end{cases}$$

The reader may check that γ is C^∞ on the whole of $\mathbf{R}$ (all derivatives are zero at $x = 0$). Since γ is C^∞, so is the function $\tau: \mathbf{R} \longrightarrow \mathbf{R}$ defined by

$$\tau(x) = \frac{\gamma(b^2 + x)\gamma(b^2 - x)}{\gamma(x - a^2) + \gamma(-x - a^2) + \gamma(b^2 + x)\gamma(b^2 - x)}.$$

We have

$$\tau(x) = 1, \qquad |x| \leqslant a^2.$$

$$\tau(x) = 0, \qquad |x| \geqslant b^2$$

$$\tau(x) \in (0, 1), \qquad a^2 < |x| < b^2.$$

The graph of τ is shown in Fig. 4.9. We now define $\Psi: \mathbf{R}^n \longrightarrow \mathbf{R}$ by

$$\Psi(x) = \tau(\| x \|_2^2).$$

Ψ clearly satisfies the conditions of the lemma. ■

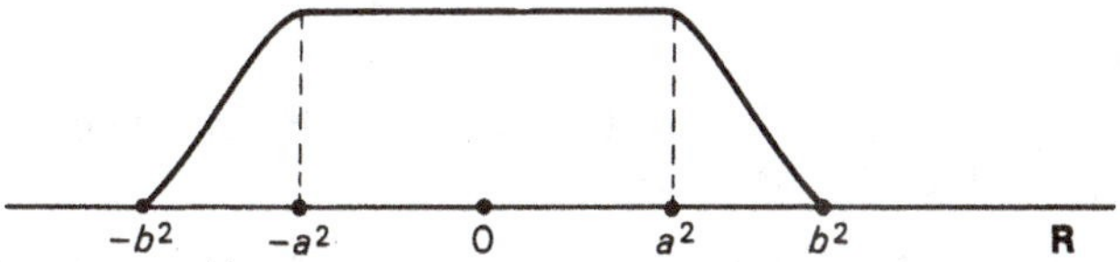

FIG. 4.9

Remarks

1. In the literature functions like Ψ or τ are usually called 'bump' functions.
2. Lemma 4.3.3 is one of the propositions of this chapter which does not generalize to arbitrary real Banach spaces. It certainly does not hold in the complex domain as analytic continuation implies that if an analytic function is zero on an open subset of a connected domain it is zero everywhere on the domain.

Lemma 4.3.4

Let M be a differential manifold with atlas $\mathscr{A}$. Let $(U, \phi) \in \mathscr{A}$ and suppose that $f : U \longrightarrow \mathbf{R}$ is differential on U (that is, $f\phi^{-1}$ is differentiable on $\phi(U) \subset \mathbf{R}^n$). Then, given any point $z \in U$, we may find an open neighbourhood W_z of z in U and a differentiable function $\tilde{f} : M \longrightarrow \mathbf{R}$ such that

$$\tilde{f} \mid W_z = f.$$

Proof. $\phi(U) \subset \mathbf{R}^n$. Without loss of generality suppose that $\phi(z) = 0$. We give $\mathbf{R}^n$ the norm $\| \; \|_2$. Choose $0 < a < b$ such that $\bar{B}_a(0) \subset \bar{B}_b(0) \subset \phi(U)$ and let $\Psi : \mathbf{R}^n \longrightarrow \mathbf{R}$ be the map defined in the preceding lemma. We define a map $g : \mathbf{R}^n \longrightarrow \mathbf{R}$ by

$$g(x) = \begin{cases} \Psi(x)(f\phi^{-1})(x), & x \in \phi(U) \\ 0, & x \in \mathbf{R}^n \setminus \phi(U). \end{cases}$$

Since $\Psi \equiv 0$ on $\mathbf{R}^n \setminus \bar{B}_b(0)$, it follows that g is identically zero on the open set $\mathbf{R}^n \setminus \bar{B}_b(0)$ and hence differentiable there. On the other hand g is certainly differentiable on $\phi(U)$ and so g is differentiable on the whole of $\mathbf{R}^n$. We now define $\tilde{f} : M \longrightarrow \mathbf{R}$ by

$$\tilde{f} \mid U = g\phi, \qquad \tilde{f} \mid M \setminus U = 0.$$

Since $\tilde{f} \equiv 0$ on an open neighbourhood of $M \setminus U$ in M it follows that $\tilde{f}$ is differentiabl on M. If we set $W_z = \phi^{-1}(B_a(0))$ it follows from our construction that

$$\tilde{f} \mid W_z = f. \quad \blacksquare$$

Remark. One can prove a much stronger result than that of Lemma 4.3.4. One can in fact show that if A and B are any two disjoint closed subsets of M, then one can find a C^∞ function on M which is zero precisely on A, equal to 1 precisely on B and takes values in $(0, 1)$ on $M \setminus (A \cup B)$. For a proof of this result we refer to [4], [13].

Suppose now that W is an open subset of M and that $\mathscr{A} = \{(U_\alpha, \phi_\alpha)\}_{\alpha \in \Lambda}$ is an atlas on M. We may define an atlas $\mathscr{A}(W)$ on W by setting

$$\mathscr{A}(W) = \{(U_\alpha \cap W, \phi_\alpha \mid (W \cap U_\alpha)) : \alpha \in \Lambda\}.$$

Since $W \cap U_\alpha$ is always open and $\bigcup_{\alpha \in \Lambda} W \cap U_\alpha = W$, $\mathscr{A}(W)$ verifies all the conditions for an atlas (on W).

We have the following basic proposition.

Proposition 4.3.5

Let $\mathscr{A}_1$ and $\mathscr{A}_2$ be two atlases on M. Then $\mathscr{A}_1 \sim \mathscr{A}_2$ if and only if $\mathscr{A}_1(W) \sim \mathscr{A}_2(W)$ for all open subsets W of M.

Proof. The non-trivial part of the implication is to show that $\mathscr{A}_1 \sim \mathscr{A}_2$ implies that $\mathscr{A}_1(W) \sim \mathscr{A}_2(W)$ for all open subsets W of M. Suppose then that $f : W \longrightarrow \mathbf{R}$ is differentiable with respect to the atlas $\mathscr{A}_1(W)$. We shall show that $\mathscr{A}_1 \sim \mathscr{A}_2$ implies that f is differentiable with respect to $\mathscr{A}_2(W)$. To do this it is clearly sufficient to show that for each $x \in W$ we can find an open neighbourhood W_x of x in W such that $f|W_x : W_x \longrightarrow \mathbf{R}$ is differentiable with respect to $\mathscr{A}_2$. That is, $f\psi^{-1} : \psi(W_x \cap V) \longrightarrow \mathbf{R}$ is differentiable for all $(V, \psi) \in \mathscr{A}_2$. But, applying Lemma 4.3.4, we can always find a differentiable function $\tilde{f} : M \longrightarrow \mathbf{R}$, which is equal to f on some neighbourhood W_x of x in W. The function $\tilde{f}$ is certainly differentiable with respect to $\mathscr{A}_2$, since $\mathscr{A}_1 \sim \mathscr{A}_2$, and so $\tilde{f}|W_x = f$ is differentiable with respect to $\mathscr{A}_2$. ■

We may now prove the main result of this section.

Theorem 4.3.6

Let $\mathscr{A}_1$ and $\mathscr{A}_2$ be two atlases on M. Then $\mathscr{A}_1 \sim \mathscr{A}_2$ if and only if $\hat{\mathscr{A}}_1 = \hat{\mathscr{A}}_2$.

Proof. It follows from Proposition 4.3.2 that we may assume that the atlases $\mathscr{A}_1$ and $\mathscr{A}_2$ are both complete. Suppose that $(U, \phi) \in \mathscr{A}_1$. Then $\phi : U \longrightarrow \mathbf{R}^m$, $m = \dim(M)$. The map ϕ is differentiable with respect to $\mathscr{A}_1(U)$ and so, by Proposition 4.3.5, ϕ is differentiable with respect to $\mathscr{A}_2(U)$. But this implies that

$$\phi\psi^{-1}: \psi(U \cap V) \longrightarrow \mathbf{R}^m$$

is differentiable for all $(V, \psi) \in \mathscr{A}_2$. Similarly, $\psi\phi^{-1}$ is differentiable for all $(V, \psi) \in \mathscr{A}_2$. Hence $(U, \phi) \in \mathscr{A}_2$. Thus $\mathscr{A}_1 \subset \mathscr{A}_2$. By the maximality of $\mathscr{A}_1$ we must therefore have $\mathscr{A}_1 = \mathscr{A}_2$. ■

Exercise

1. Let M be a topological space and suppose that we are given an open neighbourhood W_x of each point $x \in M$. Show that Proposition 4.3.5 continues to hold if we restrict attention to the family $\{W_x\}_{x \in M}$ of open subsets of M.
   This result implies that differentiability is a strictly local concept on a manifold.

4.4 Products of differential manifolds

Our aim in this section and in Sections 4.5–4.7 is to give methods for constructing differential manifolds. The formation of products of differential manifolds is the most elementary of these constructions.

Proposition 4.4.1

Let M and N be differential manifolds with atlases $\mathscr{A}(M)$ and $\mathscr{A}(N)$ respectively. Then $M \times N$ has the structure of a differential manifold with atlas $\mathscr{A}(M \times N)$ defined by

$$\mathscr{A}(M \times N) = \{(U_\alpha \times V_\beta, \phi_\alpha \times \psi_\beta): (U_\alpha, \phi_\alpha) \in \mathscr{A}(M), \quad (V_\beta, \psi_\beta) \in \mathscr{A}(N)\}.$$

$\mathscr{A}(M \times N)$ is called the *product* of the atlases $\mathscr{A}(M)$ and $\mathscr{A}(N)$.

Proof. Left to the exercises. Notice that, even if $\mathscr{A}(M)$ and $\mathscr{A}(N)$ are complete, $\mathscr{A}(M \times N)$ will not, in general, be complete. ■

Examples

1. The n-dimensional torus, denoted by T^n, is defined to be the product $S^1 \times \cdots \times S^1$ of the circle n times. In Section 4.2 we constructed an atlas on S^1 and it follows inductively from Proposition 4.4.1 that T^n may be given the structure of a differential manifold for $n \geqslant 1$. The two-dimensional torus, T^2, is of particular interest, and may be represented as the surface in $\mathbf{R}^3$ (Fig. 4.10) parametrized by the map $\omega : [0, 2\pi] \times [0, 2\pi] \longrightarrow \mathbf{R}^3$ defined by

$$\omega(\theta, \psi) = ((2 + \cos \psi) \cos \theta, (2 + \cos \psi) \sin \theta, \sin \psi).$$

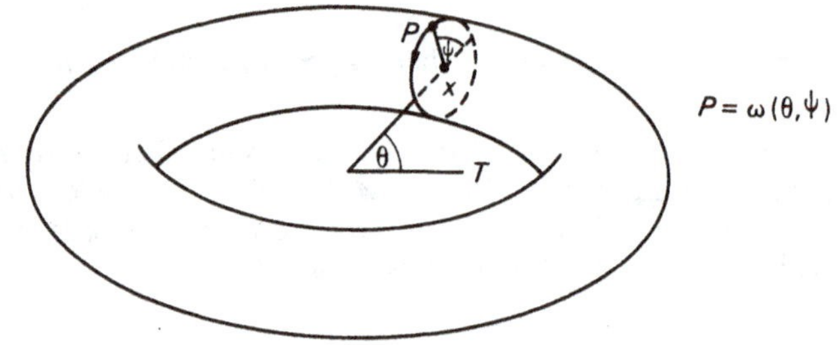

FIG. 4.10

With the notation of the diagram, OTX lies in the xy-plane and OT is part of the x-axis. As ψ increases from 0 to 2π, P moves round the surface in the direction indicated on the diagram. If ψ is held fixed and θ increased from 0 to 2π, P completes a circuit of the surface in the plane through P parallel to the xy-plane of $\mathbf{R}^3$. The parametrization ω may be used to construct an explicit homeomorphism between this surface and T^2.

2. The product $S^1 \times \mathbf{R}$ is an example of a *cylinder.* Since S^1 and $\mathbf{R}$ may be given the structure of differential manifolds, it follows from Proposition 4.4.1 that $S^1 \times \mathbf{R}$ may be given the structure of a differential manifold. Since $S^1 \subset \mathbf{R}^2$, $S^1 \times \mathbf{R}$ may be regarded as a subset of $\mathbf{R}^3$. We then have the situation illustrated in Fig. 4.11. If we take the atlas $\{(V_1, \psi_1), (V_2, \psi_2)\}$ on S^1, defined by projecting from the points (0, 1) and (0, –1), and the atlas $(\mathbf{R}, I)$ on $\mathbf{R}$, the product atlas on $S^1 \times \mathbf{R}$ is

$$\{(V_1 \times \mathbf{R}, \psi_1 \times I), (V_2 \times \mathbf{R}, \psi_2 \times I)\}.$$

Let us describe geometrically the first chart: $\psi_1 \times I$ is defined on the cylinder minus the line $\{(0, 1)\} \times \mathbf{R}$. But the cylinder minus this line is clearly homeomorphic to $\mathbf{R}^2$ (think of cutting the cylinder along this line) and the map $\psi_1 \times I$ is just the explicit homeomorphism obtained by projecting $(S^1 \setminus \{(0, 1)\}) \times \mathbf{R}$ onto the tangent plane to the cylinder, along the line $\{(0, -1)\} \times \mathbf{R}$, parallel to the xy-plane. This chart may be described intuitively as follows: Cut the cylinder along the line $\{(0, 1)\} \times \mathbf{R}$ and then 'unroll' the cylinder onto the tangent plane to the cylinder along the line $\{(0, -1)\} \times \mathbf{R}$, stretching out the cylinder as it is unrolled so that it fills up the whole of the tangent plane.

In the next section we shall show how one can use the result of this example to construct another atlas on T^2.

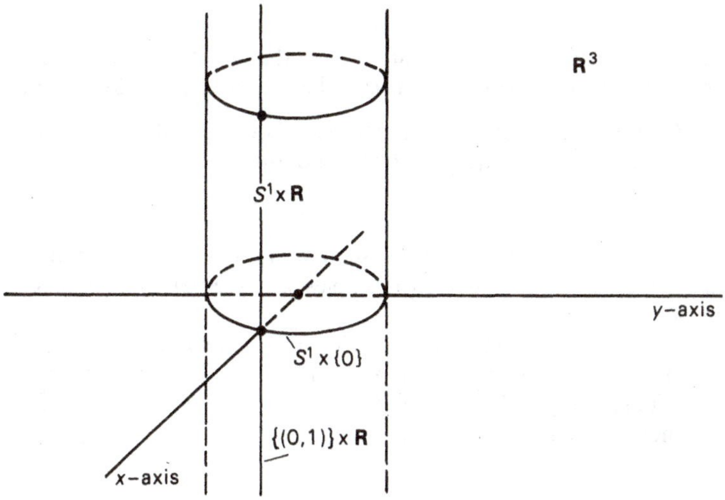

FIG. 4.11

Exercises

1. Prove that the space $S^2 \times S^2 \times T^4$ may be given the structure of a differential manifold.
2. Give an example to show that the product of complete atlases need not be complete.
3. Let $F : \mathbf{R}^2 \longrightarrow \mathbf{R}$ be a differentiable function which satisfies the condition

$$F(x + 2\pi, y + 2\pi) = F(x, y) \qquad \text{for all } x, y \in \mathbf{R}.$$

Show that F defines a *differentiable* function $\tilde{F}$ on T^2 such that

$$\tilde{F}((\cos x, \sin x), (\cos y, \sin y)) = F(x, y) \qquad \text{for all } x, y \in \mathbf{R}.$$

(F is an example of a *doubly periodic* function.)

4.5 Quotients of differential manifolds

In this section we shall study some topological methods for constructing differential manifolds and, in particular, show that real projective space may be given the structure of a differential manifold. The material of this section does, however, require a greater familiarity with topology than the rest of the chapter and since none of the results are used in the sequel the reader may pass to the next section without loss of continuity.

Recall that a map $f: M \longrightarrow N$ is a *homeomorphism* between the topological spaces M and N if f is a continuous bijection with continuous inverse. Thus f is a homeomorphism if and only if f is a continuous bijection which maps open sets to open sets.

Our first definition is about maps which locally look like homeomorphisms but globally may be far from being bijective.

Definition 4.5.1

Let M and N be topological spaces and suppose that $\pi : M \longrightarrow N$ is a continuous map. π is said to be a *local homeomorphism* if the following condition is satisfied:

For all $x \in M$ there exists an open neighbourhood U of x such that $\pi(U)$ is an open subset of N and $\pi \mid U : U \longrightarrow \pi(U)$ is a homeomorphism.

Remarks

1. Our definition of local homeomorphism does not require that π is surjective and, as we shall see from the following examples, π is in general far from being injective.
2. If $\pi : M \longrightarrow N$ is a local homeomorphism, then π maps open subsets of M onto open subsets of N. Indeed, let V be an open subset of M. Then for each $x \in V$, we may find an open subset U_x of M such that $\pi(U_x)$ is open and $\pi \mid U_x$ is a homeomorphism. Hence $\pi(U_x \cap V)$ is open. But

$$\pi(V) = \bigcup_{x \in V} \pi(U_x \cap V)$$

and so $\pi(V)$ is open.

Recall that a *base* of open neighbourhoods for the topological space X is a collection of open subsets of X, say $\mathscr{B}$, such that any open subset of X may be written as the union of elements of $\mathscr{B}$.

The following technical lemma is useful in the construction of local homeomorphisms.

Lemma 4.5.2

Let $\pi : M \longrightarrow N$ be a map between the topological spaces M and N. Suppose that $\mathscr{U}$ and $\mathscr{V}$ are bases of open sets for M and N respectively. Sufficient conditions for π to be a local homeomorphism are that

1. $\pi^{-1}(V_\beta)$ is open for all $V_\beta \in \mathscr{V}$;
2. $\pi(U_\alpha)$ is open for all $U_\alpha \in \mathscr{U}$;
3. $\pi \mid U_\alpha : U_\alpha \longrightarrow \pi(U_\alpha)$ is bijective for all $U_\alpha \in \mathscr{U}$.

Proof. (1) implies π is continuous. (2) and (3) imply $\pi \mid U_\alpha$ is a homeomorphism. ■

Examples

1. We shall define a local homeomorphism $\gamma : \mathbf{R} \longrightarrow S^1$. To do this we regard $S^1 \subset \mathbf{R}^2$ as the set of points $\{(\cos x, \sin x) : x \in [0, 2\pi)\}$ and define $\gamma : \mathbf{R} \longrightarrow S^1$ by

$$\gamma(x) = (\cos x, \sin x), \qquad x \in \mathbf{R}.$$

γ is clearly continuous. To show that γ is a local homeomorphism we consider the basis of $\mathbf{R}$ consisting of all open intervals of the form

$$I_{x,\alpha} = (x - \alpha, x + \alpha),$$

where $x \in \mathbf{R}$ and $|\alpha| < \pi/2$. $\gamma \mid I_{x,\alpha}$ is obviously a bijection onto the arc of S^1 defined by the angles $x - \alpha$ and $x + \alpha$ and $\gamma(I_{x,\alpha})$ is also an open subset of S^1. It follows from Lemma 4.5.2 that γ is a local homeomorphism. A 'pictorial' description of the map γ is shown in Fig. 4.12. $\mathbf{R}$ is represented as the infinite

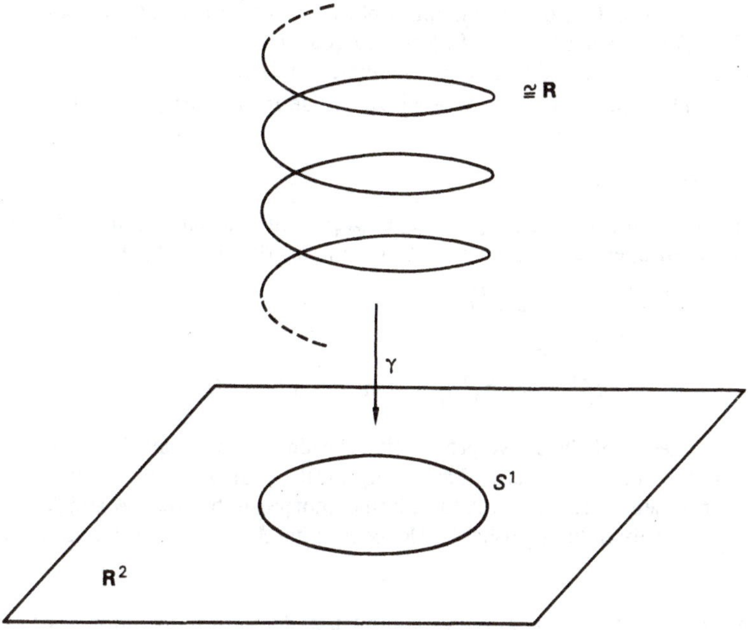

FIG. 4.12

helix $\{(\cos t, \sin t, t) : t \in \mathbf{R}\}$ contained in $\mathbf{R}^3$ and γ is just the restriction of the projection map onto the xy-plane.

2. We define a local homeomorphism $\gamma_n : \mathbf{R}^n \longrightarrow T^n, n \geqslant 1$. With the notation of the preceding example we set

$$\gamma_n = (\gamma, \ldots, \gamma) : \mathbf{R}^n \longrightarrow S^1 \times \cdots \times S^1 = T^n.$$

That γ_n is a local homeomorphism follows by looking at γ_n restricted to basic open sets of the form

$$I_{x_1, \alpha_1} \times \cdots \times I_{x_n, \alpha_n} \subset \mathbf{R}^n.$$

Notice that whilst $\mathbf{R}^n$ and T^n are locally homeomorphic, they are globally very different. For example T^n is compact and $\mathbf{R}^n$ is non-compact.

Our interest in local homeomorphisms stems from the following proposition.

Proposition 4.5.3

Let $\pi : M \longrightarrow N$ be a local homeomorphism. Then

1. If N has the structure of a differential manifold, π induces the structure of a differential manifold on M.
2. If M has the structure of a differential manifold and π is surjective, then π induces the structure of a differential manifold on N.

Proof. We shall prove (2) and leave the (similar) proof of (1) to the exercises. Let $\mathscr{A} = \{(U_\alpha, \phi_\alpha)\}$ be a complete atlas for M. We define a subset $\mathscr{A}_\pi$ of $\mathscr{A}$ by

$$\mathscr{A}_\pi = \{(U_\alpha, \phi_\alpha) \in \mathscr{A} : \pi \mid U_\alpha \text{ maps } U_\alpha \text{ homeomorphically onto } \pi(U_\alpha)\}.$$

We claim that $\mathscr{A}_\pi$ is a sub-atlas of $\mathscr{A}$. Indeed, if $x \in M$, there exists an open neighbourhood U of x such that $\pi \mid U$ is a homeomorphism. Since $\mathscr{A}$ is an atlas, there exists a chart $(U_\alpha, \phi_\alpha) \in \mathscr{A}$ such that $x \in U$. Since $\mathscr{A}$ is assumed complete the chart $(U_\alpha \cap U, \phi_\alpha | (U \cap U_\alpha))$ belongs to $\mathscr{A}$ and hence to $\mathscr{A}_\pi$.

If $(U_\alpha, \phi_\alpha) \in \mathscr{A}_\pi$, we shall set $\pi_\alpha = \pi \mid U_\alpha$ and define a chart (V_α, ψ_α) on N by the rules

$$V_\alpha = \pi_\alpha(U_\alpha); \psi_\alpha = \phi_\alpha \pi_\alpha^{-1}.$$

We claim that the set $\{(V_\alpha, \psi_\alpha): (U_\alpha, \phi_\alpha) \in \mathscr{A}_\pi\}$ is an atlas on N. Since π is surjective, the collection of open sets $\{V_\alpha\}$ covers N. On the overlap $V_\alpha \cap V_\beta$ we have

$$\begin{aligned}\psi_\alpha \psi_\beta^{-1} &= (\phi_\alpha \pi_\alpha^{-1})(\phi_\beta \pi_\beta^{-1})^{-1}\\ &= \phi_\alpha \pi_\alpha^{-1} \pi_\beta \phi_\beta^{-1}\\ &= \phi_\alpha \phi_\beta^{-1}, \text{ since } \pi_\alpha^{-1}\pi_\beta = \text{identity on } U_\alpha \cap U_\beta. \quad \blacksquare\end{aligned}$$

Remark. The result of the above proposition should be seen as intuitively obvious: A differential manifold was defined as a space locally indistinguishable from an open subset of a normed vector space. A local homeomorphism between M and N implies that they are locally indistinguishable. Hence if either M or N is a differential manifold both are.

Example. In the preceding examples we constructed a local homeomorphism $\gamma_n : \mathbf{R}^n \longrightarrow T^n$. Since $\mathbf{R}^n$ has the atlas $\mathscr{A}_n$, the above proposition implies that γ_n induces the structure of a differential manifold on T^n.

For the remainder of this section we shall show how one can put the structure of a differential manifold on the set of all lines through the origin of $\mathbf{R}^{n+1}, n \geqslant 0$.

Definition 4.5.4

We let $P^n(\mathbf{R})$ denote the set of all lines through the origin of $\mathbf{R}^{n+1}$. The set $P^n(\mathbf{R})$ is called *real projective n-space.*

Example. $P^0(\mathbf{R})$ consists of one point corresponding to $\mathbf{R}$.

Any point $x \in \mathbf{R}^{n+1}\backslash\{0\}$ determines a unique line L_x in $\mathbf{R}^{n+1}$ through x and the origin of $\mathbf{R}^{n+1}$:

$$L_x = \{tx : t \in \mathbf{R}\}.$$

Hence we have a natural map

$$q : \mathbf{R}^{n+1}\backslash\{0\} \longrightarrow P^n(\mathbf{R})$$

defined by mapping a point to the line through that point: $q(x) = L_x$.

q is clearly a surjection. Let $x, x' \in \mathbf{R}^{n+1}\backslash\{0\}$. Then $q(x) = q(x')$ if and only if $x = \lambda x'$ for some $\lambda \in \mathbf{R}\backslash\{0\}$. In coordinates this says that two non-zero $(n + 1)$-tuples $(x_1, \ldots, x_{n+1}), (x'_1, \ldots, x'_{n+1})$ determine the same point of $P^n(\mathbf{R})$ if and only if there exists $\lambda \neq 0$ such that $x_i = \lambda x'_i$, $1 \leqslant i \leqslant n$.

The $(n + 1)$-tuples $(x_1, \ldots, x_{n+1})$ are said to define *homogeneous coordinates* on $P^n(\mathbf{R})$. Two non-zero $(n + 1)$-tuples define the same point of $P^n(\mathbf{R})$ if and only if they are non-zero scalar multiples of each other.

We shall take the quotient topology on $P^n(\mathbf{R})$ induced by q. That is, a subset U of $P^n(\mathbf{R})$ is open if and only if $q^{-1}(U) \subset \mathbf{R}^{n+1}\backslash\{0\}$ is open.

Notice that if U is an open subset of $P^n(\mathbf{R})$ then $q^{-1}(U) \cup \{0\} \subset \mathbf{R}^{n+1}$ is *not* in general open. A typical open subset of $P^n(\mathbf{R})$ is defined by a *cone* or *pencil* of lines in $\mathbf{R}^{n+1}$ (Fig. 4.13).

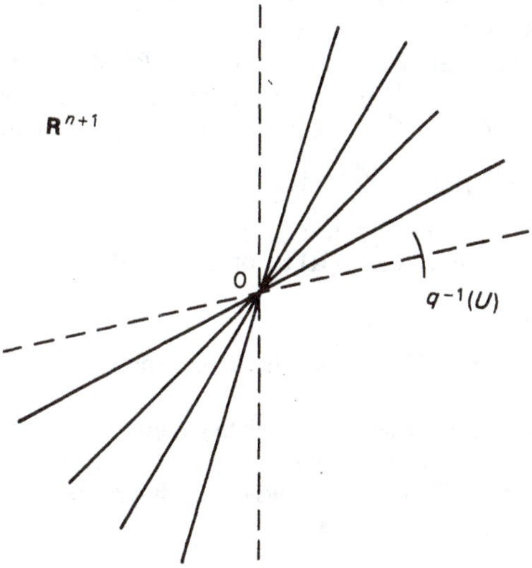

FIG. 4.13

We claim that the map $q : \mathbf{R}^{n+1} \setminus \{0\} \longrightarrow P^n(\mathbf{R})$ is an *open mapping*. That is, if $V \subset \mathbf{R}^{n+1} \setminus \{0\}$ is open, so is $q(V)$. To prove this we observe that if V is open, so is $tV, t \neq 0$. Hence

$$\tilde{V} = \bigcup_{t \neq 0} tV$$

is an open subset of $\mathbf{R}^{n+1} \setminus \{0\}$. But $\tilde{V}$ is a union of lines and so $q^{-1}(q(\tilde{V})) = \tilde{V}$. On the other hand $q(V) = q(\tilde{V})$, since any line that lies in $\tilde{V}$ passes through a point of V. Therefore, $q^{-1}(q(V))$ is open and so $q(V)$ is an open subset of $P^n(\mathbf{R})$.

Let S^n denote the unit sphere in $\mathbf{R}^{n+1}$. Each line in $\mathbf{R}^{n+1}$ meets S^n exactly twice in antipodal points (Fig. 4.14). It follows that if we restrict q to S^n, $q : S^n \longrightarrow P^n(\mathbf{R})$ is surjective and $q^{-1}(L)$ consists of precisely two points for all $L \in P^n(\mathbf{R})$.

Just as above it is easy to verify that $q : S^n \longrightarrow P^n(\mathbf{R})$ is an open mapping.

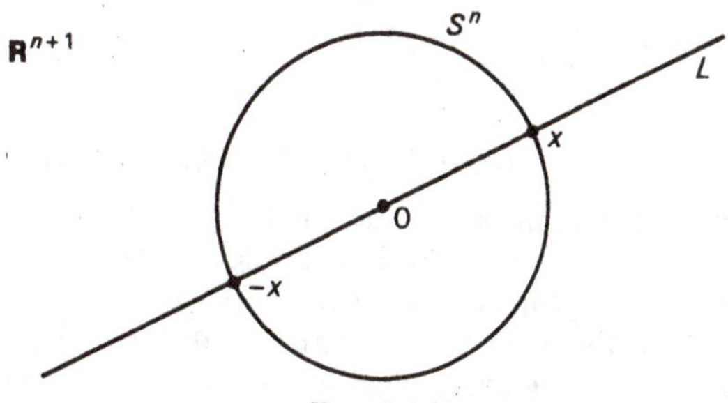

FIG. 4.14

Proposition 4.5.5

$P^n(\mathbf{R})$ is a compact Hausdorff space.

Proof. Since $q : S^n \longrightarrow P^n(\mathbf{R})$ is continuous and S^n is compact, it follows that $q(S^n) = P^n(\mathbf{R})$ is compact.

Suppose that L and L' are two distinct points of $P^n(\mathbf{R})$. Let L and L' meet S^n in the points $\{x, -x\}$ and $\{x', -x'\}$ respectively. Since S^n is Hausdorff, we may find open neighbourhoods U and U' of x and x' respectively such that

$$(U \cup (-U)) \cap (V \cup (-V)) = \emptyset.$$

It follows that $q(V) \cap q(U) = \emptyset$ and, since q is an open mapping, we have therefore separated the points L and L' of $P^n(\mathbf{R})$ by open sets. ■

Proposition 4.5.6

The map $q : S^n \longrightarrow P^n(\mathbf{R})$ is a local homeomorphism.

Proof. Let $x \in S^n$ and choose any open neighbourhood U of x such that $U \cap (-U) = \emptyset$. Then $q \mid U$ defines a bijection between U and $q(U) \subset P^n(\mathbf{R})$. Since q is an open map, $q(U)$ is an open subset of $P^n(\mathbf{R})$. It follows easily that $q \mid U$ is a homeomorphism. Hence q is a local homeomorphism. ■

Example. We showed in Section 4.2 that S^1 and S^2 could be given the structure of differential manifolds. It follows from the above proposition, together with Proposition 4.5.3, that $P^1(\mathbf{R})$ and $P^2(\mathbf{R})$ may be given the structure of differential manifolds.

We shall now construct a 'natural' atlas on $P^n(\mathbf{R}), n \geqslant 1$.

For $1 \leqslant j \leqslant n+1$, we shall let P_j denote the hyperplane of $\mathbf{R}^{n+1}$ defined by $x_j = 0$. That is

$$P_j = \{(x_1, \ldots, x_{n+1}) : x_j = 0\}.$$

Now $\mathbf{R}^{n+1} \backslash P_j$ is an open subset of $\mathbf{R}^{n+1} \backslash \{0\}$ and we shall set

$$U_j = q(\mathbf{R}^{n+1} \backslash P_j), \qquad 1 \leqslant j \leqslant n+1.$$

U_j is an open subset of $P^n(\mathbf{R})$, and consists of all lines in $\mathbf{R}^{n+1}$ through points with non-zero jth coordinate. Thus, in homogeneous coordinates,

$$U_j = \{(x_1, \ldots, x_{n+1}) : x_j \neq 0\}.$$

Since $(0, 0, \ldots, 0) \notin P^n(\mathbf{R})$, $\bigcup_{j=1}^{n+1} U_j = P^n(\mathbf{R})$.

For $1 \leqslant j \leqslant n+1$ we define a map $\phi_j : U_j \longrightarrow \mathbf{R}^n$ by

$$\phi_j(x_1, \ldots, x_{n+1}) = (x_1/x_j, \ldots, x_{j-1}/x_j, x_{j+1}/x_j, \ldots, x_{n+1}/x_j).$$

Since two $(n+1)$-tuples define the same line if and only if they are non-zero scalar multiples of one another, it is easy to check that ϕ_j is well defined by the above formula. We may describe ϕ_j geometrically as follows: Let $L \in U_j$. Then $\phi_j(L)$ is the point in $\mathbf{R}^n$ which defines the point of intersection of the line L with the affine hyperplane $x_j = 1$ of $\mathbf{R}^{n+1}$. Using this geometric description of ϕ_j, it is easy to verify that ϕ_j is a continuous open mapping. ϕ_j is obviously injective; let us show that it is surjective. If $y = (y_1, \ldots, y_n) \in \mathbf{R}^n$, we define $L_y \in P^n(\mathbf{R})$ to be the line in U_j defined by

$$L_y = \{t(y_1, \ldots, y_{j-1}, 1, y_j, \ldots, y_n) : t \in \mathbf{R}\}.$$

Clearly $\phi_j(L_y) = y$. Hence ϕ_j defines a homeomorphism between U_j and $\mathbf{R}^n$ and so (U_j, ϕ_j) is a chart on $P^n(\mathbf{R})$, $1 \leqslant j \leqslant n+1$. We claim that $\{(U_j, \phi_j)\}_{j=1}^{n+1}$ is an atlas for $P^n(\mathbf{R})$. Let $(y_1, \ldots, y_n) \in \mathbf{R}^n$. Then the homogeneous coordinates of $\phi_j^{-1}(y_1, \ldots, y_n)$ are $(y_1, \ldots, y_{j-1}, 1, y_j, \ldots, y_n)$. Without loss of generality suppose that $i > j$. Then $\phi_i\phi_j^{-1}(y_1, \ldots, y_n) =$

$$(y_1/y_{i-1}, \ldots, y_{j-1}/y_{i-1}, 1/y_{i-1}, \ldots, y_{i-2}/y_{i-1}, y_i/y_{i-1}, \ldots, y_n/y_{i-1}).$$

It follows from this explicit formula that $\phi_i\phi_j^{-1} : U_i \cap U_j \longrightarrow \mathbf{R}^n$ is C^∞. Similarly, $\phi_j\phi_i^{-1}$ is C^∞. Hence $\{(U_j, \phi_j)\}_{j=1}^{n+1}$ is an atlas on $P^n(\mathbf{R})$.

Examples

1. $P^1(\mathbf{R})$ is homeomorphic to S^1. Whilst the projection map $q : S^1 \longrightarrow P^1(\mathbf{R})$ is not a homeomorphism, since it is not injective, we can construct an explicit homeomorphism between S^1 and $P^1(\mathbf{R})$ as follows: Let $z = (\cos\theta, \sin\theta) \in S^1$, $\theta \in [0, 2\pi)$. We define $\omega : S^1 \longrightarrow P^1(\mathbf{R})$ by

   $$\omega(z) = q(\cos\theta/2, \sin\theta/2).$$

   Referring to Fig. 4.15 we see that $\omega(z)$ is the line making angle $\theta/2$ with the x-axis. As θ increases from 0 to 2π the line $\omega(z)$ rotates through angle π and as θ approaches 2π, $\omega(z)$ approaches the line $\theta = 0$. Hence ω maps S^1 bijectively onto $P^1(\mathbf{R})$. We claim that ω is a homeomorphism. This is easily seen by examining the image of an open arc of S^1 under ω. We leave details to the reader.

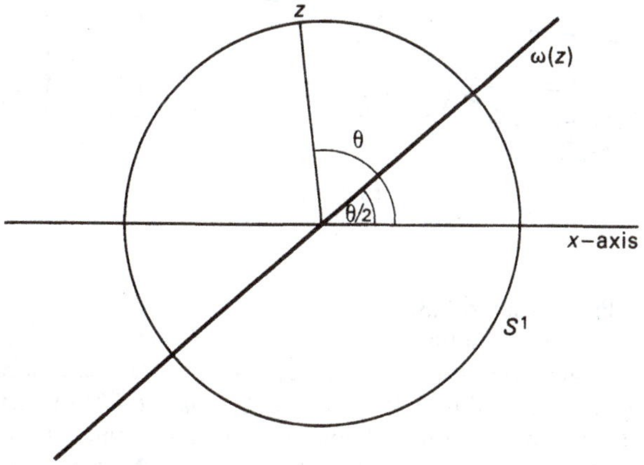

FIG. 4.15

2. $P^2(\mathbf{R})$ is not homeomorphic to S^2. We sketch two proofs of this fact.

   First proof. Let $H_1 \subset P^2(\mathbf{R})$ denote the image by the quotient map q of the plane $x_1 = 0$. Then

   $$H_1 = \{(0, x_2, x_3) : (x_1, x_2) \neq (0, 0)\}.$$

Hence H_1 is homeomorphic to $P_1(\mathbf{R})$ which, by the preceding example, is homeomorphic to S^1. On the other hand, $P^2(\mathbf{R})\backslash H_1 = U_1$ is homeomorphic to $\mathbf{R}^2$. In other words we can remove the homeomorphic image of a circle from $P^2(\mathbf{R})$ to get $\mathbf{R}^2$. Notice that $\mathbf{R}^2$ is connected. Now if $P^2(\mathbf{R})$ and S^2 were homeomorphic it would follow that if we removed the homeomorphic image of a circle from S^2 we would always obtain a connected set. But this is false by the *Jordan-curve theorem*, which states that we always get a set with *two* connected components when we remove a circle from S^2.

Second proof. S^2 minus a point is homeomorphic to $\mathbf{R}^2$. We claim that $P^2(\mathbf{R})$ minus a point is *not* homeomorphic to $\mathbf{R}^2$. In fact $P^2(\mathbf{R})$ minus a point is homeomorphic to a Möbius strip. To see this we remove the point $(0, 0, 1)$ from $P^2(\mathbf{R})$. Now $P^2(\mathbf{R})$ may be regarded as the upper hemisphere of S^2 with antipodal points identified on the equator. If we remove the point $(0, 0, 1)$ from this hemisphere it is clear, by sliding along great circles, that $P^2(\mathbf{R}) \setminus \{(0, 0, 1)\}$ is homeomorphic to the following set:

$$\{(x_1, x_2, x_3) \in S^2 : 0 \leqslant x_3 < \tfrac{1}{2}\}/\sim$$

where $\sim$ is the relation which identifies antipodal equatorial points.

The mapping from this quotient space to the Möbius strip is indicated in Fig. 4.16. Thus h takes the circle through A and B in S^2 onto the boundary of the Möbius strip. The equator of S^2 is mapped to the circle through 0 and $\{x, -x\}$, and the lines xB and $-xA$ are mapped as shown in the diagram.

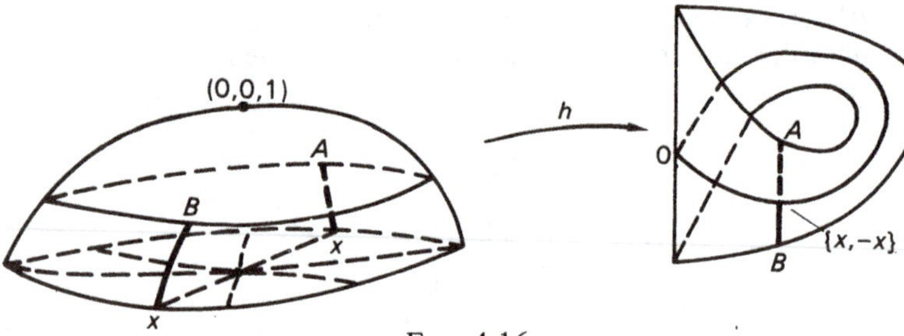

FIG. 4.16

Exercises

1. Prove (1) of Proposition 4.5.3.
2. Let P be an n-dimensional linear subspace of $\mathbf{R}^{n+1}$. Let $U_P \subset P^n(\mathbf{R})$ be defined as $q(\mathbf{R}^{n+1}\backslash P)$. Show that if $e_P \perp P$, then $\tilde{\phi}_P(x) = (x - \langle x, e_P\rangle e_P)/\langle x, e_P\rangle$, $x \in \mathbf{R}^{n+1}\setminus P$, may be used to define a chart (U_P, ϕ_P) on $P^n(\mathbf{R})$. Hence construct an atlas on $P^n(\mathbf{R})$ and show that it contains the atlas defined in the text.
3. Show that if P is an n-dimensional linear subspace of $\mathbf{R}^{n+1}$, then $q(P) \subset P^n(\mathbf{R})$ is homeomorphic to $P^{n-1}(\mathbf{R})$ and $P^n(\mathbf{R})\backslash q(P)$ is homeomorphic to $\mathbf{R}^n$. Show also that $P^n(\mathbf{R})$ is not homeomorphic to S^n, $n \geqslant 2$.

4.6 Submanifolds

In this section we wish to consider the following problem: Suppose that N is a subset of the differential manifold M. What conditions on M and N are required in order that the differential structure on M should induce a differential structure on N? In particular, if $f : M \longrightarrow \mathbf{R}$ is differentiable, what conditions on M and N do we need to ensure that $f \mid N : N \longrightarrow \mathbf{R}$ is differentiable? It should be noticed that there are *two* requirements implicit in the above demands. Firstly, N should be a differential manifold in its own right. Secondly, the differential structure on N should be compatible with that of M.

Definition 4.6.1

Let M be a differential manifold with atlas $\mathscr{A}$ and model $\mathbf{R}^m$ and suppose that N is a (connected) subset of M. We call N a *submanifold* of M, modelled on $\mathbf{R}^n$, if the following condition holds:

For all $x \in N$, there exists $(U, \phi) \in \mathscr{A}$, containing x, such that

(a) $\phi(U \cap N)$ is an open subset of the vector subspace $\mathbf{R}^n \times \{0\}$ of

$$\mathbf{R}^n \times \mathbf{R}^{m-n} = \mathbf{R}^m.$$

(b) ϕ maps $U \cap N$ homeomorphically onto $\phi(U \cap N)$.

In Fig. 4.17, ϕ maps $U \cap N$ onto the open subset $\phi(U \cap N)$ of $\mathbf{R}^n$. Since $\phi | (U \cap N)$ is a homeomorphism, $(U \cap N, \phi \mid (U \cap N))$ is a chart for N and so N 'locally looks like' an open subset of $\mathbf{R}^n$.

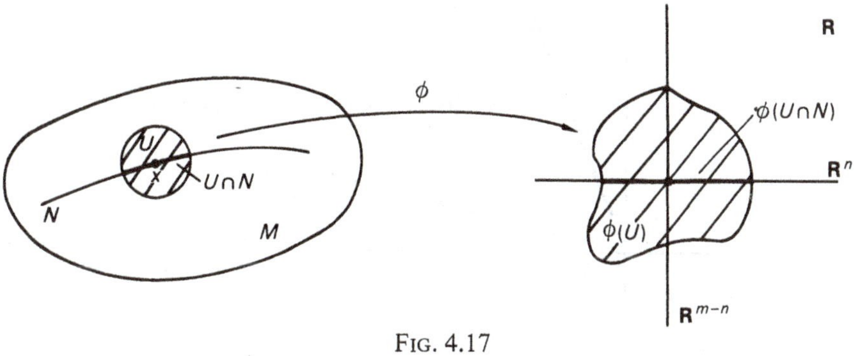

FIG. 4.17

The above picture is intended to indicate what a submanifold of M should look like. Let us look at one or two examples of subsets that are not submanifolds.

Examples

1. We take $M = \mathbf{R}^2$ and N to be the union of the x- and y-axes (Fig. 4.18). We claim that we cannot find a chart $(U, \phi) \in \mathscr{A}_2$, containing $(0, 0)$, such that $\phi(U \cap N)$ is an open subset of $\mathbf{R}$ and $\phi \mid U \cap N$ is a homeomorphism onto $\phi(U \cap N)$. Suppose the contrary. Without loss of generality we may assume that $\phi(U \cap N)$ is a connected neighbourhood of $\phi(0, 0)$. Now $(U \cap N)\backslash\{(0, 0)\}$ has at least *four* connected components and so, therefore, does its homeomorphic image $\phi(U \cap N)\backslash\{\phi(0, 0)\}$. But $\phi(U \cap N)$ is a connected open subset of $\mathbf{R}$ and so is an open interval. Removing one point from an open interval leaves a space with *two* connected components. Contradiction. Therefore, N is not a submanifold of $\mathbf{R}^2$.

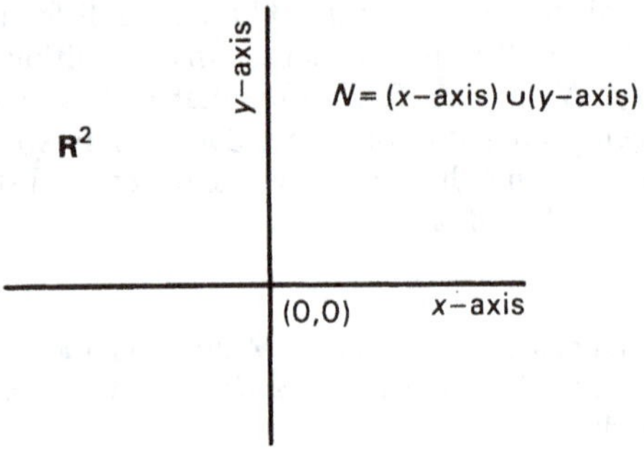

FIG. 4.18

2. We take $M = \mathbf{R}^2$ and N to be the union of the two rays $y = ax, y = bx$, where $b > a \geqslant 0$ and $x \geqslant 0$ (Fig. 4.19). We claim that we cannot find a chart $(U, \phi) \in \mathscr{A}_2$, containing $(0, 0)$, which maps $U \cap N$ homeomorphically onto an open subset of $\mathbf{R}$. Suppose the contrary. Then $\phi(U \cap N) \subset \mathbf{R} \times \{0\} \subset \mathbf{R}^2$. Restricting ϕ to the lines $y = ax$ and $y = bx$, it follows that for all $x \in \mathbf{R}$ we have

$$D\phi_{(0,0)}(x, ax) \in \mathbf{R} \times \{0\} \qquad \text{and} \qquad D\phi_{(0,0)}(x, bx) \in \mathbf{R} \times \{0\}.$$

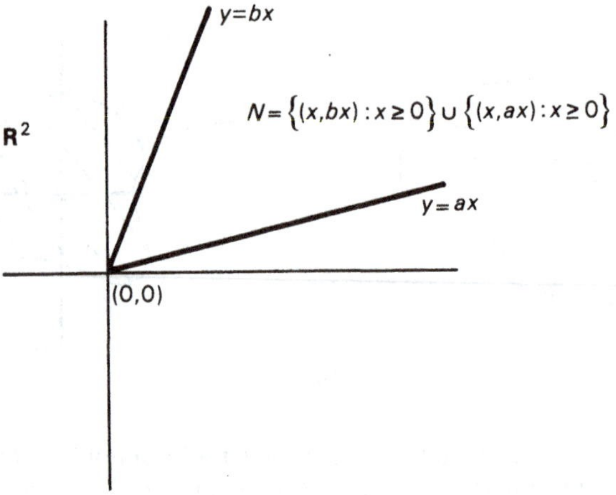

FIG. 4.19

If we let $[D\phi_{(0,0)}] = [\alpha_{ij}]$, we therefore have, for all $x \in \mathbf{R}$,

$$\alpha_{21}x + \alpha_{22}ax = 0 \qquad \text{and} \qquad \alpha_{21}x + \alpha_{22}bx = 0.$$

Since $a \neq b$, it follows that $\alpha_{21} = \alpha_{22} = 0$ and so $D\phi_{(0,0)}$ is not a linear isomorphism. Hence ϕ cannot be a diffeomorphism and this contradicts the fact that we assumed (U, ϕ) was a chart. Consequently N is not a submanifold of $\mathbf{R}^2$.

The reader should notice that whilst the subset of $\mathbf{R}^2$ in the first example cannot be given the structure of a differential manifold, the subset in the second example most certainly can, since it is homeomorphic to $\mathbf{R}$ (Proposition 4.5.3). These examples therefore emphasize the fact that a submanifold is more than just a manifold.

In the next section we shall develop powerful methods for constructing submanifolds. Here we shall content ourselves with giving just a couple of elementary examples.

Examples

1. $\mathbf{R}^n$ is a submanifold of $\mathbf{R}^{n+m}$. This is trivial: we can use the single chart $(\mathbf{R}^{n+m}, I)$ which clearly works for every point of $\mathbf{R}^n$ in Definition 4.6.1.
2. Let M and N be differential manifolds with atlases $\mathscr{M}$ and $\mathscr{N}$ respectively. Fix some point $m \in M$. We claim that $\{m\} \times N$ is a submanifold of the product manifold $M \times N$. Indeed, if we let $\mathscr{M}_m$ denote the set of charts in $\mathscr{M}$ which contain the point m, it is easy to see that any chart in $\mathscr{M}_m \times \mathscr{N}$ 'works' in Definition 4.6.1.

We now wish to show that submanifolds are the class of subspaces that solve the problems posed at the beginning of this section.

Proposition 4.6.2

Let N be a submanifold of the differential manifold M, and suppose that M and N are modelled on $\mathbf{R}^m$ and $\mathbf{R}^n$ respectively. Then N has the structure of a differential manifold modelled on $\mathbf{R}^n$.

Proof. Let us denote the complete atlas of M by $\mathscr{A}$. We define a subset $\mathscr{A}^*$ of $\mathscr{A}$ as follows: $(U, \phi) \in \mathscr{A}^*$ if and only if $U \cap N \neq \emptyset$ and (U, ϕ) satisfies the submanifold conditions of Definition 4.6.1. Since $\mathscr{A}$ is assumed to be complete and N is a submanifold, we certainly have

$$\bigcup_{(U,\phi) \in \mathscr{A}^*} (U \cap N) = N.$$

We claim that

$$\mathscr{A}_N = \{(U \cap N, \phi \mid (U \cap N)) : (U, \phi) \in \mathscr{A}^*\}$$

is an atlas for N. Let $(U, \phi), (V, \psi) \in \mathscr{A}^*$ and set $\tilde{\phi} = \phi \mid (U \cap N)$ and $\tilde{\psi} = \psi \mid (V \cap N)$. Let

$$P : \mathbf{R}^n \times \mathbf{R}^{m-n} \longrightarrow \mathbf{R}^n \qquad \text{and} \qquad J : \mathbf{R}^n \longrightarrow \mathbf{R}^n \times \mathbf{R}^{m-n}$$

respectively denote the projection and inclusion maps on the first factor. Now

$$\tilde{\psi}\tilde{\phi}^{-1} : \phi(U \cap V \cap N) \longrightarrow \psi(U \cap V \cap N)$$

may be written in terms of ψ, ϕ, P and J as

$$\tilde{\psi}\tilde{\phi}^{-1} = P\psi\phi^{-1}J.$$

This composition of maps is described in Fig. 4.20.

To prove the proposition it is clearly sufficient to show that

$$\tilde{\psi}\tilde{\phi}^{-1} : \phi(U \cap V \cap N) \longrightarrow \psi(U \cap V \cap N)$$

is C^∞. But since $\hat{\psi}\hat{\phi}^{-1} = P\psi\phi^{-1}J$, this follows immediately from the composite-mapping theorem. ■

Remark. In the sequel we shall sometimes refer to the atlas $\mathscr{A}_N$ constructed on N as the *induced atlas* on N or as the *atlas induced by $\mathscr{A}$ on N.*

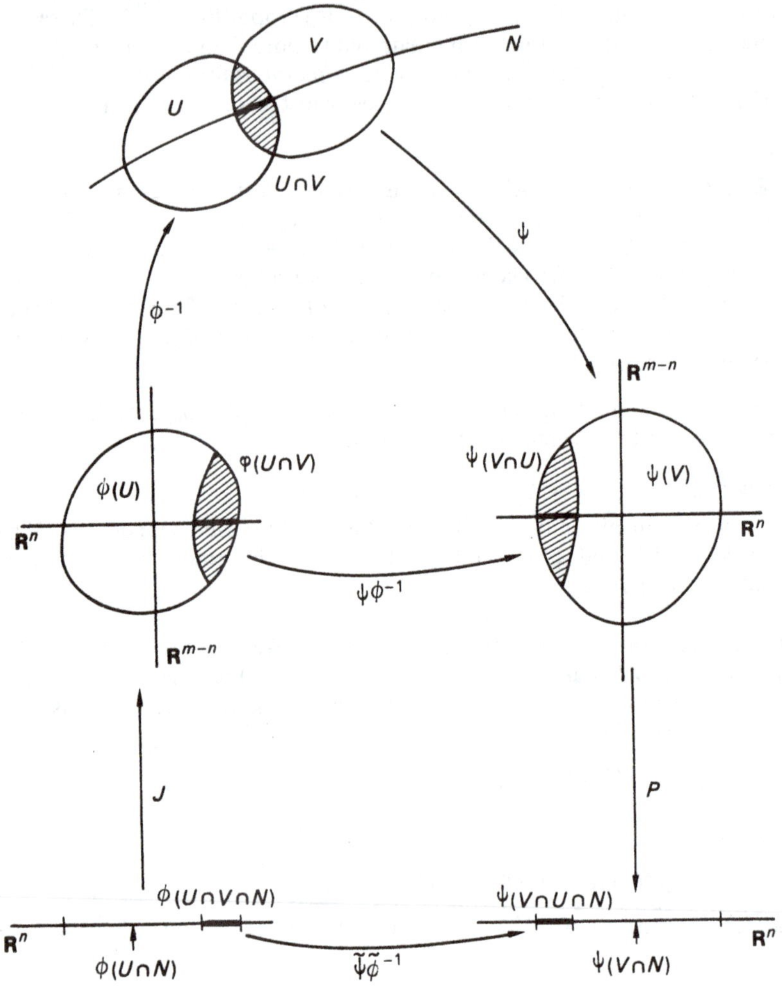

FIG. 4.20

Proposition 4.6.3

Let N be a submanifold of the differential manifold M. Then every differentiable function on M restricts to a differentiable function on N.

Proof. Let M have complete atlas $\mathscr{A}$ and let N have induced atlas $\mathscr{A}_N$. Suppose that $f: M \longrightarrow \mathbf{R}$ is differentiable. We must show that $f \mid N: N \longrightarrow \mathbf{R}$ is differentiable with respect to $\mathscr{A}_N$. Now given any chart $(V, \psi) \in \mathscr{A}_N$, we may find a chart $(U, \phi) \in \mathscr{A}$ such that $V = U \cap N$ and $\psi = \phi \mid (U \cap N)$. Hence

$$f\psi^{-1}: \psi(V) \subset \mathbf{R}^n \longrightarrow \mathbf{R}$$

may be written as the composition $f\phi^{-1}J$, where $J: \mathbf{R}^n \longrightarrow \mathbf{R}^n \times \mathbf{R}^{m-n}$ is the inclusion map. Hence, by the composite-mapping theorem, $f\psi^{-1}$ is differentiable and so, by definition, $f \mid N$ is differentiable. ■

Exercises

1. Show that the set of points $\{(x, x^2), (x, -x^2) : x \geqslant 0\}$ is not a submanifold of $\mathbf{R}^2$.
2. Let N_1 and N_2 be submanifolds of M_1 and M_2 respectively. Prove that $N_1 \times N_2$ is a submanifold of $M_1 \times M_2$.
3. Show that every open subset of a differential manifold M is a submanifold of M. Need the *boundary* of an open subset of M be a submanifold of M?

4.7 Submanifolds of $\mathbf{R}^m$

In this section we shall make a study of the most important class of submanifolds: The submanifolds of $\mathbf{R}^m$. Since this class of submanifolds is so important we shall write down their definition explicitly.

Definition 4.7.1

Let N be a subset of $\mathbf{R}^m$. Then N is a submanifold of $\mathbf{R}^m$, modelled on $\mathbf{R}^n$, if for all $x \in N$, there exists $(U, \phi) \in \mathscr{A}_m$, containing x, such that $\phi : U \longrightarrow \mathbf{R}^n \times \mathbf{R}^{m-n}$ and ϕ maps $U \cap N$ homeomorphically onto an open subset of $\mathbf{R}^n \times \{0\}$.

We have the picture shown in Fig. 4.21. Referring to Section 3.3, we see that this picture is typical of those which occur in that section, where we 'straighten out' some subset of a vector space into an open subset of another vector space. We shall now exploit the results of Section 3.3 to construct submanifolds of $\mathbf{R}^m$.

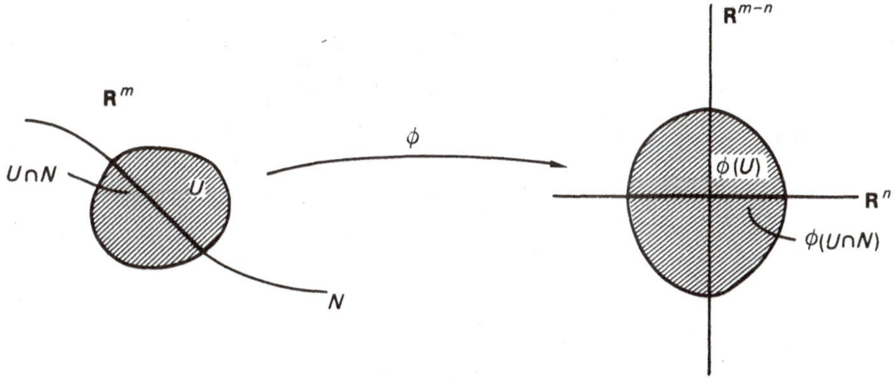

FIG. 4.21

Examples

1. Let $f : \mathbf{R}^p \longrightarrow \mathbf{R}^q$ be C^∞. Then we claim that the graph of f, $\{(x, f(x)) : x \in \mathbf{R}^p\}$, is a submanifold of $\mathbf{R}^p \times \mathbf{R}^q$. To prove this we shall consider the map $F : \mathbf{R}^p \times \mathbf{R}^q \longrightarrow \mathbf{R}^q$ defined by

$$F(x, y) = y - f(x), (x, y) \in \mathbf{R}^p \times \mathbf{R}^q$$

(See Fig. 4.22). Clearly $F^{-1}(0) = \text{graph } (f)$. We wish to apply the full force of the implicit-function theorem to F at points on the graph of f. Now

$$D_2F_{(x,y)} = I_{\mathbf{R}^q} \in GL(q; \mathbf{R}),$$

and so the conditions of Theorem 3.3.1 hold for the function F. Hence, given $x \in \mathbf{R}^p$, we may find open neighbourhoods W of $(x, f(x))$ and $V_1 \times V_2$ of $(x, 0)$ together with a map $h \in \mathrm{Diff}^\infty(V_1 \times V_2, W)$ such that $Fh = p_2$, where $p_2 \colon \mathbf{R}^p \times \mathbf{R}^q \longrightarrow \mathbf{R}^q$ is the projection map onto the second factor. We consider the chart $(W, h^{-1}) \in \mathscr{A}_m$. By construction, $h^{-1}(W \cap F^{-1}(0)) = V_1 \times \{0\} \subset \mathbf{R}^p \times \mathbf{R}^q$. Since $F^{-1}(0) = \mathrm{graph}\,(f)$, we have constructed at each point of graph (f) a chart satisfying the conditions of Definition 4.7.1. Hence graph(f) is a submanifold of $\mathbf{R}^p \times \mathbf{R}^q$.

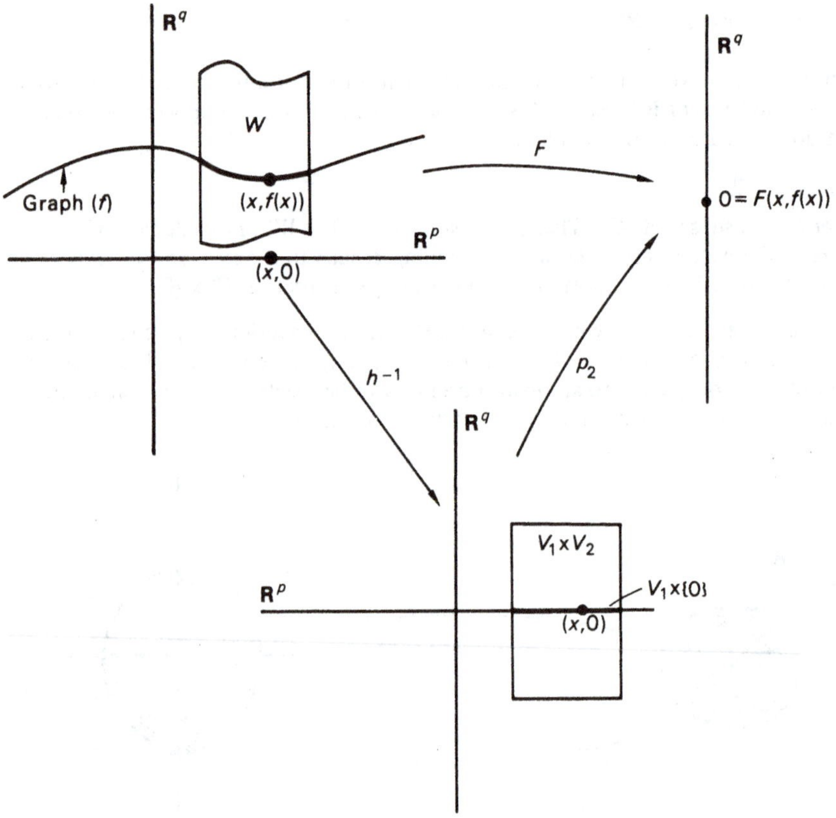

FIG. 4.22

Notice that the graph of f trivially has the structure of a differential manifold, since graph (f) is *homeomorphic* to $\mathbf{R}^p$ (Proposition 4.5.3). Indeed the graph of any continuous function from $\mathbf{R}^p$ to $\mathbf{R}^q$ may be given the structure of a differential manifold modelled on $\mathbf{R}^p$. However, we need to know that f is C^∞ before we can claim that the graph of f is a *submanifold* of $\mathbf{R}^p \times \mathbf{R}^q$.

2. Let us consider the sphere of radius r in $\mathbf{R}^{n+1}$

$$S_r^n = \{(x_1, \ldots, x_{n+1}) \colon x_1^2 + \cdots + x_{n+1}^2 = r^2\}, \qquad r > 0.$$

S_r^n is the zero set of the following function defined on $\mathbf{R}^{n+1}$:

$$f(x_1, \ldots, x_{n+1}) = x_1^2 + \cdots + x_{n+1}^2 - r^2, \qquad (x_1, \ldots, x_{n+1}) \in \mathbf{R}^{n+1}.$$

f is C^∞ and we shall apply Theorem 3.3.2 to f at points on S_r^n. First note that

$$Df_x \neq 0 \qquad \text{for all } x \in \mathbf{R}^{n+1}\backslash\{0\}.$$

Hence Df_x is surjective for every $x \in S_r^n$, and we may apply Theorem 3.3.2 to find open neighbourhoods W of $x \in S_r^n$ in $\mathbf{R}^{n+1}$ and $V_1 \times V_2$ of $(0,0) \in \mathbf{R}^n \times \mathbf{R}$, together with a map $h \in \text{Diff}^\infty(V_1 \times V_2, W)$, such that

$$fh(x,y) = y \qquad \text{for all } (x,y) \in V_1 \times V_2.$$

By construction, $(W, h^{-1}) \in \mathscr{A}_{n+1}$ and $h^{-1}(S_r^n \cap W) = V_1 \times \{0\}$. Hence (W, h^{-1}) is a submanifold chart for S_r^n. Since x was an arbitrary point of S_r^n it follows that S_r^n is a submanifold of $\mathbf{R}^{n+1}$ modelled on $\mathbf{R}^n$.

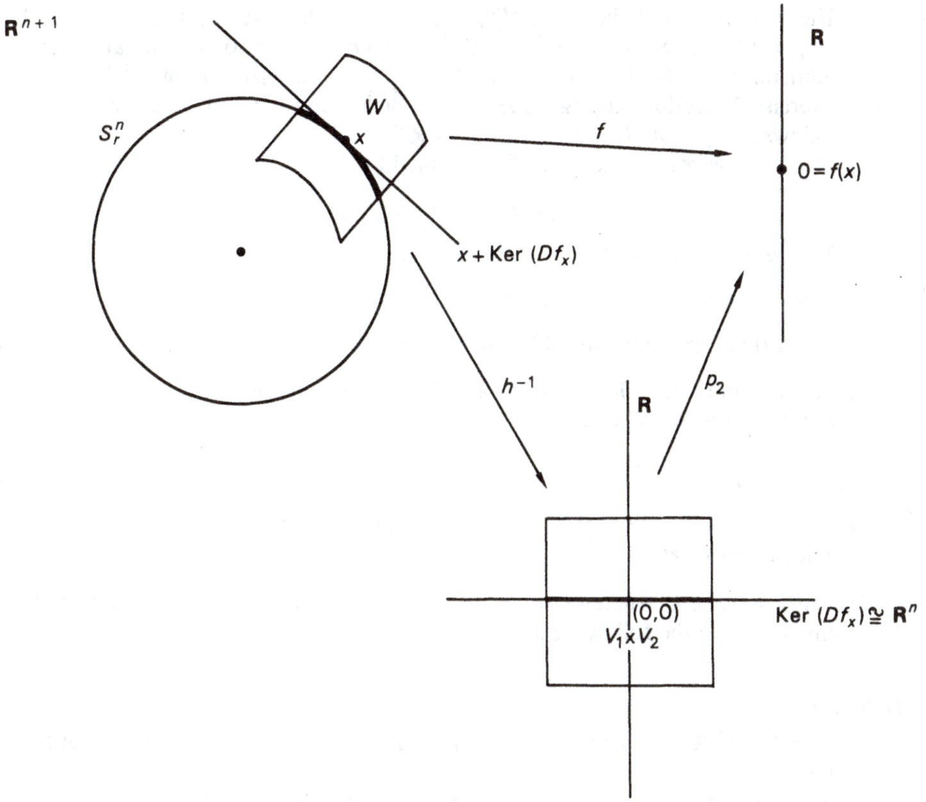

FIG. 4.23

Theorem 4.7.2

Let $f: \mathbf{R}^m \longrightarrow \mathbf{R}$ be C^∞ and let $c \in \mathbf{R}$. Suppose that $Df_x \neq 0$ for all $x \in f^{-1}(c)$. Then $f^{-1}(c)$ is a submanifold of $\mathbf{R}^m$ modelled on $\mathbf{R}^{m-1}$.

Proof. The proof, using Theorem 3.3.2, is exactly the same as the argument used in Example 2 above. ■

Theorem 4.7.2 allows us to construct many submanifolds of $\mathbf{R}^m$, modelled on $\mathbf{R}^{m-1}$.

Examples

1. Let p, q, m be positive integers satisfying the relations $0 < p + q \leqslant m + 1$. Then the subset $H^m_{p,q}$ of $\mathbf{R}^{m+1}$ defined by
$$H^m_{p,q} = \{(x_1, \ldots, x_{m+1}) : x_1^2 + \cdots + x_p^2 - x_{p+1}^2 - \cdots - x_q^2 = 1\}$$
is a submanifold of $\mathbf{R}^{m+1}$ modelled on $\mathbf{R}^m$ ($H^m_{p,q}$ is a generalized hyperboloid). We show this first in the non-degenerate case $p + q = m + 1$ by considering the function defined on $\mathbf{R}^{m+1}$ by
$$f(x_1, \ldots, x_{m+1}) = x_1^2 + \cdots + x_p^2 - x_{p+1}^2 - \cdots - x_{m+1}^2 - 1.$$
Then $H^m_{p,q}$ is the zero set of f and $Df_x \neq 0$ for $x \in H^m_{p,q}$. Hence we may apply Theorem 4.7.2 to deduce that $H^m_{p,q}$ is a submanifold of $\mathbf{R}^{m+1}$. If $p + q < m + 1$, set $p + q = r$. Then $H^m_{p,q} = H^{r-1}_{p,q} \times \mathbf{R}^{m+1-r}$. Since $H^{r-1}_{p,q}$ is non-degenerate, it is a submanifold of $\mathbf{R}^r$. It follows easily that $H^m_{p,q}$ is a submanifold of $\mathbf{R}^{m+1}$ (cf. Exercise 2, Section 4.6). Since $H^{r-1}_{p,q}$ is modelled on $\mathbf{R}^{r-1}$ and $p + q = r \geqslant 1$, it follows easily that $H^m_{p,q}$ is modelled on $\mathbf{R}^m$.
2. Consider the paraboloid P in $\mathbf{R}^{m+1}$ defined by
$$P = \{(x_1, \ldots, x_{m+1}) : x_1^2 + \cdots + x_m^2 = x_{m+1}\}.$$
Theorem 4.7.2 applied to the function defined by
$$f(x_1, \ldots, x_{m+1}) = x_1^2 + \cdots + x_m^2 - x_{m+1}$$
shows that P is a submanifold of $\mathbf{R}^{m+1}$ modelled on $\mathbf{R}^m$.

Those submanifolds of $\mathbf{R}^{m+1}$ which are of dimension m are of particular interest and we give them a special name.

Definition 4.7.3

Let N be a submanifold of $\mathbf{R}^{m+1}$ of dimension m. Then N is called a *surface* (sometimes a *hypersurface*).

The surfaces we constructed in the above examples all arose from the zero sets of functions. We make the following definition.

Definition 4.7.4

Let $f : \mathbf{R}^m \longrightarrow \mathbf{R}$ be a function. Then, if $c \in \mathbf{R}$, $f^{-1}(c)$ is called the *level set* of f through the point c.

If, in addition, f is C^∞ and $f^{-1}(c)$ is a submanifold of $\mathbf{R}^{m+1}$ of dimension m, we shall call $f^{-1}(c)$ a *level surface*.

We have a similar definition of level sets and level surfaces for a function $f : M \longrightarrow \mathbf{R}$ where M is a differential manifold modelled on $\mathbf{R}^{m+1}$.

From Theorem 4.7.2 follows immediately

Proposition 4.7.5

Let $f : \mathbf{R}^{m+1} \longrightarrow \mathbf{R}$ be a C^∞ function and suppose that for some $c \in \mathbf{R}$, $Df_x \neq 0$ for all $x \in f^{-1}(c)$. Then $f^{-1}(c)$ is a level surface.

Remark. The conditions of the statement of Proposition 4.7.5 are not *necessary* conditions for $f^{-1}(c)$ to be a level surface. They are, however, the conditions that nearly always arise in practice.

Theorem 4.7.6

Let $f: \mathbf{R}^m \longrightarrow \mathbf{R}^n$ be C^∞ and let $c \in \mathbf{R}$. Suppose that rank $(Df_x) = q$ for all $x \in f^{-1}(c)$. Then $f^{-1}(c)$ is a submanifold of $\mathbf{R}^m$ modelled on $\mathbf{R}^{m-q}$.

Proof. The result follows from the rank theorem in the same way that Theorem 4.7.2 followed from the implicit-function theorem. ■

Theorem 4.7.6 is of most interest when $q = n$. That is, when Df_x is surjective for all $x \in f^{-1}(c)$.

Example. The ellipsoid E, defined by the equation $x^2/a^2 + y^2/b^2 + z^2/c^2 = 1$, is a surface in $\mathbf{R}^3$ as is the cylinder $S^1 \times \mathbf{R}$, defined by the equation $x^2 + y^2 = 1$. We claim that if $a, b \neq 1$, the intersection of the ellipsoid and the cylinder is either empty or a submanifold of $\mathbf{R}^3$ of dimension 1:

The intersection $E \cap (S^1 \times \mathbf{R})$ is clearly the zero set of the function

$$F = (f, g) : \mathbf{R}^3 \longrightarrow \mathbf{R}^2,$$

where f and g are defined by

$$f(x, y, z) = x^2/a^2 + y^2/b^2 + z^2/c^2 - 1, \qquad g(x, y, z) = x^2 + y^2 - 1.$$

Computing derivatives, we find that

$$[DF_{(x,y,z)}] = \begin{bmatrix} 2x/a^2 & 2y/b^2 & 2z/c^2 \\ 2x & 2y & 0 \end{bmatrix}.$$

Hence the rank of $DF_{(x,y,z)}$ is 2 provided that either

$$4xy/a^2 - 4xy/b^2 \neq 0 \qquad \text{or} \qquad 4zx/c^2 \neq 0 \qquad \text{or} \qquad 4zy/c^2 \neq 0.$$

Let us suppose that $E \cap (S^1 \times \mathbf{R}) \neq \emptyset$. Then the reader may verify that if $(x, y, z) \in E \cap (S^1 \times \mathbf{R})$, then the assumption that $a, b \neq 1$ implies that at least one of the above expressions is non-zero. Hence the rank of DF is constant and equal to 2 on the intersection and we may apply Theorem 4.7.6 to deduce that the intersection is a submanifold of dimension 1. In the case when a or b equals 1, the cylinder touches the ellipsoid at the end of one of its principal axes, and the intersection degenerates to two isolated points, that is, to a submanifold of $\mathbf{R}^3$ of dimension zero.

Remark. The reader should not infer from the above example that the intersection of two submanifolds is always a submanifold. We refer the reader to the exercises at the end of this section for counterexamples.

So far in this section we have only proved that the *inverse* images of certain functions have the structure of a submanifold. In particular, we have made no use of the results in Section 3.3 which have to do with 'straightening out' the *image* of a function. For example, we have not given conditions on a curve for it to be a submanifold. The problem of finding sufficient conditions on a function in order that its image is a submanifold is a little more difficult than the problems we have been considering so far. Suppose, for example, that $f: \mathbf{R} \longrightarrow \mathbf{R}^2$ is a C^∞ function which is injective and has nowhere-vanishing derivative (see Fig. 4.24).

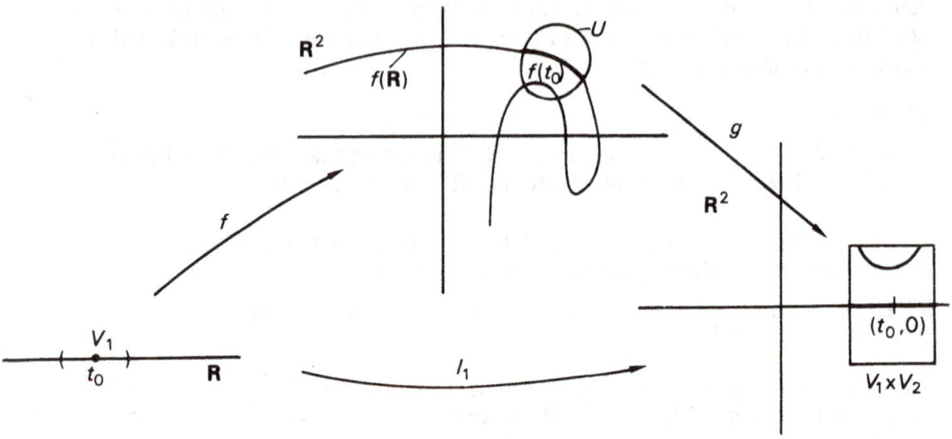

FIG. 4.24

Since $Df_t \in L(\mathbf{R}, \mathbf{R}^2)$ is obviously injective for all $t \in \mathbf{R}$, we can apply Theorem 3.3.4 to f: At each point $t_0 \in \mathbf{R}$ we may find an open neighbourhood U of $f(t_0)$ and an open neighbourhood $V_1 \times V_2$ of $(t_0, 0)$ in $\mathbf{R}^2$, together with a map

$$g \in \mathrm{Diff}^\infty(U, V_1 \times V_2),$$

such that

$$gf(t) = (t, 0) \qquad \text{for all } t \in V_1.$$

It would therefore appear that (U, g) is a good candidate for a chart in $\mathscr{A}_2$, which is also a submanifold chart for $f(\mathbf{R})$ at $f(t_0)$. However, all that our construction shows is that $f(V_1) \cap U$ is mapped by g homeomorphically onto V_1. It does *not* prove that $U \cap f(\mathbf{R})$ is mapped homeomorphically onto V_1, which is what we need if we are to prove that $f(\mathbf{R})$ (as opposed to $f(V_1)$) is a submanifold of $\mathbf{R}^2$.

It does not follow from our assumptions on f that there need *exist* an open subset U of $\mathbf{R}^2$ such that $U \cap f(\mathbf{R}) = U \cap f(V_1)$. The problem is that $f(\mathbf{R})$ can loop back and meet U infinitely many times in points that are not images of points in V_1. Let us look at a more explicit example to see the type of pathology that can occur.

Let us consider the graph of the function $f(x) = \sin(1/x)$ for $x \in (0, 1)$. Since f is C^∞, it follows by a previous example that the graph of f is a submanifold of $\mathbf{R}^2$. This graph may be represented as the image of the curve $\omega : (0, 1) \longrightarrow \mathbf{R}^2$ defined by

$$\omega(x) = (x, \sin(1/x)).$$

Notice that ω is injective, as is its derivative. We shall now extend ω to a curve $\gamma : (0, \infty) \longrightarrow \mathbf{R}^2$. Referring to Fig. 4.25, we construct a C^∞ function $\tau : (\frac{1}{2}, \infty) \longrightarrow \mathbf{R}^2$ satisfying the conditions

$$\tau \mid (\tfrac{1}{2}, 1) = \omega \mid (\tfrac{1}{2}, 1)$$

$$\tau(1, \infty) \cap \omega(0, 1) = \emptyset$$

$$\tau(x) = (0, x - 4), \qquad x \in [2, \infty).$$

We shall also suppose that τ is chosen to be injective with nowhere-vanishing derivative. (An explicit construction for τ can be given using the 'bump' functions of Section 4.3.)

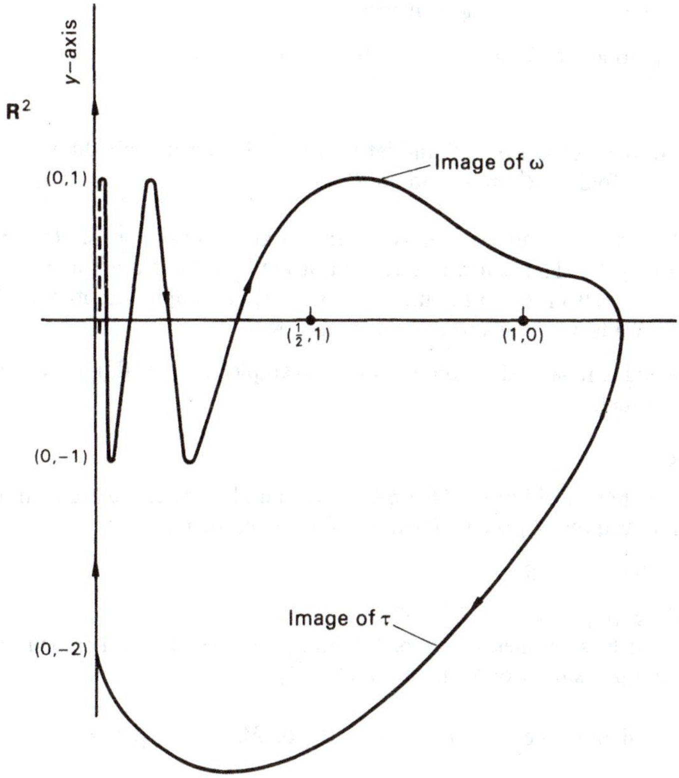

FIG. 4.25

We define $\gamma : (0, \infty) \longrightarrow \mathbf{R}^2$ by setting

$$\gamma | (0, 1) = \omega, \qquad \gamma | (\tfrac{1}{2}, \infty) = \tau.$$

Then γ is a well-defined C^∞ map which is injective with nowhere-vanishing derivative. The image of γ contains the image of ω together with that part of the y-axis above the point $(0, -2)$.

$\gamma(\mathbf{R})$ is obviously not a submanifold of $\mathbf{R}^2$ since $\gamma([3, 5]) \subset \overline{\gamma((0, 1))}$ and so it is impossible to find submanifold charts for points in $\gamma([3, 5])$.

The problem arises because the topology induced on the image of γ by $\mathbf{R}^2$ does not match up with the topology induced on the curve by γ. Otherwise said: $\gamma : (0, \infty) \longrightarrow \gamma(\mathbf{R})$ is not a homeomorphism if we give $\gamma(\mathbf{R})$ the topology induced by $\mathbf{R}^2$. This, of course, is just another example of the fact that the concept of a submanifold is a relative rather than an intrinsic property.

Motivated by the above we make the following definition.

Definition 4.7.7

Let W be an open subset of $\mathbf{R}^m$ and let $f : W \longrightarrow \mathbf{R}^n$ be C^∞. Suppose that f satisfies the following conditions:

1. Df_x is injective for all $x \in W$.
2. $f : W \longrightarrow f(W)$ is a homeomorphism, where $f(W)$ is given the topology induced by $\mathbf{R}^n$.

Then f is called an *embedding* of W into $\mathbf{R}^n$.

We have the following basic theorem about embeddings.

Theorem 4.7.8

Let W be an open subset of $\mathbf{R}^m$ and let $f: W \longrightarrow \mathbf{R}^n$ be an embedding. Then $f(W)$ is a submanifold of $\mathbf{R}^n$ modelled on $\mathbf{R}^{n-m}$.

Proof. Let V be an open subset of W. Then there exists an open subset U of $\mathbf{R}^n$ such that $U \cap f(W) = f(V)$. This is just a statement of the fact that f is a homeomorphism onto $f(W)$, where $f(W)$ is given the induced topology. We apply Theorem 3.3.4 at every point of $f(W)$. We leave details to the exercises. ■

The following lemma, apart from providing examples of embeddings, will be useful in the sequel.

Lemma 4.7.9

Let M be a submanifold of $\mathbf{R}^n$ of dimension m and let $\mathscr{B}$ denote the completion of the atlas induced by $\mathscr{A}_n$ on M. Then, if $(U, \phi) \in \mathscr{B}$, the map

$$\phi^{-1}: \phi(U) \longrightarrow \mathbf{R}^n$$

is an embedding. In particular, ϕ^{-1} is C^∞.

Conversely, if W is an open subset of $\mathbf{R}^m$ and $\gamma: W \longrightarrow \mathbf{R}^n$ is an embedding such that $\gamma(W)$ is an open subset of M, then $(\gamma(W), \gamma^{-1}) \in \mathscr{B}$.

Proof. Let $\mathscr{B}_M$ denote the atlas induced by $\mathscr{A}_n$ on M. Then $\hat{\mathscr{B}}_M = \mathscr{B}$. If $(V, \psi) \in \mathscr{B}_M$ we have

$$\phi^{-1} = \psi^{-1}(\psi\phi^{-1}) \qquad \text{on } \phi(V \cap U).$$

It follows by the composite-mapping theorem that to prove the lemma it suffices to verify it for charts in $\mathscr{B}_M$. But if $(U, \phi) \in \mathscr{B}_M$, $\exists (\tilde{U}, \tilde{\phi}) \in \mathscr{A}_n$ such that $U = \tilde{U} \cap M$ and $\phi = \tilde{\phi} \mid (\tilde{U} \cap M)$. Hence,

$$\phi^{-1} = \tilde{\phi}^{-1} J,$$

where $J: \mathbf{R}^m \longrightarrow \mathbf{R}^m \times \mathbf{R}^{n-m}$ is the inclusion map. Hence, by the composite-mapping theorem, ϕ^{-1} is C^∞. Since we already know that ϕ is a homeomorphism, this proves the first part of the lemma. The proof of the converse is similar and we leave details to the reader. ■

Example. Let $B_1(0) \subset \mathbf{R}^2$ be the unit open disc relative to the standard inner product on $\mathbf{R}^2$. We define $\phi: B_1(0) \longrightarrow \mathbf{R}^3$ by

$$\phi(x, y) = (x, y, \sqrt{(1 - x^2 - y^2)}), \qquad (x, y) \in B_1(0).$$

Clearly $\phi(B_1(0)) \subset S^2$ and ϕ maps bijectively onto the upper hemisphere

$$H_z^+ = \{(x, y, z) \in S^2 : z > 0\}.$$

ϕ is easily verified to be C^∞ and an embedding. It follows from Lemma 4.7.9 that (H_z^+, ϕ^{-1}) belongs to the completion of the atlas induced on S^2 by $\mathscr{A}_3$. A similar use of Lemma 4.7.9 enables us to check that the other atlases we have defined on S^2 and S^1 all have the same completion.

As an application of the ideas of the previous paragraphs, we shall show that the length of a smooth arc in $\mathbf{R}^n$ is independent of its parametrization (cf. Section 2.13).

First, recall that a *smooth curve* in $\mathbf{R}^n$ is defined by a C^1 map $\phi : [a, b] \longrightarrow \mathbf{R}^n$ such that $\phi' \neq 0$ on $[a, b]$. The map ϕ is said to define a *smooth arc* if, in addition, ϕ is injective. In what follows we shall always assume that ϕ is C^∞, although the main result, Corollary 4.7.11.1, is still true if ϕ is C^1.

Proposition 4.7.10

Let $\phi : [a, b] \longrightarrow \mathbf{R}^n$ be a smooth arc. Then $\phi((a, b))$ is a submanifold of $\mathbf{R}^n$ with boundary $\{\phi(a), \phi(b)\}$.

Proof. Let $t_0 \in (a, b)$. By Theorem 3.3.4 we may find open neighbourhoods $V_1 \times V_2$ of $(t_0, 0)$ in $\mathbf{R} \times \mathbf{R}^{n-1}$ and W of $\phi(t_0)$ in $\mathbf{R}^n$ together with a map $h \in \mathrm{Diff}^\infty(W, V_1 \times V_2)$ such that

$$h\phi(t) = (t, 0) \qquad \text{for all } t \in V_1.$$

(See Fig. 4.26.) Without loss of generality we may suppose that

$$V_1 = (t_0 - \epsilon, t_0 + \epsilon), \qquad V_2 = B_r(0) \subset \mathbf{R}^{n-1},$$

for strictly positive numbers ϵ and r.

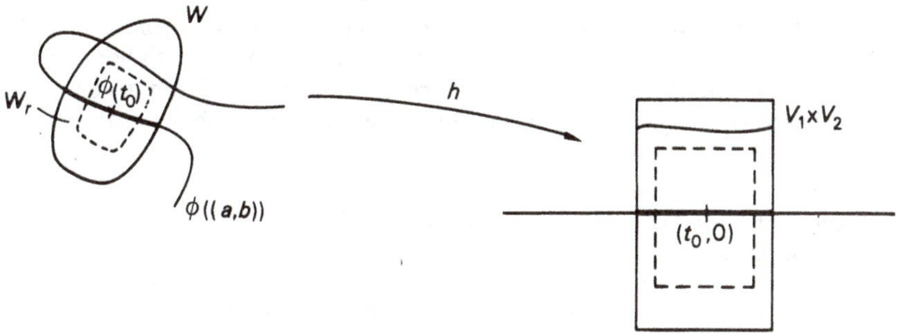

FIG. 4.26

Let us set

$$W_r = h^{-1}((t_0 - \epsilon/2, t_0 + \epsilon/2) \times B_r(0)).$$

We claim that by choosing r small enough we can ensure that

$$W_r \cap \phi((a, b)) = h^{-1}((t_0 - \epsilon/2, t_0 + \epsilon/2) \times \{0\}) \cap W_r. \tag{A}$$

Since the right-hand side of the above expression is just $\phi((t_0 - \epsilon/2, t_0 + \epsilon/2))$, it then follows that (W_r, h) is a submanifold chart for $\phi((a, b))$ containing $\phi(t_0)$.

Suppose that we cannot choose r sufficiently small to achieve (A). It will then follow that for $n = 1, 2, \ldots$ we may find a point $t_n \in (a, t_0 - \epsilon] \cup [t_0 + \epsilon, b)$ such that

$$h\phi(t_n) \in (t_0 - \epsilon/2, t_0 + \epsilon/2) \times B_{1/n}(0).$$

Since $[a, b]$ is compact, the sequence $(t_n)_{n=1}^\infty$ has a convergent subsequence, say $(s_n)_{n=1}^\infty$. Let $\lim_{n\to\infty} s_n = u$. Then $u \in [a, t_0 - \epsilon] \cup [t_0 + \epsilon, b]$. By construction

$$\phi(u) = \lim_{n\to\infty} h\phi(s_n) \in [t_0 - \epsilon/2, t_0 + \epsilon/2] \times \{0\}.$$

But this contradicts the injectivity of ϕ on $[a, b]$. Hence we can choose r so that (A) is satisfied. Since t_0 was an arbitrary point of (a, b), it follows that $\phi((a, b))$ is a submanifold of $\mathbf{R}^n$. ■

Proposition 4.7.11

Let $\phi_1 : [a, b] \longrightarrow \mathbf{R}^n$ and $\phi_2 : [c, d] \longrightarrow \mathbf{R}^n$ be two parametrizations of the smooth arc C. Then the map

$$\phi_2^{-1}\phi_1 : (a, b) \longrightarrow (c, d)$$

is a C^∞ diffeomorphism.

Proof. Let us set $C' = \phi_1((a, b)) = \phi_2((c, d))$. In the previous proposition we showed that C' was a submanifold of $\mathbf{R}^n$. By the second part of Lemma 4.7.9 it therefore follows that (C', ϕ_1^{-1}) and (C', ϕ_2^{-1}) are charts for C'. Hence $\phi_2^{-1}\phi_1$ is a C^∞ diffeomorphism. ■

Corollary 4.7.11.1

The length of a smooth arc is independent of its parametrization.

Proof. Follows from the above together with Proposition 2.13.5. ■

Remark. We could have proved the following stronger version of Corollary 4.7.11.1: Let C be a relatively compact one-dimensional submanifold of $\mathbf{R}^n$ with boundary consisting of two points. Then there exists a C^∞ parametrization of C and hence, by the above corollary, the length of C is well defined. We leave this problem to the exercises.

Exercises

1. Show that the set of solutions of the equation $x^2 + y^3 + z^4 - z^2 = 1$ is a submanifold of $\mathbf{R}^3$.
2. For what values of a is the intersection of the sphere $x^2 + y^2 + z^2 = 1$ with the parabola $1 + x^2 + y^2 = az$ a submanifold of $\mathbf{R}^3$ of dimension 1?
3. Is the set $\{(x, x \sin(1/x)) : x \in \mathbf{R}\}$ a submanifold of $\mathbf{R}^2$? What about the sets

$$\{(x, e^{-1/x} \sin(1/x)) : x > 0\} \cup \{(x, 0) : x \leqslant 0\},$$

$$\{(x, e^{-1/x^2} \sin(1/x)) : x \in \mathbf{R}\}?$$

4. Show that the two-dimensional torus T^2 can be mapped homeomorphically onto a two-dimensional submanifold of $\mathbf{R}^3$. (*Hint*: Use parametric equations.)
5. Show that the cone $z^2 = x^2 + y^2$ is not a submanifold of $\mathbf{R}^3$.
6. Show that Theorems 4.7.2 and 4.7.6 are valid if f is defined only on an open subset of $\mathbf{R}^m$.
7. Let C be a relatively compact submanifold of $\mathbf{R}^n$ of dimension 1 with boundary consisting of two points. Show that C has a C^∞ parametrization and hence that the length of C is intrinsically defined.

4.8 Differentiable maps: critical points

In this section we shall extend our definition of differentiable functions to include maps between differential manifolds.

Definition 4.8.1

Let M and N be differential manifolds with atlases $\mathscr{A}$ and $\mathscr{B}$ respectively. Suppose that $f: M \longrightarrow N$ is continuous. We shall say that f is differentiable on M if, for all charts $(U, \phi) \in \mathscr{A}$, $(V, \psi) \in \mathscr{B}$ such that $f^{-1}(V) \cap U \neq \emptyset$, the composition

$$\psi f \phi^{-1}: \phi(f^{-1}(V) \cap U) \longrightarrow \mathbf{R}^n$$

is differentiable (here $n = \dim(N)$).

We may similarly define a C^r map, $1 \leqslant r \leqslant \infty$.

Remark. If $\mathscr{A}$ and $\mathscr{B}$ are atlases on M and N respectively, then $f: M \longrightarrow N$ is differentiable with respect to $\mathscr{A}$ and $\mathscr{B}$ if and only if f is differentiable with respect to their completions $\hat{\mathscr{A}}$ and $\hat{\mathscr{B}}$. The proof of this fact is similar to the proof of Proposition 4.2.9.

Examples

1. Let W be an open subset of $\mathbf{R}^m$. Then a map $f: W \longrightarrow \mathbf{R}^n$ is differentiable in the usual sense if and only if f is differentiable with respect to the atlases $\mathscr{A}_m$ and $\mathscr{A}_n$ on $\mathbf{R}^m$ and $\mathbf{R}^n$ respectively. This follows using the charts $(\mathbf{R}^m, I) \in \mathscr{A}_m$ and $(\mathbf{R}^n, I) \in \mathscr{A}_n$ and the above remark.
2. Let M be a differential manifold modelled on $\mathbf{R}^m$ with atlas $\mathscr{A}$. Then, if $(U, \phi) \in \mathscr{A}$, the map $\phi^{-1}: \phi(U) \subset \mathbf{R}^m \longrightarrow M$ is C^∞.
3. Let N be a submanifold of M and let $i: N \longrightarrow M$ denote the inclusion map. We claim that i is C^∞: This follows by looking at charts on M which restrict to charts on N and noticing that the compositions we obtain are the restrictions of C^∞ maps.

   The reader should compare this and the previous example with Lemma 4.7.9.
4. Let $\gamma_n: \mathbf{R}^n \longrightarrow T^n$ be the projection map defined in Section 4.5. Then γ_n is C^∞. This is a consequence of the following more general result:

   Let $\pi: M \longrightarrow N$ be a local homeomorphism and suppose that the conditions of Proposition 4.5.3 are satisfied. Then π is C^∞.

   We leave details of the proof of this result to the exercises.
5. The quotient map $q: \mathbf{R}^{n+1} \setminus \{0\} \longrightarrow P^n(\mathbf{R})$ is C^∞. This amounts to showing that

   $$\phi_j q: q^{-1}(U_j) \subset \mathbf{R}^{n+1} \setminus \{0\} \longrightarrow \mathbf{R}^n$$

   is C^∞, where $\{(U_j, \phi_j)\}_{j=1}^{n+1}$ is the standard atlas for $P^n(\mathbf{R})$ constructed in Section 4.5. Now

   $$q^{-1}(U_j) = \{(x_1, \ldots, x_{n+1}) \in \mathbf{R}^{n+1} \setminus \{0\}: x_j \neq 0\}$$

   and

   $$\phi_j q(x_1, \ldots, x_{n+1}) = (x_1/x_j, \ldots, x_{j-1}/x_j, x_{j+1}/x_j, \ldots, x_{n+1}/x_j).$$

   Hence $\phi_j q$ is C^∞, $1 \leqslant j \leqslant n+1$.
6. Let $T \in L(\mathbf{R}^{n+1}, \mathbf{R}^{n+1})$ and suppose that T is an isometry of $\mathbf{R}^{n+1}$ with respect to the standard inner product on $\mathbf{R}^{n+1}$. That is, $T \in 0\,(n+1)$. We claim that T restricts to a C^∞ map $\tilde{T} = T \mid S^n: S^n \longrightarrow S^n$. Let $(U, \phi), (V, \psi) \in \mathscr{A}_{n+1}$ restrict to the charts $(\tilde{U}, \tilde{\phi})$ and $(\tilde{V}, \tilde{\psi})$ on S^n respectively. Now $\tilde{\phi}\tilde{T}\tilde{\psi}^{-1} = \phi T \tilde{\psi}^{-1}$. By Lemma 4.7.9, $\tilde{\psi}^{-1}$ is C^∞ and therefore, by the composite-mapping theorem, so is $\phi T \tilde{\psi}^{-1}$.

Example 6 is a special case of the following important proposition.

Proposition 4.8.2

Let M and N be submanifolds of $\mathbf{R}^m$ and $\mathbf{R}^n$ respectively and suppose that $f: \mathbf{R}^m \longrightarrow \mathbf{R}^n$ is C^r, $1 \leqslant r \leqslant \infty$, and that $f(M) \subset N$. Then

$$f \mid M : M \longrightarrow N$$

is C^r.

More generally, it is sufficient to assume that f is defined on an open neighbourhood of M in $\mathbf{R}^m$.

Proof. Similar to that used in Example 6, using Lemma 4.7.9. ∎

We now wish to define what the *rank* of a differentiable map $f: M \longrightarrow N$ is at a point $x \in M$.

The following technical lemma enables us to postpone the definition of the derivative of a differentiable map between manifolds.

Lemma 4.8.3

Let M and N be differential manifolds with atlases $\mathscr{A}$ and $\mathscr{B}$ respectively and suppose that $f: M \longrightarrow N$ is differentiable. Let $x \in M$ and suppose that $(U, \phi) \in \mathscr{A}$ contains x. Then, if $(V, \psi) \in \mathscr{B}$ and $x \in f^{-1}(V)$,

$$\operatorname{rank}(D(\psi f \phi^{-1})_{\phi(x)})$$

depends only on f and x and is independent of the choice of charts (U, ϕ) and (V, ψ).

Proof. Let $(\tilde{U}, \tilde{\phi}) \in \mathscr{A}$ and $(\tilde{V}, \tilde{\psi}) \in \mathscr{B}$ be any other pair of charts satisfying the conditions of the lemma. Then, on some neighbourhood of x, we have

$$\psi f \phi^{-1} = (\psi \tilde{\psi}^{-1})(\tilde{\psi} f \tilde{\phi}^{-1})(\tilde{\phi} \phi^{-1}).$$

Hence, by the composite-mapping formula,

$$D(\psi f \phi^{-1})_{\phi(x)} = D(\psi \tilde{\psi}^{-1})_z \,.\, D(\tilde{\psi} f \tilde{\phi}^{-1})_y \,.\, D(\tilde{\phi} \phi^{-1})_{\phi(x)}$$

where $y = \tilde{\phi}(x)$ and $z = \tilde{\psi} f(x)$.

But $D(\psi \tilde{\psi}^{-1})_z$ and $D(\phi \tilde{\phi}^{-1})_{\phi(x)}$ are linear isomorphisms and so the rank of $D(\psi f \phi^{-1})_{\phi(x)}$ = rank of $D(\tilde{\psi} f \tilde{\phi}^{-1})_{\tilde{\phi}(x)}$, proving the lemma. ∎

Definition 4.8.4

Let M and N be differential manifolds with atlases $\mathscr{A}$ and $\mathscr{B}$ respectively. Let $f: M \longrightarrow N$ be differentiable. Then we define the rank of f at $x \in M$ to be

$$\operatorname{rank}\,(D(\psi f \phi^{-1})_{\phi(x)}),$$

where (U, ϕ), (V, ψ) are charts satisfying the conditions of the above lemma.

Remark. It follows from the lemma that the rank of f at x is well defined. We shall use the notation 'rank $(T_x f)$' to denote the rank of f at x. The reason for this notation will become clearer in the next section. Here we remark only that we are really considering the rank of the 'derivative' of f at x rather than f at x.

Definition 4.8.5

Let M and N be differential manifolds. Suppose that $f: M \longrightarrow N$ is differentiable. We shall say that $x \in M$ is a *critical point* for f if rank $(T_x f)$ is *less* than the smaller of dim (N) and dim (M). That is, if rank $(T_x f)$ is not maximal.

If x is a critical point of f, $f(x)$ is called a *critical value* of f. If x is not a critical point, x is called a *regular point* of f. If $z \in N$ is not a critical value, we call z a regular value. We set

$$C_f = \{x \in M : x \text{ is a critical point of } f\},$$

$$R_f = \{z \in N : z \text{ is a regular value of } f\}.$$

Remarks

1. We have the following relations between C_f and R_f:

$$f(C_f) \cup R_f = N; \qquad f^{-1}(R_f) \cap C_f = \emptyset.$$

But $f^{-1}(R_f) \cup C_f$ need *not* be the whole of M.
2. In the sequel we shall be most interested in the critical points of real-valued functions on a differential manifold.

Examples

1. Let $f: \mathbf{R}^m \longrightarrow \mathbf{R}$. Then $x \in \mathbf{R}^m$ is a critical point for f according to the above definition if and only if rank $(Df_x) = 0$. That is, if and only if $Df_x = 0$. In other words, if and only if x is a critical point for f in the sense of Section 2.12. Thus, Definition 4.8.5 generalizes our previous definition of critical points.
2. Let us regard S^2 as a submanifold of $\mathbf{R}^3$ and define $f: S^2 \longrightarrow \mathbf{R}$ to be the restriction of the function $g: \mathbf{R}^3 \longrightarrow \mathbf{R}$ defined by

$$g(x, y, z) = z.$$

f measures the height of a point on S^2 above the xy-plane (see Fig. 4.27). f is obviously C^∞ (Proposition 4.8.2). We shall find the critical points of f.

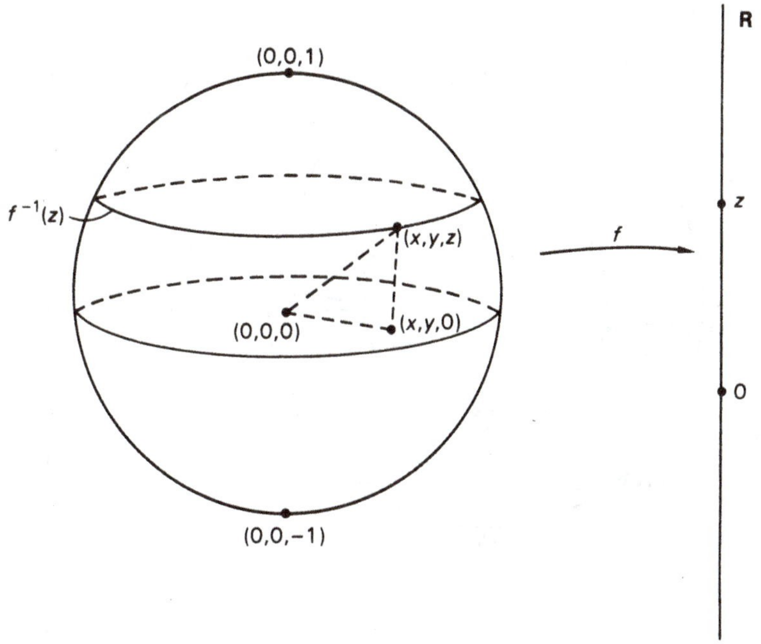

FIG. 4.27

If

$$H_z^+ = \{(x, y, z) \in S^2 : z > 0\} \qquad \text{and} \qquad \phi(x, y) = \sqrt{(1 - x^2 - y^2)},$$
$$(x, y) \in B_1(0),$$

then (H_z^+, ϕ^{-1}) is a chart for S^2. $f\phi : B_1(0) \longrightarrow \mathbf{R}$ is the function defined by

$$(f\phi)(x, y) = \sqrt{(1 - x^2 - y^2)},$$

and it is easily verified that $D(f\phi)_{(x,y)} = 0$ only when $x = y = 0$. Hence $(0, 0, 1)$ is a critical point for f. Similarly $(0, 0, -1)$ is also a critical point for f. We claim that these are the only critical points for f. This amounts to showing that there are no critical points for f on the intersection of the xy-plane with S^2 $(= S^2 \setminus (H_z^+ \cup H_z^-))$. Let

$$H_x^+ = \{(x, y, z) \in S^2 : x > 0\} \qquad \text{and} \qquad \psi(y, z) = \sqrt{(1 - y^2 - z^2)},$$
$$(y, z) \in B_1(0).$$

Then (H_x^+, ψ^{-1}) is a chart for S^2 and $f\psi(y, z) = z$. Hence $f\psi$ has no critical points and so f has no critical points on H_x^+. It follows by symmetry that f has no critical points on TH_x^+, where T is any rotation of S^2 about the z-axis. Hence $C_f = \{(0, 0, 1), (0, 0, -1)\}$.

Notice that the map $g : \mathbf{R}^3 \longrightarrow \mathbf{R}$, which induces f, has *no* critical points. Observe also that $f^{-1}(z)$ is homeomorphic to a circle whenever z is a regular point of f. At critical points of f, $f^{-1}(z)$ degenerates to a point. This dependence of $f^{-1}(z)$ on critical points is brought out more clearly by the next example.

3. Let us represent the torus T^2 as the submanifold of $\mathbf{R}^3$ parametrized by

$$(\sin\theta, (2 + \cos\theta)\sin\psi, (2 + \cos\theta)\cos\psi), \qquad \theta, \psi \in [0, 2\pi].$$

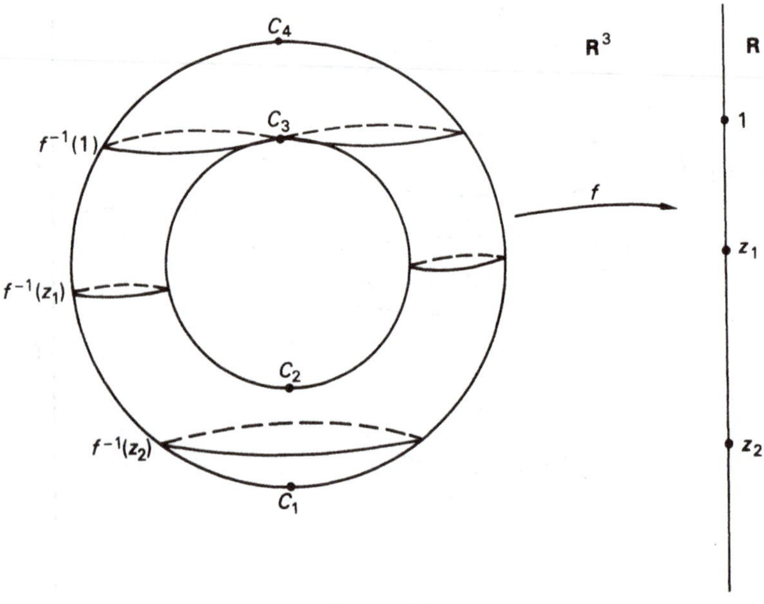

FIG. 4.28

Let $f: T^2 \longrightarrow \mathbf{R}$ denote the height function induced from the map $g: \mathbf{R}^3 \longrightarrow \mathbf{R}$ defined by $g(x, y, z) = z$ (Fig. 4.28). Thus f is C^∞. A similar analysis to that of Example 2 shows that the critical points of f are $C_1 = (0, 0, -2)$, $C_2 = (0, 0, -1)$, $C_3 = (0, 0, 1)$, $C_4 = (0, 0, 3)$.

Notice how the topological type of the level sets $f^{-1}(z)$ changes as z passes through critical values: For $-3 < z < -1$ and $1 < z < 3$, $f^{-1}(z)$ is a circle. For $-1 < z < 1$, $f^{-1}(z)$ is the disjoint union of *two* circles. At the critical points, $f^{-1}(-3)$ and $f^{-1}(3)$ are isolated points whilst $f^{-1}(-1)$ and $f^{-1}(1)$ are homeomorphic to a 'figure of eight' and not even submanifolds of T^2 (see Fig. 4.28).

This change in the topology of $f^{-1}(z)$ as z passes through critical values is of great importance in the study of the topology of differential manifolds. Its full development, *Morse theory*, is, however, beyond the scope of this text (the interested reader may consult [16]). We content ourselves with the following theorems.

Theorem 4.8.6

Let M be a differential manifold modelled on $\mathbf{R}^{m+1}$. Suppose that $f: M \longrightarrow \mathbf{R}$ is a C^∞ map. Then $f^{-1}(c)$ is a submanifold of M of dimension m for all $c \in R_f$.

Proof. Let $y \in f^{-1}(c)$ and suppose that (U, ϕ) is a chart of M which contains y. We then apply Theorem 3.3.2 to the map $f\phi^{-1}: \phi(U) \subset \mathbf{R}^{m+1} \longrightarrow \mathbf{R}$ just as in the proof of Theorem 4.7.2. This enables us to find a submanifold chart for $f^{-1}(c)$ in some neighbourhood of y. We leave details to the exercises. ■

More generally we have

Theorem 4.8.7

Let M and N be differential manifolds and $f: M \longrightarrow N$ be C^∞. Then $f^{-1}(c)$ is a submanifold of M for all $c \in R_f$.

Proof. Either Theorem 3.3.2 or Theorem 3.3.4 together with the argument of the proof of Theorem 4.8.6. ■

Remark. With the assumptions of Theorem 4.8.6, we call $f^{-1}(c)$ a *level surface* of f.

Let us now look at a more sophisticated application of some of the above ideas.

Definition 4.8.8

Let M and N be differential manifolds of the same dimension and let $f: M \longrightarrow N$ be a differentiable map. If $y \in R_f$, we let $\#f^{-1}(y)$ denote the number of points in $f^{-1}(y)$.

The number of points in $f^{-1}(y)$ is not, in general, finite. However, we have the following important result in the case when M is *compact*.

Proposition 4.8.9

Let M and N be differential manifolds of the same dimension and suppose that M is compact. If $f: M \longrightarrow N$ is C^1, then $\#f^{-1}(y)$ is finite for all $y \in R_f$. In addition, $\#f^{-1}$ is locally constant on R_f.

Proof. Since M is compact, $f(M)$ is compact and so $N \setminus f(M)$ is open. Hence if $y \notin f(M)$, there exists a neighbourhood U of y in N disjoint from $f(M)$ on which $f^{-1}(U) = \emptyset$. Hence $\#f^{-1}(z) = 0$ for all $z \in U$. Since $N \setminus f(M) \subset R_f$, this proves the result for those points where $\#f^{-1} = 0$.

Suppose that $y \in f(M) \cap R_f$. We claim that if $z \in f^{-1}(y)$, then z is an isolated point of $f^{-1}(y)$. Let (U, ϕ) be a chart on M containing z and (V, ψ) be a chart on N containing y. Then, since z is not a critical point,

$$D(\psi f \phi^{-1})_{\phi(z)} \in GL(m : \mathbf{R}),$$

where $m = \dim(M) = \dim(N)$.

Hence, by the inverse-function theorem, we may find an open neighbourhood W of $\phi(z) \in \mathbf{R}^m$ such that $\psi f \phi^{-1} | W$ is a diffeomorphism. It follows that $f | \phi^{-1}(W)$ is a homeomorphism onto an open subset of N. Therefore, z is an isolated point in $f^{-1}(y)$.

Since M is compact it follows that $\#f^{-1}(y)$ is finite for all $y \in R_f$. Finally we must show that $\#f^{-1}$ is locally constant on R_f. Let us set

$$f^{-1}(y) = \{z_1, \ldots, z_p\}.$$

As above, we may find open neighbourhoods $W_1, \ldots, W_p$ of $z_1, \ldots, z_p$ respectively such that $f | W_i$ maps W_i homeomorphically onto an open subset of N, $1 \leqslant i \leqslant p$. Set

$$V = \bigcap_{i=1}^{p} f(W_i).$$

Clearly $\#f^{-1}(y') \geqslant p$ for all $y' \in V \cap R_f$. We claim that we can find an open neighbourhood U' of y contained in U such that $\#f^{-1} = p$ on $U' \cap R_f$. If not, there exists a sequence $y_n \longrightarrow y$ such that $y_n \in U \cap R_f$ and $\#f^{-1}(y_n) > p$. In particular, therefore,

$$\exists x_n \in f^{-1}(y_n) \setminus \bigcup_{i=1}^{p} W_i.$$

Since M is compact, x_n has a subsequence converging to a point $x \in M$. Without loss of generality, let us suppose that $x_n \longrightarrow x$. Since f is continuous, $f(x_n) \longrightarrow f(x)$ and so $x \in f^{-1}(y)$. But $x \notin \{z_1, \ldots, z_p\}$ since $x_n \notin \cup_{i=1}^{p} W_i$, $n = 1, 2 \ldots$. Contradiction. ∎

Theorem 4.8.10

Let $a_0, \ldots, a_n \in \mathbf{C}$ and suppose that $a_0 \neq 0$ and $n \geqslant 1$. Then the polynomial $P(z) = a_0 z^n + \cdots + a_n$ has a root.

This result is known as the 'fundamental theorem of algebra'.

Proof (Due to Milnor [17]). Let S^2 denote the unit sphere in $\mathbf{R}^3$ (Fig. 4.29). We define an atlas on S^2 by means of stereographic projection (cf. Section 4.2). THus

$$h_1 : S^2 \setminus \{(0, 0, 1)\} \longrightarrow \mathbf{R}^2 \times \{0\} \subset \mathbf{R}^3$$

is projection from the point $(0, 0, 1)$ and

$$h_1 : S^2 \setminus \{(0, 0, -1)\} \longrightarrow \mathbf{R}^2 \times \{0\} \subset \mathbf{R}^3$$

is projection from $(0, 0, -1)$. We shall identify the xy-plane $\mathbf{R}^2 \times \{0\}$ with the complex plane $\mathbf{C}$ by mapping $(x, y) \in \mathbf{R}^2$ to the point $x + iy \in \mathbf{C}$. With this identification we may regard P as defining a map from $\mathbf{R}^2$ to $\mathbf{R}^2$. Associated with $P : \mathbf{R}^2 \longrightarrow \mathbf{R}^2$ we may define a map $\hat{P} : S^2 \longrightarrow S^2$ as follows:

$$\hat{P}(z) = \begin{cases} h_1^{-1} P h_1(z), & z \neq (0, 0, 1), \\ (0, 0, 1), & z = (0, 0, 1). \end{cases}$$

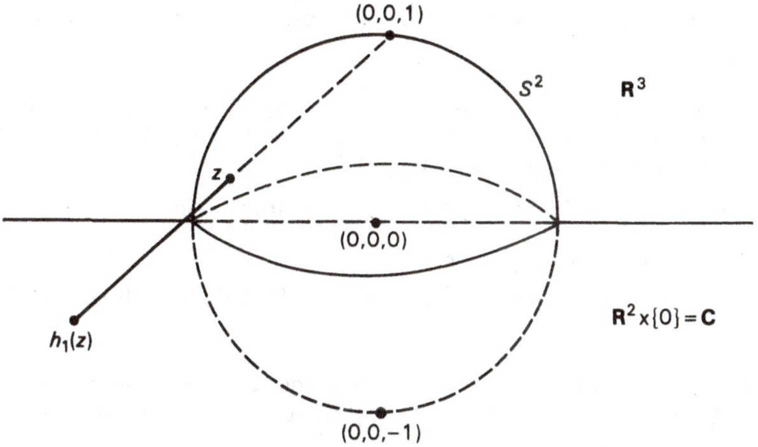

FIG. 4.29

We claim that $\hat{P}$ is C^1: Clearly $\hat{P} \mid (S^2 \setminus \{(0,0,1)\})$ is C^1 since

$$\hat{P}(S^2 \setminus \{(0,0,1)\}) \subset S^2 \setminus \{(0,0,1)\}.$$

Let $Q = h_2 \hat{P} h_2^{-1}$. It is sufficient to show that Q is C^1 in some neighbourhood of $(0,0) \in \mathbf{R}^2$ as, by the definition of differentiability, $\hat{P}$ is then differentiable as a map from S^2 to S^2. Now if $z \in \mathbf{C} \cong \mathbf{R}^2$, we have

$$h_1 h_2^{-1}(z) = z/|z|^2 \qquad \text{(as in Section 4.2)}$$
$$= 1/\bar{z}.$$

An easy computation then gives

$$Q(z) = z^n/(\bar{a}_0 + \bar{a}_1 z + \cdot \cdots + \bar{a}_n z^n).$$

Since $a_0 = 0$, it follows that Q is C^1 on a neighbourhood of 0. Hence $\hat{P}$ is differentiable.

Now $\hat{P}$ can have only a *finite* number of critical points since P fails to have maximal rank only at the zeros of $P'(z)$ of which there are only finitely many since $P' \not\equiv 0$. Therefore, the regular points of $\hat{P}$ are equal to the complement of a finite set $X \subset S^2$. This implies that $R_{\hat{P}} = S^2 \setminus \hat{P}(X)$ is *connected.* Hence, by Proposition 4.8.9, $\#\hat{P}^{-1}$ is *constant* on $R_{\hat{P}}$. If $\#\hat{P}^{-1} = 0$, it follows that Image $(\hat{P}) = \hat{P}(X)$, which is a finite set of points. Hence Image $(\hat{P})$ consists of *one* point, since $\hat{P}$ is continuous. Contradiction, since P is not constant. Therefore, $\#\hat{P}^{-1}$ is non-zero on $R_{\hat{P}}$ and so is non-zero everywhere since $S^2 \setminus R_{\hat{P}} \subset$ Image $(\hat{P})$. Therefore $\hat{P}$ is *onto* and so P must have a zero. ∎

We end this section with a few remarks about embeddings.

Definition 4.8.11

Let M and N be differential manifolds and let $f: M \longrightarrow N$ be C^∞. If f has no critical points and is a homeomorphism onto its image $f(M) \subset N$ (induced topology on $f(M)$), f is called an *embedding.*

Proposition 4.8.12

Let $f: M \longrightarrow N$ be an embedding. Then $f(M)$ is a submanifold of N.

Proof. Similar to that of Theorem 4.7.8 using Theorem 3.3.4. We leave details to the exercises. ■

Example. The inclusion $i : S^2 \longrightarrow \mathbf{R}^3$ is an embedding. More generally, if N is any submanifold of $\mathbf{R}^m$, the inclusion $i : N \longrightarrow \mathbf{R}^m$ is an embedding. The proof of this result is an immediate application of Lemma 4.7.9.

Far more generally we have the following important theorem.

Theorem (Whitney's embedding theorem)

Let M be a differential manifold of dimension m. Then there exists an embedding of M into $\mathbf{R}^{2m+1}$.

Proof. We shall not prove this result here since it requires ideas beyond the scope of this text. For a complete proof we refer to [3]. We remark that it can be shown that M may be embedded in $\mathbf{R}^{2m}$. ■

Whitney's embedding theorem shows that the theory of manifolds may be reduced to the study of submanifolds of $\mathbf{R}^n$. However, the embedding of M into $\mathbf{R}^{2m+1}$ given by Whitney's theorem is in no sense a 'natural' one (there are infinitely many such embeddings). For this reason it is often advantageous *not* to regard M as a submanifold of $\mathbf{R}^{2m+1}$ but instead study M as an abstract manifold. The situation is somewhat similar to the theory of finite-dimensional vector spaces, where one can show that a real vector space of dimension n is isomorphic to $\mathbf{R}^n$ (here we have to make an arbitrary choice of basis). We can lose insight into the structure of the vector space by making such an identification. For example, the set of solutions of $x'' = -x$ forms a vector space of dimension 2. To treat it as merely being $\mathbf{R}^2$ would divorce it from the original differential equation.

An important problem about embedding manifolds is to find the dimension of the smallest vector space into which we can embed a given manifold. For example, S^2 and T^2 both embed in $\mathbf{R}^3$ but $P^2(\mathbf{R})$ only embeds in $\mathbf{R}^4$.

We shall now give a sketch proof of a weak version of Whitney's embedding theorem.

Proposition 4.8.13

Let M be a compact differential manifold. Then M can be embedded in $\mathbf{R}^n$ for sufficiently large n.

Proof. Let dimension $M = m$. Suppose that we can find an open cover $\{V_i\}_{i=1}^p$ of M and C^∞ maps $f_i : M \longrightarrow \mathbf{R}^{m+1}$, $1 \leqslant i \leqslant p$, which satisfy the following conditions:

1. $f_i \mid V_i$ is an embedding of V_i into $\mathbf{R}^{m+1}$, $1 \leqslant i \leqslant p$;
2. $f_i(V_i) \cap f_i(M \setminus V_i) = \emptyset$, $1 \leqslant i \leqslant p$.

Then the reader may easily verify that the map

$$(f_1, \ldots, f_p): M \longrightarrow \mathbf{R}^{mp+p}$$

is an embedding. We shall use this fact to construct an embedding of M. Let $\{(U_i, \phi_i)\}_{i=1}^p$ be a cover of M by coordinate charts such that for some constants $r_i > 0$ we have

1. $\phi_i(U_i) \supset \bar{B}_{r_i}(0) \subset \mathbf{R}^m$,
2. $\{\phi_i^{-1}(B_{r_i/2}(0))\}_{i=1}^p$ is an open cover of M.

Set $V_i = \phi_i^{-1}(B_{r_i/2}(0))$, $1 \leqslant i \leqslant p$. Using bump functions (see Section 4.3) we may extend the map $\phi_i | V_i : V_i \longrightarrow \mathbf{R}^m$ to a map $g_i : M \longrightarrow \mathbf{R}^m$ such that

$$g_i | V_i = \phi_i; g_i | M \setminus U_i \equiv 0.$$

Again using bump functions we may define a C^∞ map $h_i : M \longrightarrow \mathbf{R}$ such that

$$h_i | V_i \equiv 1; h_i(x) \in (0, 1) \qquad \text{for all } x \in M \setminus V_i.$$

We define

$$f_i = (g_i, h_i): M \longrightarrow \mathbf{R}^{m+1}, \qquad 1 \leqslant i \leqslant p.$$

The reader may easily verify that $f_i | V_i$ is an embedding of V_i into $\mathbf{R}^{m+1}$ and that $f_i(V_i) \cap f_i(M \setminus V_i) = \emptyset$, $1 \leqslant i \leqslant p$. Hence $(f_1, \ldots, f_p)$ gives an embedding of M into $\mathbf{R}^{mp+p}$. ■

Exercises

1. Let $\theta \in \mathbf{R}$. Show that the map $p_\theta : S^1 \longrightarrow S^1$ defined by
$$p_\theta((\cos x, \sin x)) = (\cos(x + \theta), \sin(x + \theta))$$
is a C^∞ diffeomorphism. What about the map $\gamma_n : S^1 \longrightarrow S^1$ defined by
$$\gamma_n((\cos x, \sin x)) = (\cos nx, \sin nx), \qquad n \in Z?$$
Find the regular values of γ_n and show that $\#\gamma_n^{-1}$ is constant.
2. Let M, N and P be differential manifolds and $f : M \longrightarrow N$ and $g : N \longrightarrow P$ be C^r maps. Show that the composite $gf : M \longrightarrow P$ is C^r.
3. Let A be an $n \times n$ matrix with integer entries. Show that A induces a C^∞ map $\tilde{A} : T^n \longrightarrow T^n$ and that $\tilde{A}$ is a diffeomorphism if and only if $\det(A) \neq 0$.
4. Let $A \in GL(n + 1; \mathbf{R})$. Show that A induces a C^∞ diffeomorphism $\hat{A} : P^n(\mathbf{R}) \longrightarrow P^n(\mathbf{R})$ in the obvious way. If $A, B \in GL(n + 1; \mathbf{R})$, under what conditions do we have $\hat{A} = \hat{B}$?
5. Show that if $\pi : M \longrightarrow N$ is a local homeomorphism and M and N are differential manifolds satisfying the conditions of Proposition 4.5.3, then π is C^∞.
6. Show that if $f : M \longrightarrow \mathbf{R}$ is C^1 and M is compact, then f has at least one critical point.
7. Let $f : M \longrightarrow N$ be C^1. Show that C_f is a *closed* subset of M.
8. Complete the proofs of Theorems 4.8.6 and 4.8.7.
9. Let $f : M \longrightarrow N$ be C^∞. Suppose that f is of constant rank on $f^{-1}(c)$ for some point $c \in N$. Show that $f^{-1}(c)$ is a submanifold of M (use the rank theorem). What is the dimension of $f^{-1}(c)$ in terms of the rank of f on $f^{-1}(c)$, $\dim(M)$ and $\dim(N)$?

4.9 Tangent spaces

We introduced the theory of differentiation as a theory of linear approximation. Thus, $f : \mathbf{R}^p \longrightarrow \mathbf{R}^q$ is differentiable at x_0 if f can be approximated at x_0 by an affine linear map $g = c + Df_{x_0}$. In geometrical terms, the graph of the affine linear approximation g is the tangent plane to the graph of f at $(x_0, f(x_0))$, as shown in Fig. 4.30.

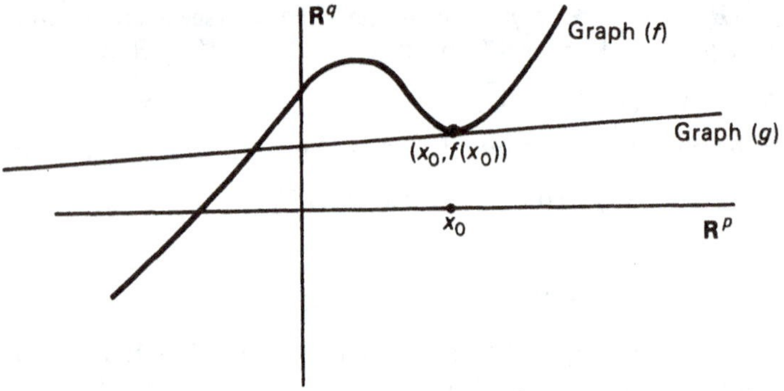

FIG. 4.30

In practice, we tend to concentrate on the linear part of g, Df_{x_0}. In this section we shall start by finding linear approximations to a submanifold N of $\mathbf{R}^m$. Our first example of this process is given by the above: graph (f) is a submanifold of $\mathbf{R}^p \times \mathbf{R}^q$ and the tangent plane, or (affine) linear approximation, to the graph of f at $(x_0, f(x_0))$ is given by the graph of g. In other words, just as a manifold locally looks like an open subset of a vector space, a submanifold of a vector space is locally like an open subset of a vector *sub*space.

Suppose that N is a submanifold of $\mathbf{R}^m$ of dimension n and that N has atlas $\mathscr{A}$. Let $x \in N$ and suppose that $(U, \phi) \in \mathscr{A}$ contains x. Then $\phi^{-1}: \phi(U) \subset \mathbf{R}^n \longrightarrow \mathbf{R}^m$ is a C^∞ embedding of the open set $\phi(U)$ onto a neighbourhood of x in N. The linear approximation to ϕ^{-1} at x is the map

$$D\phi^{-1}_{\phi(x)} : \mathbf{R}^m \longrightarrow \mathbf{R}^n.$$

The image of $D\phi^{-1}_{\phi(x)}$ will be our candidate for the tangent space to N at x. Whilst this linear map depends on the choice (U, ϕ), the following lemma will show that the *image* of this map is independent of the choice of chart.

Lemma 4.9.1

Let N be a submanifold of $\mathbf{R}^m$ of dimension n. Let (U, ϕ) be a chart for N containing the point x. Then the n-dimensional linear subspace $\operatorname{Im}(D\phi^{-1}_{\phi(x)}) = D\phi^{-1}_{\phi(x)}(\mathbf{R}^n)$ of $\mathbf{R}^m$ is independent of our choice (U, ϕ) and depends only on x.

Proof. Let $(U, \phi), (V, \psi)$ be charts on N containing the point x. Then

$$\phi^{-1} = \psi^{-1}(\psi\phi^{-1}) \qquad \text{on } \phi(V \cap U)$$

and hence, by the composite-mapping formula, we have

$$D\phi^{-1}_{\phi(x)} = D\psi^{-1}_{\psi(x)} \cdot D(\psi\phi^{-1})_{\phi(x)}.$$

Since $D(\psi\phi^{-1})_{\phi(x)} \in GL(n; \mathbf{R})$, it follows that $\operatorname{Im}(D\phi^{-1}_{\phi(x)}) = \operatorname{Im}(D\psi^{-1}_{\psi(x)})$. ∎

Definition 4.9.2

Let N be a submanifold of $\mathbf{R}^m$ of dimension n. If $x \in N$, we define the *tangent space* of N at x to be the linear subspace $\operatorname{Im}(D\phi^{-1}_{\phi(x)})$ of $\mathbf{R}^m$, where (U, ϕ) is any chart of N containing x.

We shall denote the tangent space of N at x by T_xN.

Remark. By Lemma 4.9.1, T_xN is well defined for all $x \in N$.

Figure 4.31 shows the tangent space of N at x. Notice that the tangent space T_xN is *not* necessarily tangent to N at x. (It need not pass through the point x.) The *affine* linear subspace $x + T_xN$ does pass through x and we shall refer to $x + T_xN$ as the *affine tangent space* to N at x. This distinction between the linear and affine tangent spaces is similar to the distinction between the linear and affine approximations to a differentiable map. In fact, just as in the theory of differentiation, we shall prefer to work with the tangent spaces of N as opposed to the *affine* tangent spaces to N.

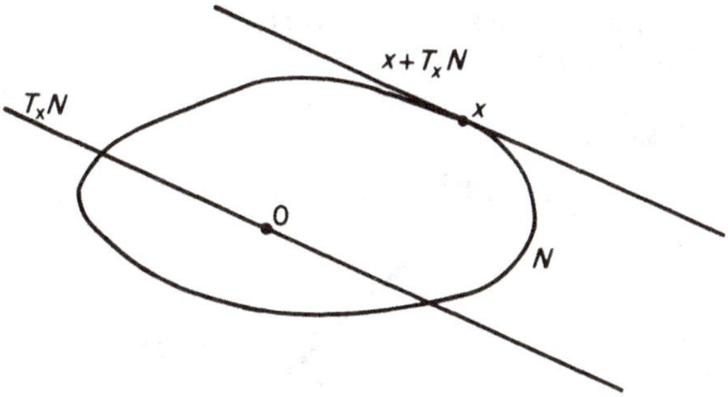

FIG. 4.31

The next two propositions enable us easily to compute the tangent spaces of a large class of submanifolds of $\mathbf{R}^n$.

Proposition 4.9.3

Let $f : \mathbf{R}^{m+1} \longrightarrow \mathbf{R}$ be C^∞, and suppose that f has no critical points on $f^{-1}(c)$, for some $c \in \mathbf{R}$. Set $N = f^{-1}(c)$. Then the tangent spaces of the level surface N are given by

$$T_xN = \text{Ker}\,(Df_x) \qquad \text{for all } x \in N.$$

Proof. Let (U, ϕ) be a chart for N containing x. Then $f\phi^{-1} : \phi(U) \longrightarrow \mathbf{R}$ is the constant function c. Hence

$$Df_x \cdot D\phi^{-1}_{\phi(x)} = 0.$$

Therefore,

$$\text{Ker}\,(Df_x) \supseteq \text{Im}\,(D\phi^{-1}_{\phi(x)}) = T_xN.$$

But since x is not a critical point of f,

$$\dim\,(\text{Ker}\,(Df_x)) = m = \dim\,(T_xN)$$

and so we have equality. ■

More generally we have

Proposition 4.9.4

Let $f : \mathbf{R}^{m+p} \longrightarrow \mathbf{R}^p$ be C^∞ and suppose that f has no critical points on $f^{-1}(c)$ for some $c \in \mathbf{R}$. If we set $N = f^{-1}(c)$, N is a dimension m submanifold of $\mathbf{R}^{m+p}$ and we have

$$T_xN = \text{Ker}\,(Df_x) \qquad \text{for all } x \in N.$$

Proof. Same argument as in the proof of Proposition 4.9.3. ∎

Examples

1. The tangent space of $\mathbf{R}^n$ at $x \in \mathbf{R}^n$ is just $\mathbf{R}^n$. Indeed, a chart for $\mathbf{R}^n$ containing x is $(\mathbf{R}^n, I)$ and so, by definition, $T_x\mathbf{R}^n = I(\mathbf{R}^n) = \mathbf{R}^n$.
2. The tangent space to $S^1 \subset \mathbf{R}^2$ at $(\cos\phi, \sin\phi)$ is the linear subspace of $\mathbf{R}^2$ defined by

$$\{(t \sin\phi, -t\cos\phi) : t \in \mathbf{R}\}.$$

S^1 is the level surface through 1 of the map $f(x,y) = x^2 + y^2$ and $[Df_{(x,y)}] = [2x, 2y]$. Hence it follows from Proposition 4.9.3 that, if $(x,y) \in S^1$, $T_{(x,y)}S^1$ is the kernel of the map $(u,v) \longmapsto 2xu + 2yv$. Thus, if $(x,y) = (\cos\phi, \sin\phi)$,

$$T_{(x,y)}S^1 = \{(u,v) \in \mathbf{R}^2 : u\cos\phi + v\sin\phi = 0\} = \{(t\sin\phi, -t\cos\phi) : t \in \mathbf{R}\}.$$

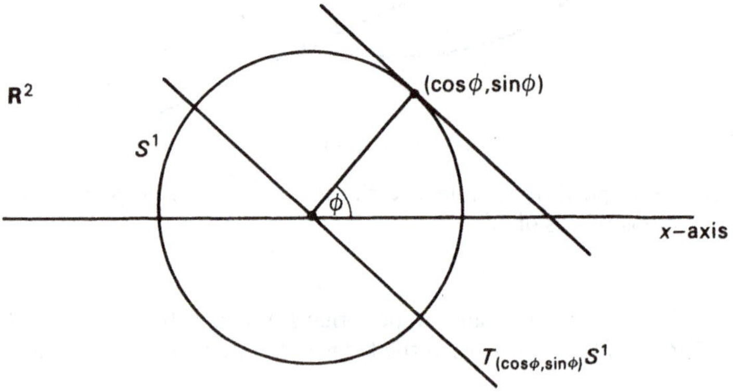

FIG. 4.32

3. The tangent space to $S^2 \subset \mathbf{R}^3$ at $(u, v, \pm\sqrt{(1-u^2-v^2)}) \in S^2$ is the plane defined by

$$\{(x,y,z) \in \mathbf{R}^3 : xu + yv + z(\pm\sqrt{(1-u^2-v^2)}) = 0\}.$$

S^2 is the level surface through 1 of the function $f(x,y,z) = x^2 + y^2 + z^2$. Therefore, by Proposition 4.9.3,

$$T_{(u,v,w)}S^2 = \operatorname{Ker}(Df_{(u,v,w)}) = \{(x,y,z) \in \mathbf{R}^3 : xu + yv + zw = 0\}.$$

Since $(u,v,w) \in S^2$, it follows that $w = \pm\sqrt{(1-u^2-v^2)}$, and substituting in the above expression we obtain the result.
4. In Section 4.7 we showed that the graph of a C^∞ function $f : \mathbf{R}^p \longrightarrow \mathbf{R}^q$ was a p-dimensional submanifold of $\mathbf{R}^{p+q}$. Let us verify that the tangent space of graph (f) at $(x_0, f(x_0))$–as defined in Definition 4.9.2–is given by the graph of the derivative of f at x_0. Recall that graph (f) is the zero set of the map $F : \mathbf{R}^p \times \mathbf{R}^q \longrightarrow \mathbf{R}^q$ defined by $F(x,y) = y - f(x)$ and that F has no critical points. Applying Proposition 4.9.4 we obtain

$$T_{(x_0, f(x_0))}(\text{graph}(f)) = \operatorname{Ker}(DF_{(x_0, f(x_0))}) = \operatorname{Ker}(p_2 - Df_{x_0} \cdot p_1).$$

But $(x,y) \in \operatorname{Ker}(p_2 - Df_{x_0} \cdot p_1)$ if and only if $y = Df_{x_0}(x)$. That is, if and only if $(x,y) \in \text{graph}(Df_{x_0})$.

At this point it is worthwhile recalling the 'arrow' notation for vectors first introduced in Section 2.13. Suppose that x is a fixed point in $\mathbf{R}^n$. Then by $x \longrightarrow y$ is meant the point y in the vector space isomorphic to $\mathbf{R}^n$ but with origin at x. We denote this vector space by $\mathrm{Arr}_x\,(\mathbf{R}^n)$. (See Fig. 4.33.)

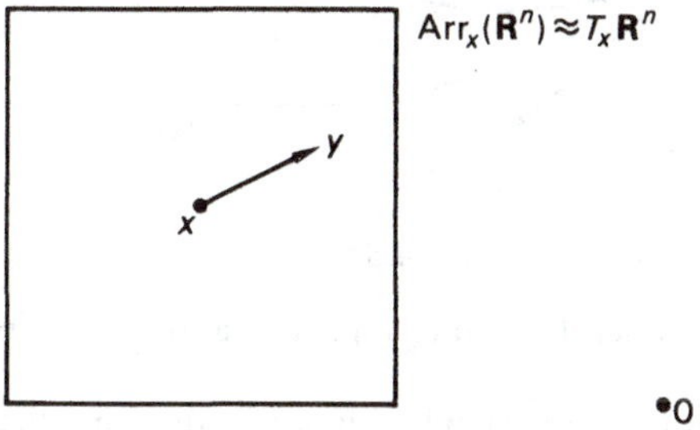

FIG. 4.33

In Example 1 above we showed that $T_x\mathbf{R}^n = \mathbf{R}^n$. It is natural to identify $T_x\mathbf{R}^n$ with $\mathrm{Arr}_x\,(\mathbf{R}^n)$. That is, $y \in T_x\mathbf{R}^n$ is identified with $x \longrightarrow y \in \mathrm{Arr}_x\,(\mathbf{R}^n)$. In this way we may think of $y \in T_x\mathbf{R}^n$ as representing a tangent vector to $\mathbf{R}^n$ at x. Equivalently, we may think of $T_x\mathbf{R}^n$ as being a vector space which is tangent to $\mathbf{R}^n$ at x.

Suppose that N is a submanifold of $\mathbf{R}^m$ and let $x \in N$. A point $y \in T_xN$ is called a *tangent vector* to N at x (Fig. 4.34).

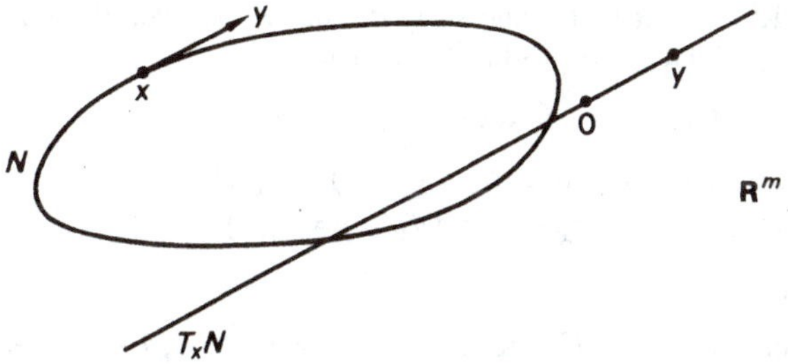

FIG. 4.34

If we regard $T_xN \subset T_x\mathbf{R}^m \approx \mathrm{Arr}_x\,(\mathbf{R}^m)$, we may represent $y \in T_xN$ by an arrow $x \longrightarrow y$ as in Fig. 4.34. Conversely, all arrows tangent to N at x represent elements of T_xN. If we let $\mathrm{Arr}_x\,(N)$ denote the vector subspace of $\mathrm{Arr}_x\,(\mathbf{R}^m)$ formed by arrows tangent to N at x, then $\mathrm{Arr}_x\,(N)$ is naturally isomorphic to T_xN. This device of representing tangent vectors by arrows is very useful as a diagrammatic convention. We emphasize, however, that $\mathrm{Arr}_x\,(N)$ should *not* be thought of as an affine subspace of $\mathbf{R}^m$. It is a vector subspace of the vector space $\mathrm{Arr}_x\,(\mathbf{R}^m)$ whose origin is at x.

The device of using arrows to represent tangent vectors is of particular use in the following situation. Let $f: \mathbf{R}^n \longrightarrow \mathbf{R}^n$. We may represent the map f by the set of all arrows $x \longrightarrow f(x)$. That is, a diagram such as Fig. 4.35 may be used to describe f. At each point in x we draw the arrow $x \longrightarrow f(x)$. The set of all arrows $x \longrightarrow f(x)$ uniquely determines f, and conversely.

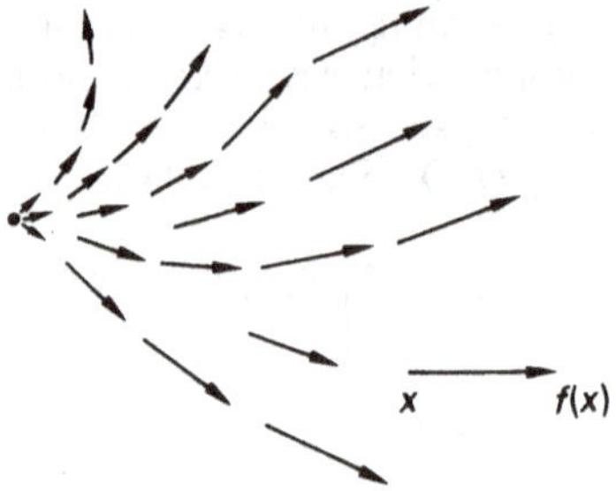

FIG. 4.35

This type of diagram is frequently used to represent a force or velocity field which is independent of time.

Let us suppose that $\mathbf{R}^n$ is given the standard inner product and that $\omega : \mathbf{R}^{n*} \longrightarrow \mathbf{R}^n$ denotes the canonical isomorphism given by the Riesz representation theorem. Then ω is characterized by the relation

$$\langle \omega(e^*), f \rangle = e^*(f) \qquad \text{for all } f \in \mathbf{R}^n \text{ and } e^* \in \mathbf{R}^{n*}$$

Definition 4.9.5

Let U be an open subset of $\mathbf{R}^n$ and let $f: U \longrightarrow \mathbf{R}$ be C^1. The *gradient* of f at x, grad $f(x)$, is the point in $\mathbf{R}^n$ defined by

$$\operatorname{grad} f(x) = \omega(Df_x),$$

where we regard $Df_x \in L(\mathbf{R}^n, \mathbf{R}) = \mathbf{R}^{n*}$.

Example. Let $f: \mathbf{R}^n \longrightarrow \mathbf{R}$ be C^1 and suppose that $\mathbf{R}^n$ is given the standard orthonormal basis $(e_i)_{i=1}^n$. Relative to this basis we have

$$\operatorname{grad} f(x) = (f_{;1}(x), \ldots, f_{;n}(x)) = \sum_{i=1}^{n} f_{;i}(x) e_i.$$

This follows since the ith coordinate of grad $f(x)$ is equal to

$$\langle \operatorname{grad} f(x), e_i \rangle = \langle \omega(Df_x), e_i \rangle = Df_x(e_i) = f_{;i}(x).$$

Proposition 4.9.6

Let U be an open subset of $\mathbf{R}^n$ and let $f: U \longrightarrow \mathbf{R}$ be C^r, $1 \leqslant r \leqslant \infty$. Then

1. grad $f: U \longrightarrow \mathbf{R}^n$ is C^{r-1}.
2. x is a critical point for f if and only if grad $f(x) = 0$.
3. $\langle \operatorname{grad} f(x), h \rangle = Df_x(h)$, $x \in U$, $h \in \mathbf{R}^n$.
4. grad $f(x) \perp \operatorname{Ker} Df_x$, where we regard $Df_x \in L(\mathbf{R}^n, \mathbf{R})$.

Proof. (1) and (2) follow immediately from the definition of grad f. (3) is just the definition of ω. (4) follows trivially from (3). ∎

Corollary 4.9.6.1

Let U be an open subset of $\mathbf{R}^{n+1}$ and let $f: U \longrightarrow \mathbf{R}$ be C^∞. Suppose that f has no critical points on $f^{-1}(c)$ for some point $c \in \mathbf{R}$. Let N denote the hypersurface $f^{-1}(c)$. Then

1. grad $f(x) \perp T_x N$ for all $x \in N$;
2. $\{t\,(\operatorname{grad} f(x)) : t \in \mathbf{R}\} \oplus T_x N = T_x \mathbf{R}^{n+1} = \mathbf{R}^{n+1}$ for all $x \in N$.

Proof. By Proposition 4.9.3, $T_xN = \text{Ker } Df_x, x \in N$. Hence, by Proposition 4.9.6, $\text{grad } f(x) \perp T_xN$ for all $x \in N$. This proves (1). (2) follows since $\dim (T_xN) = n$ and x is not a critical point of f. ∎

Example. S_r^1 is the level surface through $r > 0$ of the function $f(x_1, x_2) = x_1^2 + x_2^2$. Let $x = (x_1, x_2) \in S_r^1$. Then $\text{grad } f(x) = (2x_1, 2x_2) = 2x$ and we have the picture shown in Fig. 4.36. Notice that grad f always points in the direction of f *increasing.* For this reason grad $f(x_1, x_2)$ is said to define an *outward*-pointing normal to S^1 at (x_1, x_2). More generally we may make the following definition.

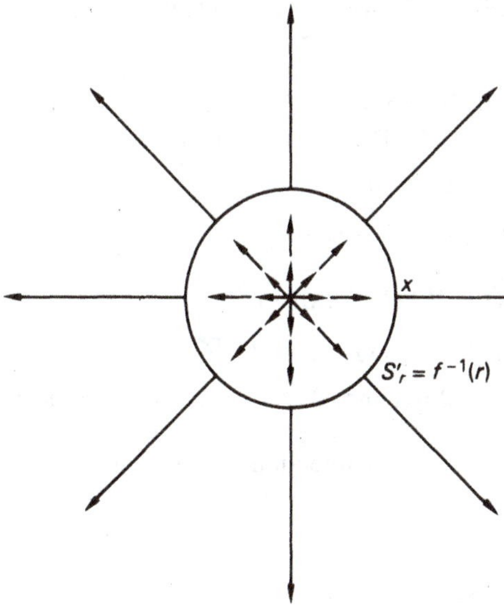

FIG. 4.36

Definition 4.9.7

Let U be an open subset of $\mathbf{R}^{n+1}$ and let $f: U \longrightarrow \mathbf{R}$ be C^1 and have no critical points on the level set $f^{-1}(c)$. Then, if $x \in f^{-1}(c)$, the vector

$$n^+(x) = \text{grad } f(x)/\| \text{grad } f(x) \|_2$$

is called the unit outward normal to the level set $f^{-1}(c)$ at x. Similarly, we call the vector $n^-(x) = -n^+(x)$ the unit inward normal to $f^{-1}(c)$ at x.

Remark. Notice that $n^+(x)$ and $n^-(x)$ are well defined for all $x \in f^{-1}(c)$, as f has no critical points on $f^{-1}(c)$ and so $\text{grad } f(x) \neq 0$ for all $x \in f^{-1}(c)$. We remind the reader that, according to our definitions, $f^{-1}(c)$ is not a *submanifold* of $\mathbf{R}^{n+1}$ unless f is C^∞. This is why we used the term level *set* rather than level *surface* in our definition.

Example. The ellipse $x^2/a^2 + y^2/b^2 = 1, a \geqslant b > 0$, is the level surface through 1 of the function $f(x, y) = x^2/a^2 + y^2/b^2$. Computing the gradient we find that

$$\text{grad } f(x, y) = (2x/a^2, 2y/b^2), \qquad (x, y) \in \mathbf{R}^2.$$

Hence the unit outward normal to the ellipse at the point (x,y) is the vector

$$n^+(x,y) = (x/a^2, y/b^2)/\sqrt{(x^2/a^4 + y^2/b^4)}.$$

In parametric form, taking $x = a\cos\theta, y = b\sin\theta$, we have

$$n^+(x,y) = ((\cos\theta)/a, (\sin\theta)/b)/\sqrt{((\cos^2\theta)/a^2 + (\sin^2\theta)/b^2)}.$$

The tangent line of the ellipse at the point $(a\cos\theta, b\sin\theta)$ is the line orthogonal to this vector. That is, the line

$$\{t((\sin\theta)/b, -(\cos\theta)/a) \colon t \in \mathbf{R}\}.$$

To justify the term 'outward' normal we have the following result.

Proposition 4.9.8

Let U be an open subset of $\mathbf{R}^{n+1}$ and let $f\colon U \longrightarrow \mathbf{R}$ be C^1. Suppose that grad $f(x_0) \neq 0$ for some point $x_0 \in U$. Then $\exists\, \epsilon > 0$ such that

$$f(x_0 + \lambda n^+(x_0)) \begin{cases} > f(x_0), & 0 < \lambda < \epsilon. \\ < f(x_0), & -\epsilon < \lambda < 0. \end{cases}$$

Proof. $f(x_0 + \lambda n^+(x_0)) = f(x_0) + Df_{x_0}(\lambda n^+(x_0)) + o(\lambda)$

$$= f(x_0) + \lambda \,\|\, \text{grad}\, f(x_0) \,\|_2 + o(\lambda).$$

Since grad $f(x_0) \neq 0$, the result follows immediately from this expression. ■

In Fig. 4.37, $n^+(x)$ always points in the direction of f *increasing.*

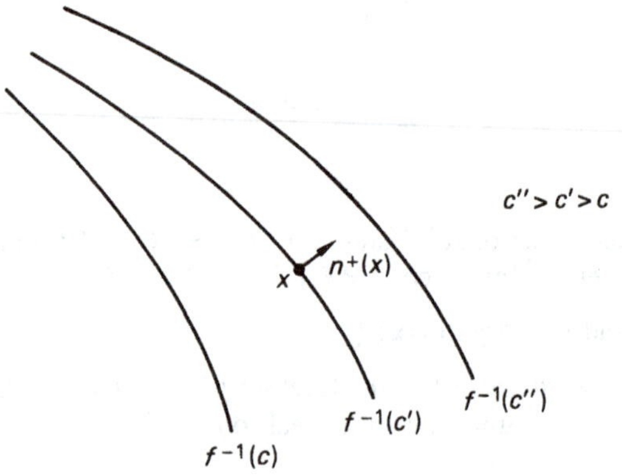

FIG. 4.37

Proposition 4.9.9

Let U be an open subset of $\mathbf{R}^{n+1}$ and let $f\colon U \longrightarrow \mathbf{R}$ be C^1. Suppose that f has no critical points on the level set $f^{-1}(c)$. Then

$$\|\, \text{grad}\, f(x) \,\|_2 = D_{n^+} f_x \qquad \text{for all } x \in f^{-1}(c),$$

where $D_{n^+} f_x$ denotes the directional derivative of f at x in the direction $n^+(x)$.

Proof. By definition

$$D_{n^+}f_x = Df_x(n^+(x)) = \langle \text{grad } f(x), n^+(x)\rangle$$
$$= \| \text{grad } f(x) \|_2, \qquad \text{by definition of } n^+(x). \quad \blacksquare$$

Remark. In the literature, the directional derivative $D_{n^+}f_x$ is usually denoted by $\partial f/\partial n$. Thus, with this notation,

$$\text{grad } f(x) = (\partial f/\partial n)n^+(x).$$

We prefer the notation $f_{;n}(x)$ for $\partial f/\partial n$. With this notation we have

$$\text{grad } f(x) = f_{;n}(x)n^+(x).$$

Proposition 4.9.10

Let U be an open subset of $\mathbf{R}^{n+1}$ and let $f: U \longrightarrow \mathbf{R}$ be C^1. Suppose that grad $f(x) \neq 0$ for some point $x \in U$. Then $n^+(x)$ is the unique unit vector in $\mathbf{R}^{n+1}$ that maximizes the directional derivative $D_u f_x$. Moreover, if $u \in \mathbf{R}^{n+1}$ is a unit vector

$$D_u f_x = (\cos\theta)f_{;n}(x) = \langle \text{grad } f(x), u\rangle,$$

where θ is the angle between the lines defined by $n^+(x)$ and u. That is, $\cos\theta = \langle u, n^+(x)\rangle$.

Proof. Left to the exercises. ∎

Example. Let $f: \mathbf{R}^3 \longrightarrow \mathbf{R}$ be defined by $f(x, y, z) = x^3 - xyz + z^2 + 3x$. Then $(1, 4, 2) \in f^{-1}(0)$ and the reader may easily verify that

$$\text{grad } f(1, 4, 2) = (-2, -2, 0).$$

Hence the unit outward normal at $(1, 4, 2)$ to the level set $f^{-1}(0)$ is

$$(-1/\sqrt{2}, -1/\sqrt{2}, 0).$$

Set $n^+ = (-1/\sqrt{2}, -1/\sqrt{2}, 0)$. Then $f_{;n^+}(1, 4, 2) = 2\sqrt{2}$ and

$$\text{grad } f(1, 4, 2) = (2\sqrt{2})n^+.$$

Let us compute the directional derivative of f at $(1, 4, 2)$ in the direction $(0, 1, 0)$. From Proposition 4.9.10 we have

$$D_{(0,1,0)}f_{(1,4,2)} = \langle (0, 1, 0), n^+\rangle f_{;n^+}(1, 4, 2)$$
$$= (-1/\sqrt{2})(2\sqrt{2}) = -2.$$

Notice how knowledge of the directional derivative $f_{;n}(x)$ and the unit outward normal at x enables us easily to compute all directional derivatives of f at x.

'grad' is an example of one of the classical differential operators defined on an inner-product space. We shall now define two other such operators: The divergence and Laplace operator. We shall not, however, make any use of these operators in the sequel.

Let U be an open subset of $\mathbf{R}^n$ and suppose that $g: U \longrightarrow \mathbf{R}^n$ is differentiable. The *divergence* of g, Div g, is the function Div $g: U \longrightarrow \mathbf{R}$ defined by

$$\text{Div } g(x) = \text{Trace } (Dg_x), \qquad x \in U.$$

That is, $Dg: U \longrightarrow L(\mathbf{R}^n, \mathbf{R}^n)$ and we take the trace of Dg_x at each point x of U. If g is C^r, Div g is obviously C^{r-1}. In coordinates, if $g = (g_1, \ldots, g_n)$,

$$\text{Div } g(x) = \sum_{i=1}^{n} g_{i;i}(x) = \sum_{i=1}^{n} \partial g_i/\partial x_i.$$

To define the *Laplace* operator we proceed as follows. Suppose that $f: U \longrightarrow \mathbf{R}$ is at least C^2. The *Laplacian* of f, Δf, is the function $\Delta f: U \longrightarrow \mathbf{R}$ defined by

$$\Delta f(x) = \text{Div grad}\,(x), \qquad x \in U.$$

In coordinates, we have

$$f(x) = \sum_{i=1}^{n} f_{;ii}(x) = \sum_{i=1}^{n} \partial^2 f/\partial x_i^2.$$

Any function f satisfying the equation $\Delta f = 0$ is called *harmonic*. Harmonic functions are of great importance and have many remarkable properties. For example, every harmonic function is necessarily C^∞.

For the next few paragraphs we shall continue to concentrate our attention on surfaces and, in particular, on the question of when we can continuously define a normal at each point of the surface. As we shall make some use of projective spaces the reader may omit this material if he wishes.

Suppose then that S is a submanifold of $\mathbf{R}^{m+1}$ of dimension m. That is, S is a surface We take the standard inner product on $\mathbf{R}^{m+1}$. At each point $x \in S$ we have a unique tangent space T_xS (Fig. 4.38). Since $\dim(T_xS) = m$, the orthogonal complement to T_xS in $\mathbf{R}^{m+1}$ is a line. For all $x \in N$ let us set

$$N_x = T_xS^\perp.$$

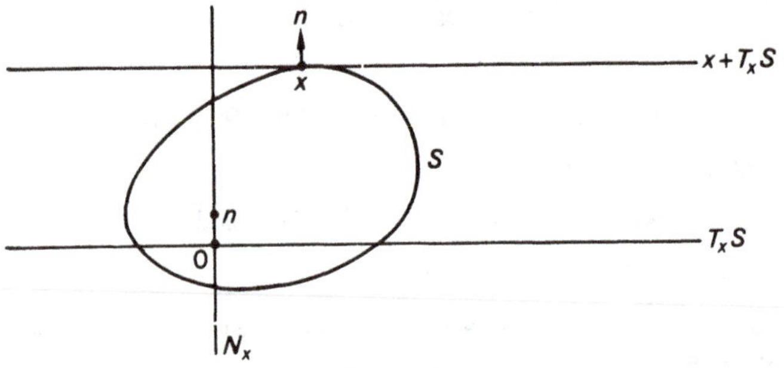

FIG. 4.38

We shall call N_x the *normal line* of S at x. Clearly

$$T_xS \oplus N_x = T_x\mathbf{R}^{m+1} = \mathbf{R}^{m+1} \qquad \text{for all } x \in N.$$

Since N_x is a line through the origin of $\mathbf{R}^{m+1}$ we have a map

$$q_S : S \longrightarrow P^m(\mathbf{R})$$

defined by: $q_S(x) = N_x$ for all $x \in N$.

Proposition 4.9.11

Let S be a surface in $\mathbf{R}^{m+1}$. The map $q_S : S \longrightarrow P^m(\mathbf{R})$ is C^∞.

Proof. Let us first assume that S is the level surface through 0 of the C^∞ function $f: U \longrightarrow \mathbf{R}$, where U is an open neighbourhood of S, and that f has no critical points on S. It follows from Corollary 4.9.6.1. that $q_S(x) = q(\text{grad}\, f(x))$, $x \in S$, where

$$q : \mathbf{R}^{m+1} \setminus \{0\} \longrightarrow P^m(\mathbf{R})$$

is the natural projection map onto $P^m(\mathbf{R})$. Since q and grad f are both C^∞ the result follows in this special case. We shall prove the general case by showing that if $x \in S$, we can find an open neighbourhood U of x in $\mathbf{R}^{m+1}$ such that $U \cap S$ is the level surface of a function $g: U \longrightarrow \mathbf{R}$. Then we can apply the special case. Let $(U, \phi) \in \mathscr{A}_{m+1}$ contain x and restrict to a chart on S. Thus $\phi(U \cap S) \subset \mathbf{R}^m \times \{0\} \subset \mathbf{R}^m \times \mathbf{R}$. We define $g: U \longrightarrow \mathbf{R}$ by

$$g(z) = p_2\phi(z),$$

where $p_2: \mathbf{R}^m \times \mathbf{R} \longrightarrow \mathbf{R}$ is the projection on the second factor. Clearly $g^{-1}(0) = U \cap S$ and we leave it to the reader to verify that g is C^∞ and has no critical points on $S \cap U$. ∎

The proof of the preceding proposition suggests the following question: Can every surface in $\mathbf{R}^{m+1}$ be represented as the level surface of a function (without critical points)? To obtain a satisfactory answer to this question we introduce the idea of an *orientable surface.*

Definition 4.9.12

Let S be a connected surface in $\mathbf{R}^{m+1}$. We shall say that S is *orientable* if the map $q_S: S \longrightarrow P^m(\mathbf{R})$ can be factored through the unit sphere in $\mathbf{R}^{m+1}$. That is, if there exists a C^∞ map $\tilde{n}: S \longrightarrow S^m \subset \mathbf{R}^{m+1}$ making the following diagram commute:

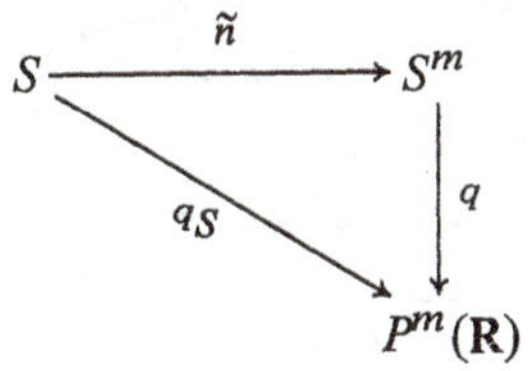

(The vertical map q is the restriction of the natural projection

$$q: \mathbf{R}^{m+1} \setminus \{0\} \longrightarrow P^m(\mathbf{R}) \text{ to } S^m.$$

The geometrical content of this definition is easy to describe: Let $x \in S$. Then $\tilde{n}(x) \in S^m$ is a unit vector. Since $q(\tilde{n}(x)) = q_S(x)$, it follows that $\tilde{n}(x) \in N_x$. In other words, $\tilde{n}(x)$ is a unit normal to S at x for all $x \in S$. Therefore, the map $\tilde{n}$ assigns in a differentiable way a unit normal to each point of S.

Notice that if $\tilde{n}(x) \in S^m, -\tilde{n}(x) \in S^m$. Therefore, if there exists one map $\tilde{n}$ making the above diagram commute, there exist at least two such maps: $\tilde{n}$ and $-\tilde{n}$. It is not hard to show that if S is connected there can exist at most two continuous maps making the above diagram commute and that any continuous map making the diagram commute is necessarily C^∞.

Proposition 4.9.13

Let U be an open subset of $\mathbf{R}^{m+1}$ and suppose that S is a level surface of the C^∞ map $f: U \longrightarrow \mathbf{R}$. Assume that f has no critical points on S. Then S is orientable.

Proof. We define $\tilde{n}: S \longrightarrow S^m$ by

$$\tilde{n}(x) = n^+(x), \qquad x \in S. \quad ∎$$

Remark. The condition that f does not have any critical points on S cannot be dropped from Proposition 4.9.13 unless we make further assumptions about S (for example, S is compact).

Examples

1. $S^1, S^2, \ldots, S^n, \ldots; H^m_{p,q}$ are all orientable surfaces. The torus T^2 is an orientable surface in $\mathbf{R}^3$.
2. The Möbius strip is not orientable: We give the Möbius strip the following parametrization as a surface $M \subset \mathbf{R}^3$:

$$M = \{(\cos\phi(2 + r\cos(\phi/2)), \sin\phi\,(2 + r\cos(\phi/2)), r\sin(\phi/2)) : \phi \in [0, 2\pi), r \in (-1, 1)\}.$$

Set

$$M_\phi = \{(\cos\phi(2 + r\cos(\phi/2)), \sin\phi(2 + r\cos(\phi/2)), r\sin(\phi/2)) : r \in (-1, 1)\}.$$

Then M is formed by rotating the line segment M_ϕ through π as ϕ increases from 0 to 2π (see Fig. 4.39).

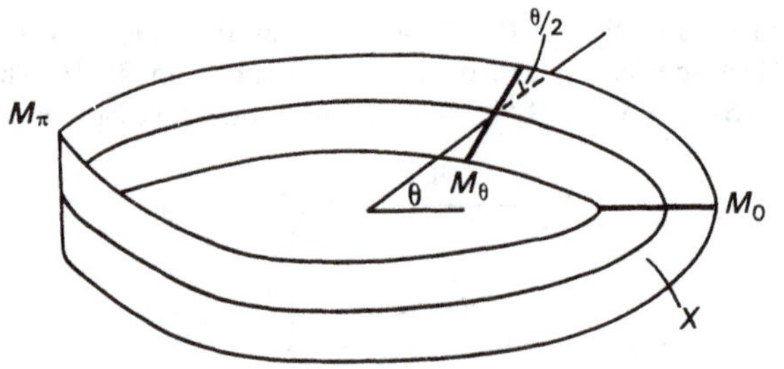

FIG. 4.39

The intersection of M with the xy-plane is a circle of radius 2. Let us denote this circle by X. A unit normal for M at the point $(1, 0, 0)$ is the vector $(0, 0, 1)$. If we move this normal once round the circle X, always keeping it orthogonal to M, it arrives back at $(1, 0, 0)$ as the vector $(0, 0, -1)$. Hence there can be no *continuous* map $\tilde{n}: M \longrightarrow S^2$ such that $q_M = q\tilde{n}$ and so the Möbius strip is not orientable.

We have the following characterization of compact orientable surfaces in $\mathbf{R}^{m+1}$.

Proposition 4.9.14

Let $S \subset \mathbf{R}^{m+1}$ be a compact surface. Then S is orientable if and only if S is the level surface of a C^∞ function, defined on some neighbourhood of S in $\mathbf{R}^{m+1}$, with no critical points on S.

Proof. We have already proved one implication (Proposition 4.9.13). Suppose, therefore that S is orientable. Then there exists a C^∞ map $\tilde{n}: S \longrightarrow S^m$ making the following diagram commute:

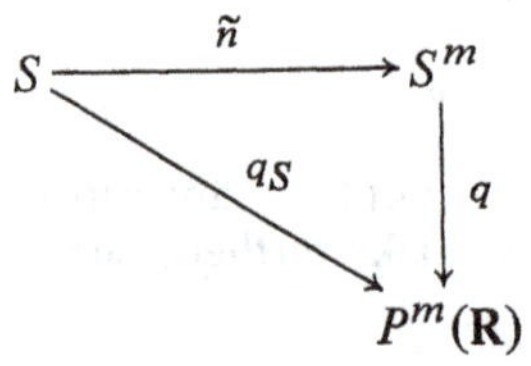

Consider the map $F: S \times \mathbf{R} \longrightarrow \mathbf{R}^{m+1}$ defined by

$$F(x, t) = x + t\tilde{n}(x).$$

F is certainly C^∞ and we claim that F has no critical points on the set $S \times \{0\} \subset S \times \mathbf{R}$: Indeed, using charts on S, we may suppose that S (locally) is an open subset of $\mathbf{R}^m$. Then, if $x \in S$, $D_1F_{(x,0)} \in \mathrm{Iso}\,(\mathbf{R}^m, T_xS)$ and $D_2F_{(x,0)}: \mathbf{R} \longrightarrow N_x$ is the isomorphism $t \longmapsto t\tilde{n}(x)$. Hence F has no critical points on $S \times \{0\}$.

Fix $x \in S$. It follows from the inverse-function theorem that we can find open neighbourhoods V of x in S and I of 0 in $\mathbf{R}$ such that $F|(V \times I)$ is a diffeomorphism onto an open neighbourhood of x in S. See Fig. 4.40. Doing this for each point of S and using the compactness of S it follows easily that there exists $a > 0$ such that

$$F: S \times (-a, +a) \longrightarrow \mathbf{R}^{m+1}$$

is a diffeomorphism onto an open neighbourhood of S in $\mathbf{R}^{m+1}$. We set $U = F(S \times (-a, +a))$ and define $f: U \longrightarrow \mathbf{R}$ by

$$f = p_2F^{-1},$$

where $p_2: \mathbf{R}^m \times \mathbf{R} \longrightarrow \mathbf{R}$ is projection onto the second factor. We leave it to the reader to check that f satisfies the conditions of the proposition. ∎

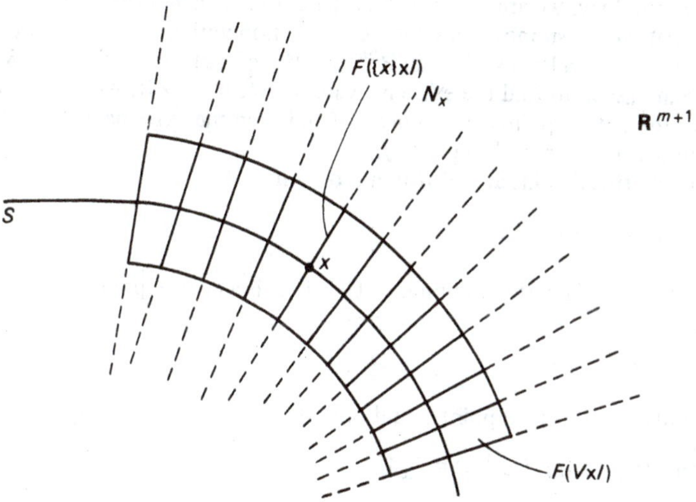

FIG. 4.40

Remarks

1. Using bump functions it is not difficult to show that we may always assume that S is defined as the level surface of a function defined on the *whole* of $\mathbf{R}^{m+1}$.
2. The neighbourhood U of S constructed in the above proposition is of a type extremely important in applications. It is usually called the *normal bundle* of S in $\mathbf{R}^{m+1}$ on account of the way in which it is constructed from the set of normal lines to S. Notice that since $F: S \times (-a, +a) \longrightarrow U$ is a diffeomorphism which maps $S \times \{0\}$ onto S, we have a natural coordinate system on U induced from the product $S \times (-a, +a)$. That is, each point in U may be uniquely represented as a pair (x, t), where $x \in S$ and $t \in (-a, +a)$. It follows from the existence of the

normal bundle that every differentiable function $f: S \longrightarrow \mathbf{R}^k$ may be regarded as the restriction of a differentiable function defined on some open neighbourhood of S in $\mathbf{R}^{m+1}$. Indeed, we may define $\tilde{f}: U \longrightarrow \mathbf{R}^k$, extending f, by setting $\tilde{f} = f\pi_1$, where π_1 is induced from the projection $S \times (-a, +a) \longrightarrow S$.

3. Although we have only constructed the normal bundle of a compact orientable hypersurface it is not hard to generalize the construction to all closed submanifolds of $\mathbf{R}^m$. More precisely, if N is a closed submanifold of $\mathbf{R}^m$ we obtain an open neighbourhood U of N in $\mathbf{R}^m$ which is the union of (translates) of open discs in the normal spaces $T_xN^\perp$. However, U is *not*, in general, diffeomorphic to a *product* $N \times B_r(0)$. As a consequence of the existence of normal bundles it may be shown that a differentiable map defined on a closed submanifold N of $\mathbf{R}^m$ may always be represented as the restriction of a differentiable function defined on an open neighbourhood of N in $\mathbf{R}^m$. (See the exercises at the end of the section.)

As an example of the insight that can be gained by working with the tangent spaces to a submanifold let us look at the problem of finding the extremal values of a function subject to a constraint. Such problems are classically formulated in the following way:

Let $f: \mathbf{R}^m \longrightarrow \mathbf{R}$ be a differentiable function. Find the extremal values of f subject to the constraint $g(x) = 0$, where $g: \mathbf{R}^m \longrightarrow \mathbf{R}$ is another function.

In our terminology, we are asked to find the extremal values of f restricted to the level set $g^{-1}(0)$. Let us suppose that g is a C^∞ function with no critical points on $g^{-1}(0)$. Then $g^{-1}(0)$ is a level surface in $\mathbf{R}^m$. Set $M = g^{-1}(0)$ and let $\tilde{f} = f|M$. With this notation we are asked to find the extremal values of $\tilde{f}: M \longrightarrow \mathbf{R}$. As a necessary first step we find the critical points and values of $\tilde{f}$ and then the extremal values of $\tilde{f}$ will be contained in the set of critical values of $\tilde{f}$.

Now $\tilde{f}$ has a critical point at $x \in M$ if and only if

$$D(\tilde{f}\phi^{-1})_{\phi(x)} = 0,$$

where (U, ϕ) is any M chart containing x. That is, x is a critical point for $\tilde{f} = f|M$ if and only if

$$Df_x \,.\, D\phi^{-1}_{\phi(x)} = 0.$$

In other words, x is a critical point if and only if

$$\operatorname{Ker}(Df_x) \supseteq \operatorname{Im}(D\phi^{-1}_{\phi(x)}) = T_xM.$$

Hence x is a critical point for $\tilde{f}$ if and only if

$$\operatorname{grad} f(x) \perp T_xM.$$

Two possibilities may occur:

1. $\operatorname{Dim}(\operatorname{Ker}(Df_x)) = m$. In this case x is also a critical point for the map $f: \mathbf{R}^m \longrightarrow \mathbf{I}$ Notice that this does *not* imply that $f^{-1}(f(x))$ and M have common tangent spaces at x. Indeed, since x is a critical point for f, $f^{-1}(f(x))$ need not even be a manifold.
2. $\operatorname{Dim}(\operatorname{Ker}(Df_x)) = m - 1$. This is by far the most important case in applications. We now have $\operatorname{Ker}(Df_x) = T_xM$. Since x is not a critical point of f, it follows that $f^{-1}(f(x))$ is a level surface—at least near x—and so the tangent space to the level surface of f through $f(x)$ at x *coincides* with that of M at x. Let us set $f(x) = y$.

Then

$$T_xM = T_x(f^{-1}(y)).$$

But this equality occurs if and only if

$$\text{grad } f(x) = \lambda \text{ grad } g(x) \qquad \text{for some } \lambda \neq 0.$$

That is, the tangent spaces coincide if and only if the normal lines coincide. But we have this equality if and only if the function

$$f + \lambda g: \mathbf{R}^m \longrightarrow \mathbf{R}$$

has a critical point for some non-zero value of λ. (If it has a critical point for $\lambda = 0$, this is covered by case 1.) If we write this out in coordinate form, the problem is therefore reduced to finding points x and corresponding non-zero values of λ such that

$$f_{;j}(x) + \lambda g_{;j}(x) = 0, \qquad 1 \leqslant j \leqslant m.$$

Any such point x will then be a critical point of $f|M$. This method of finding critical points of functions subject to constraints is known as the method of *Lagrange multipliers.* It may easily be generalized to submanifolds of $\mathbf{R}^m$ other than surfaces. (See the exercises at the end of this section.)

Example. Find the maximum value of the function

$$f(x, y, z) = x^2 + y - z^2$$

subject to the constraint

$$x^2 + y^2 + z^2 = 1.$$

That is, we are asked to find the maximum value of f on the unit sphere S^2. Consider the function $F: \mathbf{R}^3 \longrightarrow \mathbf{R}$ defined by

$$F(x, y, z) = (x^2 + y - z^2) + \lambda(x^2 + y^2 + z^2).$$

Writing down the partial derivatives of F we must solve the equations

$$x(1 + \lambda) = 0, \qquad 1 + 2\lambda y = 0, \qquad z(\lambda - 1) = 0$$

subject to the constraint $x^2 + y^2 + z^2 = 1$. Let us examine the various possibilities.

$x = 0$. We must have $y^2 + z^2 = 1$. If $z = 0$, $y = \pm 1$ and $\lambda = \pm 1/2$, giving the possible solutions $(0, \pm 1, 0)$. If $z \neq 0$, $\lambda = 1$ and so $y = -1/2$. It follows easily that $z = \pm\sqrt{3}/2$, giving the possible solutions $(0, -1/2, \pm\sqrt{3}/2)$.

$x \neq 0$. $\lambda = -1$ and so $y = 1/2$ and $z = 0$. It follows that $x = \pm\sqrt{3}/2$ and so the remaining solutions are $(\pm\sqrt{3}/2, 1/2, 0)$.

Since S^2 is compact, f does attain both maximum and minimum values on S^2 and these values are a subset of the critical values of f. Inspection shows that

$$\text{Maximum value of } f = 5/4$$

$$\text{Minimum value of } f = -5/4.$$

Exercises

1. Find the tangent space of the hyperboloid
$$x_1^2 + x_2^2 + x_3^2 - x_4^2 = 2$$
at the point (1, 1, 1, 1).
2. Find the gradient of the function
$$f(x_1, x_2, x_3, x_4) = x_1 x_2 x_3 + x_3^2 x_4^3$$
at the point (1, 1, −1, 1).
3. Draw the level surfaces of the function $f(x, y) = x^2 - y^2$. What are the critical points of f? Find the unit outward normal at the point (1, 2).
4. Complete the proof of Proposition 4.9.11.
5. Using the explicit parametrization given for the Möbius strip as a surface in $\mathbf{R}^3$, construct a normal bundle for the Möbius strip and hence show that the Möbius strip may be represented as the zero set of a C^∞ function defined on a neighbourhood of the Möbius strip and with all its critical points lying on the Möbius strip. (*Hint:* map $x + t\tilde{n}(x)$ to t^2. See also Exercise 6.)
6. (Existence of normal bundles.) Let P be a p-dimensional compact submanifold of $\mathbf{R}^n$. Let $N_x(P) = T_x P^\perp$ denote the normal hyperplane to P at x and let $N_x^s(P)$ denote the disc radius $s > 0$ in $N_x(P)$. Show that $\exists\ r > 0$ such that the following conditions are satisfied.

   (a) $N^r(P) \bigcup_{x \in N} x + N_x^r(P)$ is an open neighbourhood of P in $\mathbf{R}^n$
   (b) $N_x^r(P) \cap N_y^r(P) \neq \emptyset$ if and only if $x = y$.
   (c) Given $x_0 \in N$, we can find a chart $(U, \phi) \in \mathscr{A}_n$ which restricts to a chart (V, ψ) for N containing x_0 such that there exists a C^∞ diffeomorphism
   $$F: V \times B_r(0) \longrightarrow N^r(P \cap V)$$
   satisfying the condition
   $$F(x, t) \in N_x^t(P) \qquad \text{for all } x \in V \text{ and } t \in B_r(0).$$
   Here $B_r(0)$ denotes the open disc of radius r in $\mathbf{R}^{n-p}$.
   $N^r(P)$ is called the *normal bundle* of P in $\mathbf{R}^{n-p}$.

   (*Hint:* Start by proving (c) and use the compactness of P to deduce (a). To prove (a) proceed as follows: Let $(W, \gamma) \in \mathscr{A}_n$ be a chart containing $x_0 \in N$ which restricts to a chart on N. Now
   $$\gamma: W \longrightarrow \gamma(W) \subset \mathbf{R}^p \times \mathbf{R}^{n-p}$$
   and we let $\tau: W \longrightarrow \mathbf{R}^{n-p}$ denote the C^∞ map $p_2\gamma$. Then $W \cap N = \tau^{-1}(0)$. Set $\tau = (\tau_1, \ldots, \tau_{n-p})$. Show that grad $\tau_j(x) \perp T_x N$ for all $x \in W \cap N$, $1 \leqslant j \leqslant n - p$. Show also that (grad $\tau_1(x), \ldots,$ grad $\tau_{n-p}(x)$) is a basis for $T_x N^\perp$ for all $x \in W \cap$. Using this fact, together with the inverse-function theorem, construct the map F whose existence is asserted in (c).)

   Show, by example, that we cannot require the map F of condition (c) to be defined on the whole of P.
7. Show, using the results of Exercise 6, that every differentiable function defined on a compact submanifold P of $\mathbf{R}^n$ is the restriction of a differentiable function defined on an open neighbourhood of P in $\mathbf{R}^n$.
8. (For discussion.) Let Gr $(p; n)$ denote the set of p-dimensional linear subspaces of $\mathbf{R}^n$. Show that associated with any $(n - p)$-dimensional submanifold S of $\mathbf{R}^n$

we have a natural map $q_S : S \longrightarrow \mathrm{Gr}(p; n)$. How might one attempt to define the notion of orientability for S?

9. Find the minimum value of the function $f(x, y, z) = x^4 + y^4 - xyz$ on the ellipsoid $x^2 + y^2 + 4z^2 = 1$.
10. Show how the method of Lagrange multipliers may be generalized to find the extremal values of a differentiable map $f : M \subset \mathbf{R}^n \longrightarrow \mathbf{R}$, where M is a submanifold of $\mathbf{R}^n$ which is defined as the level surface through zero of a C^∞ map $g : \mathbf{R}^n \longrightarrow \mathbf{R}^p$, $p > 1$. Hence find the maximum value of the function

$$f(x_1, x_2, x_3, x_4) = x_2x_3 - x_4x_1^3 + x_4$$

on the torus $x_1^2 + x_2^2 = 1$, $x_3^2 + x_4^2 = 1$.

4.10 The tangent bundle of a submanifold

In this section we shall return to the problem of defining what is meant by the derivative of a map $f : M \longrightarrow N$ between the differential manifolds M and N. For simplicity, we shall always assume that our manifolds are represented as submanifolds of Euclidean space. However, this is not a necessary restriction and we indicate how to remove it in the exercises at the end of the section.

Definition 4.10.1

Let N be a submanifold of $\mathbf{R}^m$. We define the *tangent bundle* of N, denoted by TN, to be the (disjoint) union of the set of tangent spaces to N. That is

$$TN = \bigcup_{x \in N} T_xN.$$

Remarks

1. For the moment we may think of TN as being a set of vector subspaces of $\mathbf{R}^m$ parametrized by $x \in N$.
2. If we recall the definition of the tangent space to a submanifold N, it is clear that TN represents the set of all 'linear approximations' to the submanifold N.

Examples

1. As a set we claim that $T\mathbf{R}^m = \mathbf{R}^m \times \mathbf{R}^m$. In fact since $T_x\mathbf{R}^m = \mathbf{R}^m$ for all $x \in \mathbf{R}^m$, we may define a bijection between $T\mathbf{R}^m$ and $\mathbf{R}^m \times \mathbf{R}^m$ by mapping $y \in T_x\mathbf{R}^m$ to the point $(x, y) \in \mathbf{R}^m \times \mathbf{R}^m$. In the sequel, we shall always use this bijection to identify the sets $T\mathbf{R}^m$ and $\mathbf{R}^m \times \mathbf{R}^m$. We stress that we do *not* regard $T\mathbf{R}^m$ as a vector space: The only vector space structure 'in' $T\mathbf{R}^m$ is that on the tangent spaces $T_x\mathbf{R}^m$, $x \in \mathbf{R}^m$.
2. Let U be an open subset of $\mathbf{R}^m$. Then $T_xU = \mathbf{R}^m$ for all $x \in U$ and so, restricting the bijection defined in the above example to U, we have

$$TU = U \times \mathbf{R}^m.$$

In particular, if $(a, b) \subset \mathbf{R}$ is an open interval,

$$T(a, b) = (a, b) \times \mathbf{R}.$$

If TN is the tangent bundle of the submanifold N of $\mathbf{R}^m$, we have a natural projection map

$$\pi_N : TN \longrightarrow N$$

defined by mapping points in the tangent space of N at x to x. That is

$$\pi_N(y) = x \qquad \text{for all } y \in T_xN.$$

Notice that $\pi_N^{-1}(x) = T_xN$ for all $x \in N$.

Suppose that N is a submanifold of $\mathbf{R}^m$. Then T_xN is a vector subspace of $T_x\mathbf{R}^m$ for all $x \in N$. Hence, in a natural way,

$$TN \subset T\mathbf{R}^m.$$

We shall topologize TN by giving it the topology induced from the usual product topology on $\mathbf{R}^m \times \mathbf{R}^m = T\mathbf{R}^m$. With this topology on TN, $\pi_N : TN \longrightarrow N$ is obviously continuous.

Far more important than the topological structure on TN is the existence of a differential structure.

Proposition 4.10.2

Let N be a submanifold of $\mathbf{R}^m$. Then TN has the natural structure of a submanifold of $T\mathbf{R}^m$.

Proof. Let us start by defining a special atlas for $T\mathbf{R}^m$. If $(U, \phi) \in \mathscr{A}_m$, we shall define a chart $(\hat{U}, \hat{\phi}) \in \mathscr{A}_{2m}$ by

$$\hat{U} = U \times \mathbf{R}^m, \qquad \hat{\phi} = (\phi, D\phi),$$

where $\hat{\phi}: U \times \mathbf{R}^m \longrightarrow \phi(U) \times \mathbf{R}^m$ is the C^∞ map defined by

$$\hat{\phi}(x, e) = (\phi(x), D\phi_x(e)) \qquad \text{for all } (x, e) \in U \times \mathbf{R}^m.$$

Observe that $\hat{U} = TU$. We claim that

$$\{(\hat{U}, \hat{\phi}) : (U, \phi) \in \mathscr{A}_m\}$$

defines an atlas on $T\mathbf{R}^m$. This is trivial since $\widehat{\psi\phi^{-1}} = \hat{\psi}\hat{\phi}^{-1}$. Let us denote this atlas that we have constructed on $T\mathbf{R}^m$ by $T\mathscr{A}_m$.

Let $\mathscr{A}^*$ denote the set of charts in $\mathscr{A}_m$ which restrict to charts on N. We shall show that the set of charts

$$T\mathscr{A}^* = \{(\hat{U}, \hat{\phi}) : (U, \phi) \in \mathscr{A}^*\}$$

restricts to charts on TN and therefore gives TN the structure of a submanifold of $T\mathbf{R}^m$.

Suppose that $(U, \phi) \in \mathscr{A}^*$ restricts to the chart (V, ψ) on N. That is, $N \cap U = V$, $\psi = \phi | (N \cap U)$, $\phi : U \longrightarrow \phi(U) \subset \mathbf{R}^n \times \mathbf{R}^{m-n}$ and $\phi(V) \subset \mathbf{R}^n \times \{0\}$. For all $x \in V$ we have (by definition)

$$T_xN = D\phi^{-1}_{\phi(x)}(\mathbf{R}^n \times \{0\}).$$

We may rewrite this statement equivalently as

$$D\phi_x(T_xN) = \mathbf{R}^n \times \{0\} \subset \mathbf{R}^m \qquad \text{for all } x \in V,$$

or, in terms of $\hat{\phi}$ as

$$\hat{\phi}(TN \mid V) = \phi(V) \times (\mathbf{R}^n \times \{0\}).$$

If we set $\hat{\psi} = \hat{\phi} \mid (TN \mid V)$ it follows immediately that $(TN \mid V, \hat{\psi})$ is a chart for TN.

Hence every chart in $\mathscr{A}^*$ restricts to a chart for TN and so TN is a submanifold of $T\mathbf{R}^m$. ■

Remarks

1. We shall denote the atlas constructed for TN in the above proposition by $T\mathscr{A}_N$.
2. We do not take the completion of the atlases that we have defined on $T\mathbf{R}^m$ and TN. Notice that open sets in the atlas we have constructed for TN are always of the form $U \times \mathbf{R}^n$, where U is an open subset of N and $\mathbf{R}^n$ is the model space for N.

Proposition 4.10.3

Let N be a submanifold of $\mathbf{R}^m$ with atlas $\mathscr{A}_N$ and let $(\hat{V}, \hat{\psi}) \in T\mathscr{A}_N$ be induced from the chart $(V, \psi) \in \mathscr{A}_N$. Then the map

$$\hat{\psi}: TN \mid V \longrightarrow \psi(V) \times \mathbf{R}^n$$

is linear in the second variable. That is, for all $x \in V$, we have

$$\hat{\psi}(x, te + f) = t\hat{\psi}(x, e) + \hat{\psi}(x, f) \qquad \text{for all } e, f \in T_xN \text{ and } t \in \mathbf{R}.$$

Proof. Left to the exercises. ■

Examples

1. TS^1 is homeomorphic to $S^1 \times \mathbf{R}$. In fact we have

$$TS^1 = \{(\cos\phi, \sin\phi), (t\sin\phi, -t\cos\phi) : t \in \mathbf{R}, \phi \in [0, 2\pi)\}$$

and we define a homeomorphism (actually diffeomorphism) $F: TS^1 \longrightarrow S^1 \times \mathbf{R}$ by

$$F((\cos\phi, \sin\phi), t(\sin\phi, -\cos\phi)) = ((\cos\phi, \sin\phi), t) \in S^1 \times \mathbf{R}.$$

We remark that $F \mid T_xS^1 : T_xS^1 \longrightarrow \mathbf{R}$ is linear for all $x \in S^1$ and so the map F preserves the linear structure on the tangent spaces of S^1.

2. TS^2 is not homeomorphic to $S^2 \times \mathbf{R}^2$. The proof of this fact is beyond the scope of this text and requires some algebraic topology.

Let U be an open subset of $\mathbf{R}^m$ and suppose that $f: U \longrightarrow \mathbf{R}^n$ is differentiable. Then f induces a map

$$Tf: TU \longrightarrow T\mathbf{R}^n$$

defined, for all $x \in U$ and $e \in T_xU = T_x\mathbf{R}^m = \mathbf{R}^m$, by

$$Tf(x, e) = (f(x), Df_x(e)).$$

If f is C^r, $1 \leqslant r \leqslant \infty$, it is easy to check that $Tf: TU \longrightarrow T\mathbf{R}^n$ is C^{r-1} (as a map between manifolds). The map Tf will be our model for the derivative of a map between submanifolds of $\mathbf{R}^m$ and $\mathbf{R}^n$. Notice that

$$Tf \mid T_x\mathbf{R}^m : T_x\mathbf{R}^m \longrightarrow T_{f(x)}\mathbf{R}^n$$

is *linear* for all $x \in U$. In fact we have

$$Tf \mid T_x\mathbf{R}^m = Df_x \qquad \text{for all } x \in U.$$

Also notice that the chart $(\hat{U}, \hat{\phi}) \in T\mathscr{A}_m$ associated with the chart $(U, \phi) \in \mathscr{A}_m$ is equal to $(TU, T\phi)$ according to the above definition.

Lemma 4.10.4

Let M and N be submanifolds of $\mathbf{R}^p$ and $\mathbf{R}^q$ respectively and suppose that $f: M \longrightarrow N$ is differentiable. Then, for each $x \in M$, we have a well-defined linear map

$$T_x f: T_x M \longrightarrow T_{f(x)} N$$

defined as follows:

Let $(U, \phi) \in \mathscr{A}_p$ and $(V, \psi) \in \mathscr{A}_q$ restrict to charts on M and N respectively and suppose that $x \in U$ and $f(x) \in V$. Then

$$T_x f = D\psi^{-1}_{\psi(f(x))} \cdot D(\psi f \phi^{-1})_{\phi(x)} \cdot D\phi_x \mid T_x M.$$

Proof. Let the dimensions of M and N be m and n respectively. We first remark that $T_x f(T_x M) \subseteq T_{f(x)} N$. Indeed, $T_x M = D\phi^{-1}_{\phi(x)}(\mathbf{R}^m \times \{0\})$, and so

$$D(\psi f \phi^{-1})_{\phi(x)} \cdot D\phi_x(T_x M) \subseteq \mathbf{R}^n \times \{0\} = D\psi_{f(x)}(T_{f(x)} N).$$

The argument used to show that $T_x f$ is well defined, independent of the choice of charts, is standard and best left to the reader. That $T_x f$ is linear is obvious from the definition. ■

Definition 4.10.5

Let M and N be submanifolds of $\mathbf{R}^p$ and $\mathbf{R}^q$ respectively and let $f: M \longrightarrow N$ be differentiable. The *derivative* of f is the map

$$Tf: TM \longrightarrow TN$$

defined by

$$Tf \mid T_x M = T_x f \qquad \text{for all } x \in M.$$

Remark. The derivative of f is perhaps best thought of in the following way: At each point $x \in M$ we have linearizations $T_x M$ and $T_{f(x)} N$ of M and N respectively. f induces a map on these linearizations (with the notation of Lemma 4.10.4, the map $\psi f \phi^{-1}$) and the derivative of f at x is defined be the linearization of the induced map (that is, the linear map $D(\psi f \phi^{-1})_{\phi(x)}$). We then take the set of all these linearizations to f at all points of M to form the derivative map Tf.

Suppose that W is an open subset of $\mathbf{R}^p$ and that $f: W \longrightarrow \mathbf{R}^q$ is differentiable. The reader may check that Tf defined according to Definition 4.10.5 coincides with Tf as defined in the paragraph before Lemma 4.10.4. That is,

$$Tf(x, e) = (f(x), Df_x(e)) \qquad \text{for all } x \in W \text{ and } e \in \mathbf{R}^p.$$

Using this fact we may obtain a more explicit local representation of Tf in the general case. Suppose then that M and N are submanifolds of $\mathbf{R}^p$ and $\mathbf{R}^q$ respectively and let $\mathscr{A}_M$ and $\mathscr{A}_N$ denote the respective atlases induced on M and N. Let $f: M \longrightarrow N$ be differentiable. If $(U, \phi) \in \mathscr{A}_M$, then $(TU, T\phi)$ defines a chart for TM, and

$$T\phi(T(U)) = \phi(U) \times \mathbf{R}^m.$$

Similarly for charts in $\mathscr{A}_N$. Suppose that $(V, \psi) \in \mathscr{A}_N$ and that $f(U) \subset V$. We have the following commutative diagram of maps:

$$\begin{array}{ccc} TM\,|\,U & \xrightarrow{\quad Tf \quad} & TN\,|\,V \\ \downarrow{\scriptstyle T\phi} & & \downarrow{\scriptstyle T\psi} \\ \phi(U) \times \mathbf{R}^m & \xrightarrow{\quad T\tilde{f} \quad} & \psi(V) \times \mathbf{R}^n \end{array}$$

Here $\tilde{f} = \psi f \phi^{-1} : \phi(U) \subset \mathbf{R}^m \longrightarrow \mathbf{R}^n$ and, by the above remarks

$$T\tilde{f}(x, e) = (\tilde{f}, D\tilde{f}_x(e)) \qquad \text{for all } x \in \phi(U) \text{ and } e \in \mathbf{R}^m.$$

In other words, Tf can be written (locally) as the composite

$$Tf = (T\psi)^{-1}(T\tilde{f})(T\phi).$$

This, of course, follows immediately from the definition of $T_x f$, $x \in U$. From this local representation of Tf as a composite of maps follows immediately

Proposition 4.10.6

Let $f: M \longrightarrow N$ be a C^r map between the submanifolds M and N of $\mathbf{R}^p$ and $\mathbf{R}^q$ respectively. Then $Tf: TM \longrightarrow TN$ is C^{r-1}.

We leave to the exercises the proof of the following

Proposition 4.10.7

Let U be an open neighbourhood of the submanifold M of $\mathbf{R}^p$ and let N be a submanifold of $\mathbf{R}^q$. Suppose that $f: U \longrightarrow \mathbf{R}^q$ is a C^r map and that $f(M) \subset N$. Then $Tf: TU \longrightarrow T\mathbf{R}^q$ restricts to a C^{r-1} map

$$Tf: TM \longrightarrow TN.$$

Moreover, for all $x \in M$ we have

$$T_x f \,|\, T_x M = Df_x \,|\, T_x M.$$

Example. Let us consider the function $F: S^2 \longrightarrow \mathbf{R}$ defined as the restriction of the map

$$f(x_1, x_2, x_3) = x_1^3 - 2x_1 x_2 + x_3 x_1.$$

Let us find $T_x F$ for $x = (1, 0, 0) \in S^2$. Computing $T_x f$ at $(1, 0, 0)$ we find that

$$[T_x f] = [3, -2, 1].$$

Now the tangent space of S^2 at $(1, 0, 0)$ is the plane $x_1 = 0$. Therefore, by the previous proposition we have

$$[T_x F] = [T_x f] \,|\, \{x_1 = 0\} = [-2, 1].$$

To end this section we remark that the composite-mapping formula takes a rather more natural form when written in terms of the derivative maps Tf. We have

Proposition 4.10.8 (Composite-mapping formula)

Let M, N and P be submanifolds of $\mathbf{R}^s$, $\mathbf{R}^t$ and $\mathbf{R}^u$ respectively and let $f: M \longrightarrow N$ and $g: N \longrightarrow P$ be differentiable maps. Then the composite gf is differentiable and we have

$$T(gf) = (Tg)(Tf).$$

Proof. Left to the exercises. ∎

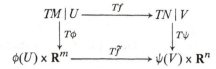

Exercises

1. Prove Proposition 4.10.3.
2. Prove that the tangent bundle of the torus T^2 is diffeomorphic to $T^2 \times \mathbf{R}^2$.
3. Prove Propositions 4.10.7 and 4.10.8.
4. Let M and N be submanifolds of $\mathbf{R}^q$ and suppose that N is a submanifold of M. Let $i: N \longrightarrow M$ denote the inclusion map. Show that $Ti: TN \longrightarrow TM$ is C^∞ and that $Ti(TN) = TN$, regarded as a submanifold of $\mathbf{R}^q$.
5. Let S be a p-dimensional submanifold of $\mathbf{R}^n$ with induced atlas $\mathscr{A}_S$. Construct the atlas $T(T\mathscr{A}_S)$ and hence show that $T(TS)$ is a submanifold of $T(T\mathbf{R}^n) = \mathbf{R}^n \times \mathbf{R}^n$ $\mathbf{R}^n \times \mathbf{R}^n$ and that open sets in $T(T\mathscr{A}_S)$ are of the form $(V \times \mathbf{R}^p) \times (\mathbf{R}^p \times \mathbf{R}^p)$.
6. In this example we show how one may define the tangent bundle of an arbitrary differential manifold M. Let M have atlas $\{(U_\alpha, \phi_\alpha)\}_{\alpha \in \Lambda}$ and be modelled on $\mathbf{R}^n$. We form the disjoint union

$$X = \bigcup_{\alpha \in \Lambda} U_\alpha \times \mathbf{R}^n.$$

We define an equivalence relation $\sim$ on X by

$$(\alpha, x, e) \sim (\beta, y, f)$$

if and only if $x = y$ and $D(\phi_\alpha \phi_\beta^{-1})_{\phi_\beta(y)}(f) = e$. We set $TM = X/\sim$, and let $\pi_M: TM \longrightarrow M$ denote the map induced from the projection of X onto M defined by mapping (α, x, e) to x. We shall call TM the tangent bundle of M. The next few problems show that TM has the correct properties and, in particular, generalizes the constructions given in the text for submanifolds of $\mathbf{R}^n$.

(a) Show that for all $\alpha \in \Lambda$ we have a natural map

$$\psi_\alpha: \pi_M^{-1}(U_\alpha) = TM \mid U_\alpha \longrightarrow U_\alpha \times \mathbf{R}^n$$

and that ψ_α is a homeomorphism.

(b) If $x \in M$ we shall set $T_x M = \pi_M^{-1}(x)$. Let $x \in U_\alpha$. We define

$$\psi_{\alpha x} = \psi_\alpha \mid T_x M: T_x M \longrightarrow \mathbf{R}^n.$$

Show that if $x \in U_\alpha \cap U_\beta$, the map

$$\psi_{\alpha x} \psi_{\beta x}^{-1}: \mathbf{R}^n \longrightarrow \mathbf{R}^n$$

is a *linear* isomorphism.

Hence show that if we define

$$e + tf = \psi_{\alpha x}^{-1}(\psi_{\alpha x}(e) + t\psi_{\alpha x}(f))$$

for all $e, f \in T_x M$ and $t \in \mathbf{R}$, then this definition gives each space $T_x M$ the structure of a vector space and that this structure does not depend on the choice of chart (U_α, ϕ_α).

(c) Set

$$\hat{\phi}_\alpha = (\phi_\alpha \times I)\psi_\alpha : TM \mid U_\alpha \longrightarrow \phi_\alpha(U_\alpha) \times \mathbf{R}^n.$$

Show that $\{(TM \mid U_\alpha, \hat{\phi}_\alpha) : \alpha \in \Lambda\}$ gives TM the structure of a differential manifold modelled on $\mathbf{R}^n \times \mathbf{R}^n$.

(d) Show that this definition of the tangent bundle generalizes that of the tangent bundle of a submanifold of $\mathbf{R}^m$. In particular, show that the tangent bundle as constructed above is isomorphic to the tangent bundle constructed i the text, by a diffeomorphism which restricts to linear isomorphisms on the tangent spaces $T_x M$.

4.11 Differential equations on manifolds

To conclude this chapter we wish to show how the theory of ordinary differential equations has a natural generalization to differential manifolds. This section is only intended to be introductory and we refer the reader to references [1], [2] and [13] for more comprehensive treatments.

Let us start by making a brief resumé of the theory of first-order ordinary differential equations on open subsets of $\mathbf{R}^n$.

Suppose that U is an open subset of $\mathbf{R}^n$ and that

$$F = (F_1, \ldots, F_n): U \longrightarrow \mathbf{R}^n$$

is a continuous function. Then the set of first-order ordinary differential equations

$$\frac{\mathrm{d}x_1}{\mathrm{d}t} = F_1(x_1, \ldots, x_n)$$

$$\cdots\cdots\cdots\cdots$$

$$\cdots\cdots\cdots\cdots$$

$$\frac{\mathrm{d}x_n}{\mathrm{d}t} = F_n(x_1, \ldots, x_n)$$

is said to define a *first-order ordinary differential equation on U.* We may write this equation more concisely in vector notation as

$$\frac{\mathrm{d}x}{\mathrm{d}t} = F(x).$$

An *integral curve* for the differential equation $\mathrm{d}x/\mathrm{d}t = F(x)$ consists of a differentiable function

$$c:(a, b) \longrightarrow U$$

that solves the differential equation, U being an open interval of the real line. That is, if we set $c = (c_1, \ldots, c_n)$, we have at all points of the interval (a, b)

$$\frac{\mathrm{d}c_i}{\mathrm{d}t} = F_i(c_1, \ldots, c_n), \qquad 1 \leqslant i \leqslant n$$

Or, emphasizing the dependence on t and using vector notation, we have

$$c'(t) = F(c(t)) \qquad \text{for all } t \in (a, b).$$

Remark. Other names frequently met in place of 'integral curve' are 'solution curve' or 'trajectory'.

In what follows we shall always assume that integral curves are defined on an open interval containing zero.

An integral curve $c:(a, b) \longrightarrow U$ is said to have *initial condition* $x_0 \in U$ if

$$c(0) = x_0.$$

c may be referred to as an integral curve *through the point* x_0.

Example. Let us consider the first-order differential equation on $\mathbf{R}^2$ defined by

$$\mathrm{d}x/\mathrm{d}t = y, \qquad \mathrm{d}y/\mathrm{d}t = -x.$$

An integral curve for this system through the point (x_0, y_0) is given by the function

$$S_0(t) = (x_0 \cos t, y_0 \sin t), \qquad t \in \mathbf{R}.$$

Figure 4.41 shows the integral curves of the differential equation. The solution curves are all circles (degenerating to a point at the origin). We may think of each curve as parametrizing the motion of a particle in $\mathbf{R}^2$ whose velocity at the point (x, y) is $(-y, x)$.

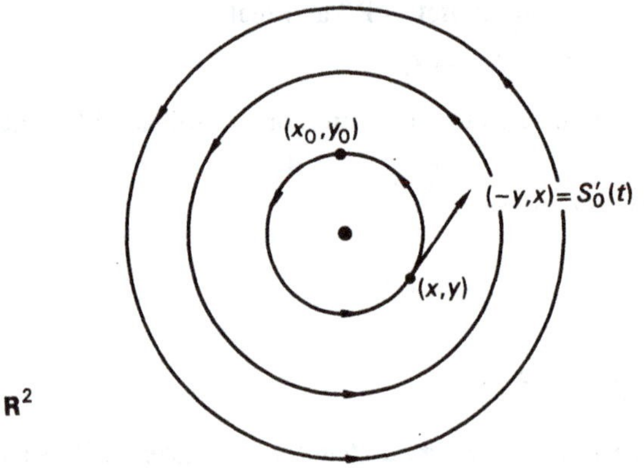

FIG. 4.41

We wish to emphasize the above interpretation of an integral curve as parametrizing the motion of a particle. Referring to Fig. 4.42, suppose that $c:(a, b) \longrightarrow U$ is an integral curve through x_0 of the differential equation $\mathrm{d}x/\mathrm{d}t = F(x)$. We shall think of c as parametrizing the motion of a particle whose velocity at the point $x \in U$ is equal to $F(x)$. That is, at each point of the curve, the velocity, $c'(t)$, is determined by the relation

$$c'(t) = F(c(t)).$$

Now the velocity, $c'(t)$, may be thought of as defining a tangent vector to the curve defined by c. Hence it is natural to regard $c'(t) \in T_{c(t)}U$. Since $c'(t) = F(c(t))$, this suggests that we should consider F as defining a *tangent vector* at each point of U.

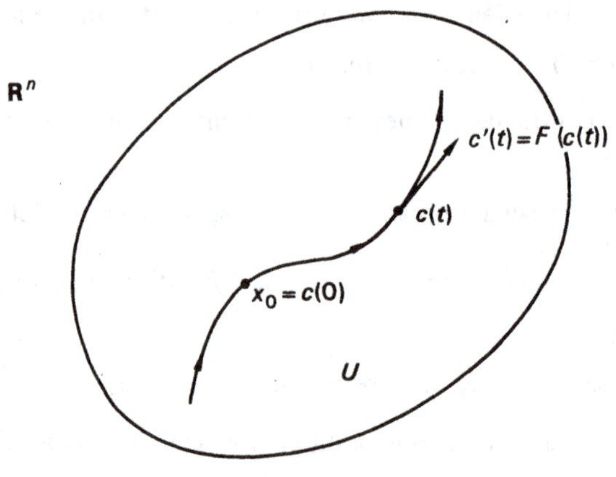

FIG. 4.42

That is, we shall consider F as a map from U to TU which maps $x \in U$ to $F(x) \in T_x U$ for all $x \in U$.

Referring to Fig. 4.43, F defines a tangent vector (or 'arrow') at each point of U. The problem of finding integral curves is the problem of finding smooth curves whose tangents are given by F. Intuitively: Think of U as being fluid in motion with velocity at $x \in U$ equal to $F(x)$. A 'light' object placed at the point $x_0 \in U$ starts moving with velocity $F(x_0)$, and if we allow it to 'flow' with the direction and velocity given by F it will trace out an integral curve for the differential equation $x' = F(x)$.

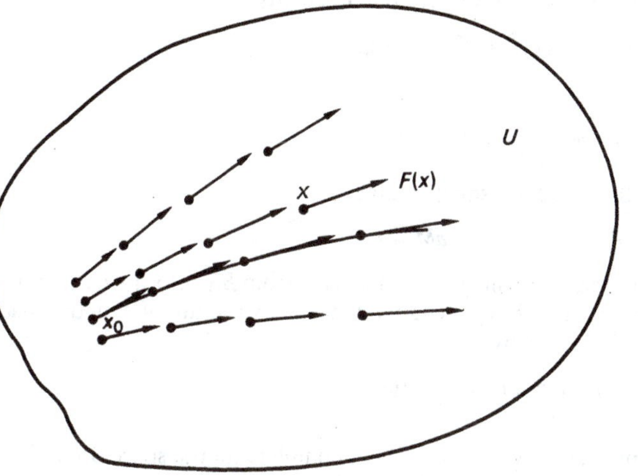

FIG. 4.43

It is this interpretation of an ordinary differential equation as a 'field of tangent vectors' that we shall exploit in the sequel. However, before defining differential equations on submanifolds of $\mathbf{R}^n$, we wish to state and prove a basic existence and uniqueness theorem for ordinary differential equations.

Theorem 4.11.1

Let U be an open subset of $\mathbf{R}^n$ and suppose that $F: U \longrightarrow \mathbf{R}^n$ is C^r, $r \geqslant 1$. Then for each point $x_0 \in U$, there exist an open neighbourhood U_0 of x_0 in U, an open interval J_0 containing zero and a continuous function

$$\phi: U_0 \times J_0 \longrightarrow U$$

such that for all $x \in U_0$ the map $\phi_x: J_0 \longrightarrow U$, defined by $\phi_x(t) = \phi(x, t)$, is an integral curve through x for the differential equation $x' = F(x)$.

Furthermore, integral curves for the differential equation are *unique* in the sense that if $c_1: (a_1, b_1) \longrightarrow U$ and $c_2: (a_2, b_2) \longrightarrow U$ are integral curves through x_0, then

$$c_1 \mid (a_1, b_1) \cap (a_2, b_2) = c_2 \mid (a_1, b_2) \cap (a_2, b_2).$$

Finally, every integral curve for the differential equation is of class at least C^{r+1}.

Proof. We first remark that if $c: (a, b) \longrightarrow U$ is any differentiable map satisfying $c'(t) = F(c(t))$ on (a, b), then c is necessarily of class C^{r+1}. Indeed, if c is differentiable, then c is continuous and so $c' = F(c)$ is continuous and, therefore, c is C^1. Proceeding inductively, suppose that we have shown that c is of class C^s, $s \leqslant r$. $c' = F(c)$ is C^s by the composite-mapping theorem and so c is of class C^{s+1}.

The proof of the existence and uniqueness of integral curves will be a straightforward application of the contraction-mapping lemma. Let $\| \quad \|$ be a fixed norm on $\mathbf{R}^n$. Choose $r > 0$ so that $\bar{B}_{2r}(x_0) \subset U$. We set

$$K = \sup_{x \in \bar{B}_{2r}(x_0)} \| DF_x \|.$$

$K < \infty$ since F is C^1 and $\bar{B}_{2r}(x_0)$ is compact. It follows from the mean-value theorem that for all $x, y \in \bar{B}_{2r}(x_0)$ we have the inequality

$$\| F(x) - F(y) \| \leqslant K \| x - y \|.$$

Let us set

$$M = \sup_{x \in \bar{B}_{2r}(x_0)} \| F(x) \|.$$

Choose $a > 0$ sufficiently small so that

$$Ka \leqslant \tfrac{1}{2} \qquad \text{and} \qquad aM \leqslant r.$$

Let X denote the set of all continuous maps from $\bar{B}_r(x_0) \times [-a, a]$ to $\bar{B}_{2r}(x_0)$ which ma $(x, 0)$ to x for all $x \in \bar{B}_r(x_0)$. The set X has the structure of a metric space with metric defined for all $f, g \in X$ by

$$d(f, g) = \sup_{x,t} \| f(x, t) - g(x, t) \|.$$

The reader may easily verify that X is a complete metric space. Indeed, with the notation of Section 1.9, X is a closed subset of the Banach space $C(\bar{B}_r(x_0) \times [-a, a]; \mathbf{R}^n$

Let $f \in X$. We define a map $T(f): \bar{B}_r(x_0) \times [-a, a] \longrightarrow \mathbf{R}^n$ by

$$T(f)(x, t) = x + \int_0^t F(f(x, s))\, ds.$$

(We refer to Section 2.13 for the theory of integration of vector-valued functions.) Fix $x \in \bar{B}_r(x_0)$. Then

$$T(f)(x, t) - x_0 = \int_0^t F(f(x, s))\, ds + (x - x_0)$$

and so for all $t \in [-a, a]$, we have

$$\| T(f)(x, t) - x_0 \| \leqslant \left\| \int_0^t F(f(x, s))\, ds \right\| + \| x - x_0 \|$$

$$\leqslant \int_0^t \| F(f(x, s)) \|\, ds + r \leqslant M \int_0^t ds + r \leqslant Ma + r \leqslant 2r.$$

Hence

$$T(f): \bar{B}_r(x_0) \times [-a, a] \longrightarrow \bar{B}_{2r}(x_0) \qquad \text{for all } f \in X.$$

Since $T(f)$ is obviously continuous and $T(f)(x, 0) = x$ for all $x \in \bar{B}_r(x_0)$, it follows that $T: X \longrightarrow X$. We shall now show that T is a contraction mapping. Let $f, g \in X$.

Then

$$d(T(f), T(g)) = \sup_{x,t} \left\| \int_0^t (F(f(x, s)) - F(g(x, s)))\, ds \right\|$$

$$\leqslant \sup_{x,t} \int_0^t \|F(f(x, s)) - F(g(x, s))\|\, ds$$

$$\leqslant \sup_{x,t} \int_0^t K \|f(x, s) - g(x, s)\|\, ds \leqslant Kad(f, g).$$

Since $Ka \leqslant \frac{1}{2}$ it follows that T is a contraction mapping. By the contraction-mapping lemma (see Section 3.2), T has a unique fixed point $\phi \in X$. That is,

$$\phi: \bar{B}_r(x_0) \times [-a, a] \longrightarrow \bar{B}_{2r}(x_0)$$

is continuous and satisfies the integral equation

$$\phi(x, t) = x + \int_0^t F(\phi(x, s))\, ds \qquad \text{for all } x \in \bar{B}_r(x_0) \text{ and } t \in [-a, a].$$

By the fundamental theorem of the integral calculus, the right-hand side of this equation is differentiable with respect to t on $(-a, a)$. Hence ϕ is differentiable in the second variable. Differentiating we find that, for fixed $x \in \bar{B}_r(x_0)$, we have

$$\phi_{;2}(x, t) = F(\phi(x, t)) \qquad \text{on } (-a, a).$$

In other words, the map $\phi_x : (-a, a) \longrightarrow U$ is an integral curve through x for the differential equation $x' = F(x)$. Taking $U_0 = B_r(x_0)$ and $J_0 = (-a, a)$, this proves the existence part of the theorem.

Suppose that $\gamma: (\alpha, \beta) \longrightarrow U$ is any other integral curve through $x \in U_0$. We claim that

$$\gamma \mid (\alpha, \beta) \cap (-a, a) \equiv \phi_x \mid (\alpha, \beta) \cap (-a, a).$$

If $(\alpha, \beta) \supset (-a, a)$, this is an immediate consequence of the uniqueness part of the contraction-mapping lemma. We leave details of the proof in the case when $(\alpha, \beta) \not\supset (-a, a)$ to the reader. This argument shows that the integral curve through x is unique at least in some neighbourhood of 0. Before completing the proof of uniqueness we shall prove an elementary lemma.

Lemma 4.11.2

Let $\gamma: (\alpha, \beta) \longrightarrow U$ be an integral curve through x for the differential equation $x' = F(x)$. Then, if $s \in (\alpha, \beta)$, the function

$$\gamma(+s): (\alpha - s, \beta - s) \longrightarrow U,$$

defined by $\gamma(+s)(t) = \gamma(s + t)$, is an integral curve through the point $\gamma(s)$.

Proof. $\gamma(+s)$ is the composite of γ and the translation $t \longmapsto t + s$. Hence, by the composite-mapping formula, $\gamma(+s)$ is differentiable and

$$D\gamma(+s)_t = D\gamma_{t+s}.$$

It follows that

$$\gamma(+s)'(t) = \gamma'(s + t) = F(\gamma(s + t)) = F(\gamma(+s)(t)). \quad \blacksquare$$

Let $c_1:(a_1, b_1) \longrightarrow U$ and $c_2:(a_2, b_2) \longrightarrow U$ be integral curves through x_0. Let $(\alpha, \beta) = (a_1, b_1) \cap (a_2, b_2)$ and let $Y \subset (\alpha, \beta)$ denote the set of points where $c_1 = c_2$. By the remarks preceding the above lemma, Y contains an open neighbourhood of 0 and so is not empty. The continuity of c_1 and c_2 implies that Y is closed. Let $s \in Y$. By the above lemma, $c_1(+s)$ and $c_2(+s)$ are integral curves through the point $x_s = c_1(s) = c_2(s)$. Again by the remarks preceding the lemma, c_1 and c_2 are equal on an open neighbourhood of s. Hence Y is open. Since (α, β) is connected it follows that $Y = (\alpha, \beta)$ and the uniqueness is proved. ■

Remark. The above theorem is relatively weak. For example, it can be shown that ϕ is C^r (we have only shown that ϕ is C^0). Also the conditions on F may be weakened. For much more extensive expositions of the existence and uniqueness theory for ordina differential equations we refer the reader to [5], [18].

Let us now return to the problem of defining differential equations on submanifolds of $\mathbf{R}^n$. Motivated by the discussion preceding the existence and uniqueness theorem we make the following definition.

Definition 4.11.3

Let N be a submanifold of $\mathbf{R}^m$. A C^r map $X:N \longrightarrow TN$ is called a C^r *vector field* on N if

$$X(x) \in T_xN \qquad \text{for all } x \in N.$$

Equivalently, X is a C^r vector field if X is C^r and $\pi_N X$ = identity map on N.

A vector field on N therefore assigns a tangent vector to each point of N.

Remark. In the sequel we shall always assume that vector fields are at least C^1.

Suppose that N is a submanifold of $\mathbf{R}^m$ and that (a, b) is an open interval containing zero. Recall that the tangent bundle of (a, b) is equal to $(a, b) \times \mathbf{R}$. If

$$c:(a, b) \longrightarrow TN$$

is a differentiable map then

$$Tc:(a, b) \times \mathbf{R} \longrightarrow TN.$$

We shall define a map $c':(a, b) \longrightarrow TN$ by

$$c'(t) = Tc(t, 1), \qquad t \in (a, b).$$

Since $N \subset \mathbf{R}^m$, $TN \subset T\mathbf{R}^m = \mathbf{R}^m \times \mathbf{R}^m$ and it follows from Proposition 4.10.7 that

$$Tc(t, 1) = (c(t), Dc_t(1)) \in \mathbf{R}^m \times \mathbf{R}^m,$$

where Dc_t is the derivative of c (regarded as an $\mathbf{R}^m$-valued map). Now $Dc_t(1)$ is what we previously defined to be $c'(t)$ in Chapter 2. The above equation gives the relation between our two definitions of c'. We use the same notation, because if we regard $T_{c(t)}N$ as a linear subspace of $T_{c(t)}\mathbf{R}^m = \mathbf{R}^m$, then the point $Tc(t, 1) \in T_{c(t)}N$ corresponds to the point $Dc_t(1) \in \mathbf{R}^m$. The context will always make it clear which interpretation of c' is intended.

As in the case of differential equations defined on open subsets of $\mathbf{R}^n$ we may make the following definition.

Definition 4.11.4

Let $X:N \longrightarrow TN$ be a C^r vector field on the submanifold N of $\mathbf{R}^m$. A differentiable map $c:(a, b) \longrightarrow N$ is said to be an *integral curve* for X if

$$c'(t) = X(c(t)) \qquad \text{for all } t \in (a, b).$$

If, in addition, $c(0) = x_0 \in N$, c is said to have *initial condition* x_0 and to be an integral curve for X *through the point* x_0.

Example. Let N be a surface in $\mathbf{R}^3$ and suppose that a particle is constrained to move on N in such a way that its velocity depends only on its position on N. That is, the velocity of the particle is given by some function $F:N \longrightarrow \mathbf{R}^3$. Suppose the particle is moving along the smooth curve $x:(a, b) \longrightarrow N$. Regarding $N \subset \mathbf{R}^3$, the function $x':(a, b) \longrightarrow \mathbf{R}^3$ gives the velocity of the particle at time t. That is

$$x'(t) = F(x(t)) \qquad \text{for all } t \in (a, b).$$

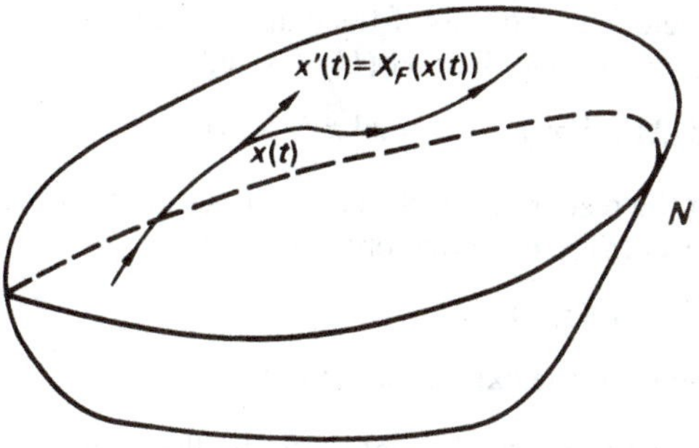

FIG. 4.44

On the other hand, since the particle is constrained to move on N, its velocity is always *tangential* to N. That is $F(x) \in T_xN$ for all $x \in N$. Hence F defines a vector field X_F on N. If the particle is moving along the curve $x:(a, b) \longrightarrow N$, we may regard the velocity of the particle on the curve as defining a map

$$x':(a, b) \longrightarrow TN$$

and we have, for all $t \in (a, b)$, the equation

$$x'(t) = X_F(x(t)).$$

In other words, an integral curve for the vector field X_F corresponds to a solution of the first-order ordinary differential equation $x' = F(x)$. Later we shall show that any integral curve for the differential equation $x' = F(x)$, with initial condition on N, defines an integral curve for X_F.

The above example suggests that to find integral curves for a vector field we have to solve first-order ordinary differential equations. The next two propositions make this relationship between vector fields and ordinary differential equations rather more precise.

Proposition 4.11.5

Let U be an open subset of $\mathbf{R}^n$ and let $F: U \longrightarrow \mathbf{R}^n$ be C^r, $r \geqslant 1$. Let $X_F: U \longrightarrow TU$ denote the C^r vector field on U defined by $X_F(x) = (x, F(x))$ for all $x \in U$. Then a differentiable map $c:(a, b) \longrightarrow U$ is an integral curve for the differential equation $x' = F(x)$ if and only if c is an integral curve for the vector field X_F.

Proof. Suppose that c is an integral curve for the differential equation $x' = F(x)$. Then we have

$$c'(t) = Dc_t(1) = F(c(t)) \qquad \text{on } (a, b).$$

That is, with the alternative interpretation of c' as a map into TU,

$$c'(t) = (c(t), Dc_t(1)) = (c(t), F(c(t))) = X_F(c(t)) \qquad \text{on } (a, b).$$

Hence c defines an integral curve for X_F. Reversing the argument gives the converse. ∎

Lemma 4.11.6

Let X be a C^r vector field on the n-dimensional submanifold N of $\mathbf{R}^m$. Suppose that (U, ϕ) is a chart for N. Then a differentiable map $c:(a, b) \longrightarrow U \subset N$ is an integral curve for the vector field X if and only if the map $\phi c:(a, b) \longrightarrow \phi(U) \subset \mathbf{R}^n$ is an integral curve for the vector field X_ϕ on $\phi(U)$ defined by

$$X_\phi(x) = (T\phi)X(\phi^{-1}(x)) \qquad \text{for all } x \in \phi(U).$$

Proof. We leave it to the reader to check that X_ϕ does define a vector field on $\phi(U)$. The proposition follows easily once one observes that

$$(\phi c)'(t) = (T\phi)(Tc)(t, 1) = T\phi(c'(t))$$

and we shall leave details to the exercises. ∎

With the notation of Lemma 4.11.6, let us define the C^r map $S_\phi: \phi(U) \longrightarrow \mathbf{R}^n$ by

$$X_\phi(x) = (x, S_\phi(x)), \qquad x \in \phi(U).$$

The next proposition follows immediately from Proposition 4.11.5 and Lemma 4.11.6.

Proposition 4.11.7

Let X be a C^r vector field on the n-dimensional submanifold N of $\mathbf{R}^m$. Suppose that (U, ϕ) is a chart for N. Then a differentiable map $c:(a, b) \longrightarrow U \subset N$ is an integral curve for the vector field X if and only if $\phi c:(a, b) \longrightarrow \phi(U) \subset \mathbf{R}^n$ is an integral curve for the first-order ordinary differential equation $x' = S_\phi(x)$.

Proposition 4.11.7 shows that vector fields are 'globalizations' of first-order ordinary differential equations. Our study of the integral curves of a vector field will make much use of the 'local' result on existence and uniqueness of solutions to ordinary differential equations (Theorem 4.11.1) and our main result will be a global form of this theorem for vector fields defined on submanifolds.

Proposition 4.11.8

Let X be a C^r vector field on the submanifold N of $\mathbf{R}^m$. Suppose that $c:(a, b) \longrightarrow N$ is an integral curve for X through the point x_0. Then

1. c is of class C^{r+1}.

2. For any $s \in (a, b)$, the curve $c(+s): (a - s, b - s) \longrightarrow N$ is an integral curve for X through $c(s)$.
3. Integral curves are unique in the sense that if

$$c_1 : (a_1, b_1) \longrightarrow N \qquad \text{and} \qquad c_2 : (a_2, b_2) \longrightarrow N$$

are integral curves for X through x_0, then

$$c_1 \mid (a_1, b_1) \cap (a_2, b_2) = c_2 \mid (a_1, b_1) \cap (a_2, b_2).$$

Proof. Let us start by proving (2). $c(+s)$ is the composite of c and the translation $t \longmapsto t + s$. Hence $T(c(+s)) = TcT(+s)$ and so

$$c'(+s)(t) = Tc(s + t, 1) = c'(s + t), \text{ proving (2).}$$

Now Theorem 4.11.1, together with Proposition 4.11.7, show that $c(+s)$ is C^{r+1} in a neighbourhood of $c(s)$ for all $s \in (a, b)$. Hence c is C^{r+1}. Theorem 4.11.1 implies that integral curves for X are locally unique and (3) then follows immediately using (2). ∎

Definition 4.11.9

Let X be a C^r vector field on the submanifold N of $\mathbf{R}^m$. Suppose that $c: (a, b) \longrightarrow N$ is an integral curve for X through x_0. The map c is said to be a *maximal integral* curve (through x_0) if (a, b) contains the domains of all other integral curves through x_0.

c is said to be *complete*, or a complete integral curve, if the domain of c is the whole of the real line.

Example. Maximal integral curves need not be complete as the following example shows. Let N be the submanifold of the real line consisting of the real line minus the origin. Take the vector field on N corresponding to velocity +1 in the positive direction. That is, define $X: N \longrightarrow TN$ by $X(t) = (t, 1)$. A point starting to the left of the origin 'drops off' N as it reaches the origin:

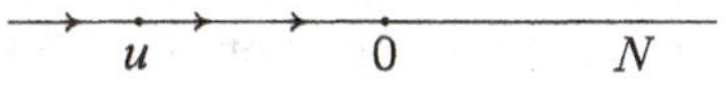

More explicitly, let $u < 0$. Then the maximal integral curve through u is given by the function $c: (-\infty, -u) \longrightarrow N$ defined by

$$c(t) = u + t.$$

Similarly, integral curves through points to the right of the origin are not defined for all negative values of t.

Proposition 4.11.10

Let X be a C^r vector field on the submanifold N of $\mathbf{R}^m$. Then there exists a unique maximal integral curve through each point of N. Furthermore, if $\gamma_x : (\alpha, \beta) \longrightarrow N$ is the maximal integral curve for X through x, then $\gamma_x(+s): (\alpha - s, \alpha - s) \longrightarrow N$ is the maximal integral curve for X through $\gamma_x(s)$.

Proof. The uniqueness of maximal integral curves follows immediately from Proposition 4.11.8. We shall prove the existence. Let (α, β) denote the union of all domains of all integral curves for X through the point $x \in N$. We shall define a map $\gamma_x : (\alpha, \beta) \longrightarrow N$. Let $t \in (\alpha, \beta)$. Then there exists an integral curve $c: (a, b) \longrightarrow N$ through x such that $t \in (a, b)$. We define $\gamma_x(t) = c(t)$. Since integral curves are unique

$\gamma_x(t)$ is defined independently of our choice c. Since $\gamma_x \mid (a, b) = c$, it follows that γ_x is differentiable at t and that $\gamma_x'(t) = X(\gamma_x(t))$. Since t was chosen arbitrarily, it follows that $\gamma_x : (\alpha, \beta) \longrightarrow N$ is an integral curve for X through x. Obviously, γ_x is the maximal integral curve through x. We leave the remaining part of the proposition to the reader. ■

Notation. Let X be a vector field on the submanifold N. For the rest of this section we shall always let $\gamma_x : D_x \longrightarrow N$ denote the maximal integral curve for X through x. If it is necessary to display the end points of D_x, we shall write $D_x = (\alpha_x, \beta_x)$. If γ_x is complete, we shall set $D_x = \mathbf{R}$ and write $\gamma_x : \mathbf{R} \longrightarrow N$ instead.

Corollary 4.11.10.1

Let X be a C^r vector field on the submanifold N of $\mathbf{R}^m$. Then, for all $x, y \in N$, either

$$\gamma_x(D_x) = \gamma_y(D_y)$$

or

$$\gamma_x(D_x) \cap \gamma_y(D_y) = \emptyset.$$

In particular, the integral curves of X define a partition of N.

Proof. Left to the exercises. ■

The set of maximal integral curves of a vector field X is usually referred to as the *phase portrait* of X. It is not our intention in this text to make any detailed analysis of the phase-portrait structure of a vector field. We refer to the references for comprehensive treatments. However, we give by Fig. 4.45 an indication of the variety of maximal integral curves that can occur for a vector field defined on an open subset of the plane.

A, C, R_1, R_2, S_1, S_2 are points where the vector field is zero and the corresponding maximal integral curves through the points are constant maps. Such points are called *zeros* or *critical points* of the vector field.

A is an example of an *attractor* (or *sink*)
R_1 and R_2 are examples of *repellers* (or *sources*)
S_1 and S_2 are examples of *saddles*
C_1 is an example of a *centre.*

The closed curve L is an example of an (attracting) *limit cycle* and the corresponding maximal integral curves are complete and periodic. That is, there exists a least strictly positive number T (the *period* of L) such that $\gamma_x(t + T) = \gamma_x(t)$ for all $t \in \mathbf{R}$ and any point $x \in L$. Similarly, we have a nested family of closed or periodic orbits about the centre point C. All other integral curves shown in the above phase portrait are non-compact.

We shall now state and prove the global existence and uniqueness theorem which generalizes Theorem 4.11.1 to vector fields.

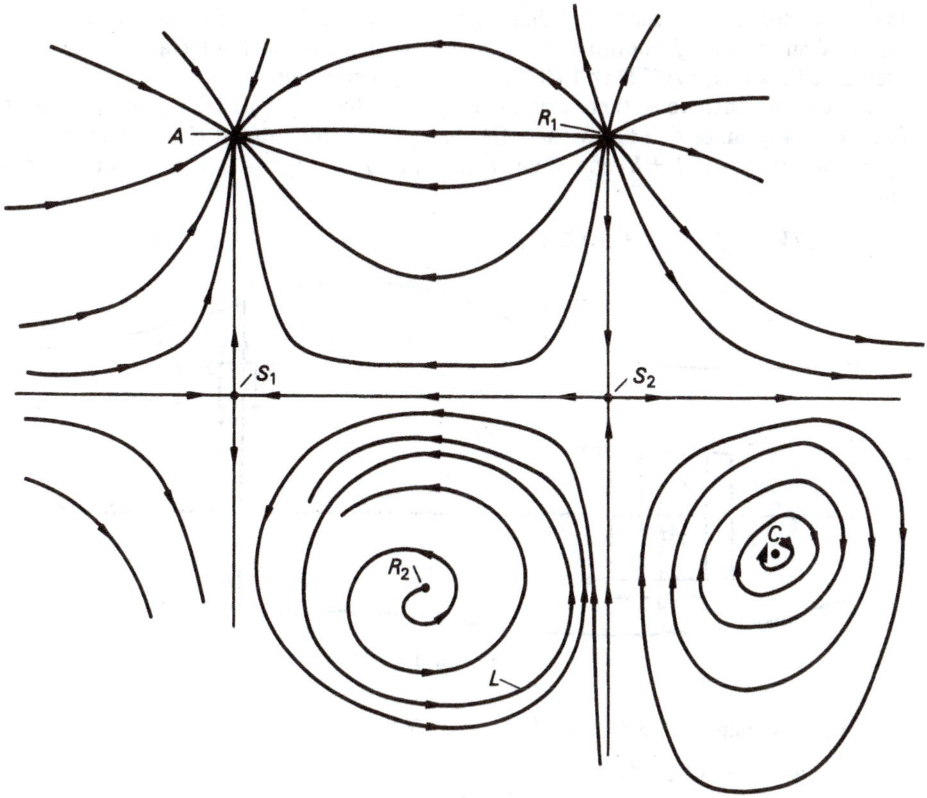

FIG. 4.45

Theorem 4.11.11

Let X be a C^r vector field on the submanifold N of $\mathbf{R}^m$. Then there exists an open neighbourhood D of $N \times \{0\}$ in $N \times \mathbf{R}$ and a continuous function $\Gamma: D \longrightarrow N$ such that if we set $D_x = \{t: (x, t) \in D\}$ for each $x \in N$, then

1. The map $\gamma_x: D_x \longrightarrow N$, defined by $\gamma_x(t) = \Gamma(x, t)$, is the maximal integral curve for X through x.
2. If N is compact, $D = N \times \mathbf{R}$. In particular, all maximal integral curves for X are complete.

Proof. Let $\gamma_x: D_x \longrightarrow N$ denote the maximal integral curve through x for X. We set

$$D = \bigcup_{x \in N} \{x\} \times D_x$$

and define $\Gamma: D \longrightarrow N$ by

$$\Gamma(x, t) = \gamma_x(t) \qquad \text{for all } (x, t) \in D.$$

We have to prove that Γ is continuous, D is open and that $D = N \times \mathbf{R}$ if N is compact.

Fix $x \in N$. Define the subset Y_x of D_x by the condition that $t \in Y_x$ if there exist open neighbourhoods V of x and J of t such that $V \times J \subset D$ and $\Gamma \mid V \times J$ is continuous. Y_x is not empty since $0 \in Y_x$ by Theorem 4.11.1. Y_x is certainly open. We shall show that Y_x is closed in D_x and hence that $Y_x = D_x$. Let $t_0 \in \bar{Y}_x$. By Theorem 4.11.1 we may find an open neighbourhood V_0 of $x_0 = \gamma_x(t_0)$ and an open interval $(-a, +a)$ such that $V_0 \times (-a, +a) \subset D$ and $\Gamma | V_0 \times (-a, +a)$ is continuous.

Since γ_x is continuous, there exists $t' \in Y_x$ such that $x' = \gamma_x(t') \in V_0$ and $|t' - t_0| <$ (see Fig. 4.46). Since $t' \in Y_x$, we may find an open neighbourhood V' of x and an open interval $(t' - b, t' + b)$ containing t' such that $\Gamma | V' \times (t' - b, t' + b)$ is continuous and

$$\Gamma(V' \times (t' - b, t' + b)) \subset V_0.$$

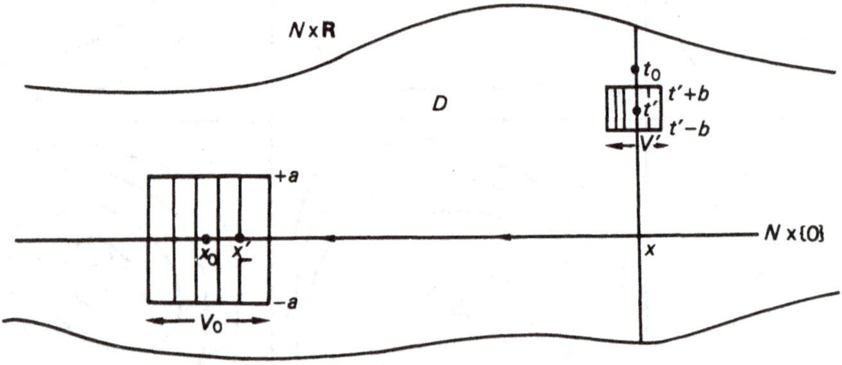

FIG. 4.46

We define the map $\tilde{\Gamma}: V' \times (t' - a, t' + a) \longrightarrow N$ by

$$\tilde{\Gamma}(x, t) = \Gamma(\Gamma(x, t'), t - t').$$

Since $\tilde{\Gamma}$ is a composite of continuous maps, $\tilde{\Gamma}$ is continuous. Also $\tilde{\Gamma}$ is differentiable in the t variable by the composite-mapping formula. Let $\tilde{\gamma}_x : (t' - a, t' + a) \longrightarrow N$ be the map defined, for $x \in V'$, by

$$\tilde{\gamma}_x(t) = \tilde{\Gamma}(x, t).$$

Now

$$\gamma_x'(t) = \Gamma_{;2}(\Gamma(x, t'), t - t') = \gamma_{x'}'(t - t')$$
$$= X(\gamma_{x'}(t - t')) = X(\Gamma(x, t'), t - t').$$

Hence $\tilde{\gamma}_x$ defines an integral curve for X for all $x \in V'$. We also have

$$\tilde{\gamma}_x(t') = \Gamma(\Gamma(x, t'), 0) = x' = \gamma_x(t')$$

and so, by the uniqueness of integral curves, $\tilde{\gamma}_x$ is equal to γ_x on $(t' - a, t' + a) \subset D_x$. Therefore, $\tilde{\Gamma} = \Gamma$. Since $\tilde{\Gamma}$ is continuous and $V' \times (t' - a, t' + a) \subset D$, it follows that $t_0 \in Y_x$ and hence $D_x = Y_x$. Since x was chosen arbitrarily, D is open and Γ is continuous on D.

To complete the proof, we must show that $D = N \times \mathbf{R}$ if N is compact. Let $x \in N$ and suppose that $D_x = (\alpha_x, \beta_x) \neq \mathbf{R}$. Without loss of generality suppose that $\beta_x < \infty$. Since N is compact we may find an increasing sequence $(t_n)_{n=1}^{\infty}$ of points of D_x converging to β_x such that the sequence $(\gamma_x(t_n))_{n=1}^{\infty}$ converges to some point $y \in N$. By Theorem 4.11.1 we can find an open neighbourhood V of y and $a > 0$ such that

$$D_z \supset (-a, +a) \qquad \text{for all } z \in V.$$

Choose n large, so that

$$\beta_x - t_n < a \qquad \text{and} \qquad \gamma_x(t_n) \in V.$$

Set $x_n = \gamma_x(t_n)$. It follows that the maximal integral curve through x_n meets the maximal integral curve through x. Hence, as sets, the two curves are equal. It follows from Proposition 4.11.10 that

$$(\alpha_{x_n}, \beta_{x_n}) = (\alpha_x - t_n, \beta_x - t_n).$$

This is absurd since $a \in (\alpha_{x_n}, \beta_{x_n})$ and $\beta_x - t_n < a$. Hence $\beta_x = \infty$. ∎

Remarks

1. It may be shown that the map Γ defined in Theorem 4.11.11 is C^r as a map from D to N.
2. If N is compact, $\Gamma: N \times \mathbf{R} \longrightarrow N$ and for all $x \in N$ and $t, t' \in \mathbf{R}$ we have the identity

   $$\Gamma(\Gamma(x, t), t') = \Gamma(x, t + t').$$

   Let $\Gamma_t: N \longrightarrow N$ denote the map defined for all $t \in \mathbf{R}$ by

   $$\Gamma_t(x) = \Gamma(x, t), \qquad x \in N.$$

   It follows from the above identity that for all $t, t' \in \mathbf{R}$ we have

   $$\Gamma_t \Gamma_{t'} = \Gamma_{t+t'} \qquad \text{and} \qquad \Gamma_0 = \text{identity map on } N.$$

   In particular, taking $t = -t'$, Γ_t is a homeomorphism of N for all $t \in \mathbf{R}$. (Using Remark 1, Γ_t is actually a C^r diffeomorphism of N.) Any C^r map $\Gamma: N \times \mathbf{R} \longrightarrow N$ satisfying the above relations is called a *flow* or *one-parameter group of diffeomorphisms* of N. Notice that if we are given a C^r flow on N we can construct a (C^{r-1}) vector field X on N by setting $X(x)$ = time derivative of Γ at $(x, 0)$.

To end this section we describe an important way of constructing vector fields on submanifolds of Euclidean space.

Let N be a submanifold of $\mathbf{R}^m$ and suppose that $\mathbf{R}^m$ is given the standard Euclidean inner product $\langle \, , \rangle$. Since $T_xN \subset \mathbf{R}^m$ as a linear subspace, $\langle \, , \rangle$ induces an inner product $\langle \, , \rangle_x$ on T_xN for all $x \in N$. Suppose that $f: N \longrightarrow \mathbf{R}$ is a differentiable map. Then

$$T_xf: T_xN \longrightarrow \mathbf{R}$$

is a linear map for all $x \in N$. Using the inner product $\langle \, , \rangle_x$ on T_xN we may define a unique point grad $f(x) \in T_xN$ by the rule

$$T_xf(e) = \langle \operatorname{grad} f(x), e \rangle_x, \qquad e \in T_xN.$$

This construction defines a map grad $f: N \longrightarrow TN$. Since grad $f(x) \in T_xN$ for all $x \in N$, grad f is a vector field on N. We shall call grad f the *gradient* of f or the *gradient vector field* of f.

Lemma 4.11.12

Let N be a submanifold of $\mathbf{R}^m$ and $f: N \longrightarrow \mathbf{R}$ be a C^r function. Then grad f is a C^{r-1} vector field on N.

Proof. Let $\phi: T\mathbf{R}^m \longrightarrow \mathbf{R}^m \times \mathbf{R}^{m*}$ be the C^∞ map defined by

$$\phi(x, e) = (x, \gamma(e)),$$

where $\gamma: \mathbf{R}^m \longrightarrow \mathbf{R}^{m^*}$ is the linear isomorphism given by the Riesz representation theorem. Since ϕ is a C^∞ diffeomorphism, $\phi(TN)$ is a submanifold of $\mathbf{R}^m \times \mathbf{R}^{m^*}$. Now $\phi(T_xN) = \gamma(T_xN) = T_xN^*$ (the dual space of T_xN). Hence $\phi(TN)$ is the disjoint union of the set of dual tangent spaces of N. We set $\phi(TN) = TN^*$. Now since $T_xf: T_xN \longrightarrow \mathbf{R}$ is linear for all $x \in N$, it follows that Tf induces a C^{r-1} map $\mathrm{d}f: N \longrightarrow TN^*$ defined by

$$\mathrm{d}f(x) = Tf_x \in L(T_xN, \mathbf{R}) = T_xN^*.$$

grad f is equal to the composite $\phi^{-1}\,\mathrm{d}f$ and, by the composite-mapping theorem, is C^{r-1}. ∎

Characteristic of the properties of the gradient of a function is the following result.

Proposition 4.11.13

Let N be a submanifold of $\mathbf{R}^m$ and suppose that $f: N \longrightarrow \mathbf{R}$ is a C^∞ map. Let $c \in \mathbf{R}$ be a regular value of f. Then grad $f(x) \perp T_x(f^{-1}(c))$ for all $x \in f^{-1}(c)$.

Proof. $T_x(f^{-1}(c)) = \mathrm{Ker}\,(T_xf) \subset T_xN$. ∎

Example. Let $f: S^2 \longrightarrow \mathbf{R}$ denote the restriction of the height function $g(x, y, z) = z$ to S^2, illustrated in Fig. 4.47. (See p. 265.)

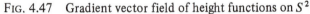

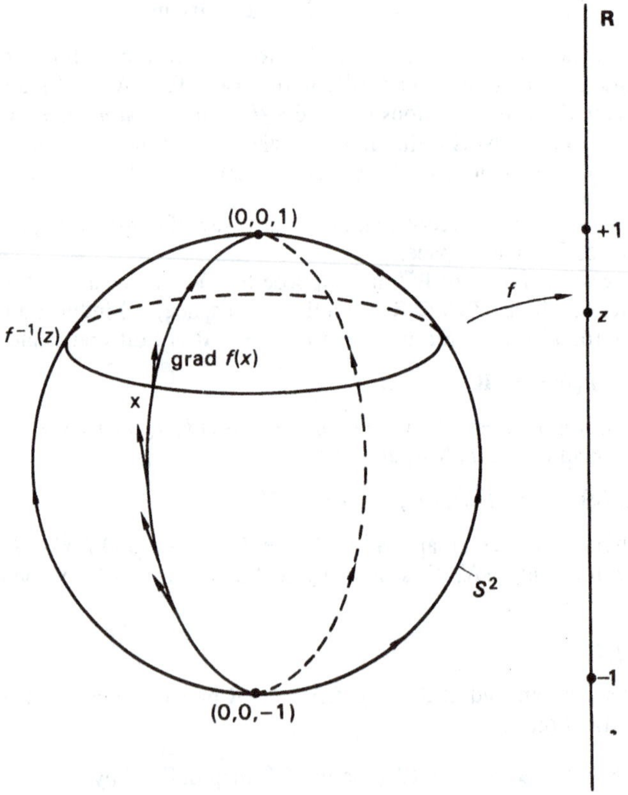

FIG. 4.47 Gradient vector field of height functions on S^2

grad $f(x) \in T_xS^2$ for all $x \in S^2$. The only critical points for f are (0, 0, 1) and (0, 0, −1). Therefore, $f^{-1}(z)$ is a submanifold of S^2 of dimension 1 for $z \in (-1, 1)$. From Proposition 4.11.13 we have

$$\text{grad } f(x) \perp T_x(f^{-1}(f(x)))$$

and grad $f(x)$ is the tangent vector to S^2 at x pointing in the direction of the greatest rate of increase of f. That is, in the direction of steepest 'slope' or 'ascent' (remember f is a height function). This explains the use of the term 'gradient'. It is easy to see what the integral curves of the gradient vector field are: The points (0, 0, 1) and (0, 0, −1) are respectively a sink and a source. If x is any other point on S^2, the integral curve through x is the great circle through x linking (0, 0, 1) and (0, 0, −1). Notice that the integral curves are orthogonal to the level surfaces of f.

Exercises

1. Let $X: N \longrightarrow TN$ be a vector field on the submanifold N of $\mathbf{R}^m$. Show that if N is an open subset of $\mathbf{R}^m$ it is always possible to find X without any zeros. Can one find a vector field on S^2 with no zeros? (Proof not required; draw phase portraits.) On T^2?
2. Complete the proof of Lemma 4.11.6.
3. Prove Corollary 4.11.10.1.
4. Let N be a submanifold of $\mathbf{R}^p$ and suppose that $f: N \longrightarrow \mathbf{R}$ is differentiable. Let $\phi: (a, b) \longrightarrow N$ be a trajectory of the vector field grad f. Show that

   (a) If ϕ is not a constant trajectory, f increases strictly along the curve defined by ϕ.
   (b) If ϕ is not a constant trajectory, there do not exist points $t, t' \in (a, b)$ such that $\phi(t) = \phi(t')$. (A gradient vector field does not have any periodic orbits.)
5. Show, by means of examples, that not every vector field is a gradient vector field.
6. Show that a point in TS^2 naturally defines the position and velocity of a point on S^2. For discussion: Suppose that a point moves on S^2 under a force depending on position and velocity according to Newton's laws of motion. Show that the differential equations of motion define a vector field on TS^2, that is a map $F: TS^2 \longrightarrow T(TS^2)$ satisfying $\pi_{TS^2} F =$ identity map on S^2. Show also that $(T\pi_{S^2})F =$ identity map on S^2. Show that vector fields on TS^2 satisfying the latter condition are naturally associated with force fields on S^2 depending only on position and velocity.

Bibliography

[1] R. Abraham and J. E. Marsden, *Foundations of mechanics*, Benjamin, New York, 1967.

[2] R. Abraham and J. Robbin, *Transversal mappings and flows*, Benjamin, New York, 1967.

[3] L. Auslander and I. Mackensie, *Introduction to differentiable manifolds*, McGraw-Hill, New York, 1963.

[4] F. Brickell and R. S. Clark, *Differentiable manifolds: An introduction*, Van Nostrand Reinhold, London, 1970.

[5] R. V. Churchill, *Fourier series and boundary value problems*, 2nd ed. McGraw-Hill, 1963.

[6] E. Coddington and N. Levinson, *Theory of ordinary differential equations*, McGraw-Hill, New York, 1955.

[7] J. Dieudonné, *Foundations of modern analysis*, Academic Press, New York and London, 1960.

[8] R. E. Edwards, *Fourier series, a modern introduction*, Vol. I, Holt, Rinehart and Winston, 1967.

[9] P. Halmos, *Finite dimensional vector spaces*, 2nd ed., Van Nostrand, London, 1958.

[10] M. W. Hirsch and S. Smale, *Differential equations, dynamical systems and linear algebra*, Academic Press, New York and London, 1974.

[11] J. Kelley, *General topology*, Van Nostrand; London, 1963.

[12] J. La Salle and S. Lefschetz, *Stability by Liapunov's direct method with applications*, Academic Press, New York and London, 1961.

[13] S. Lang, *Differential manifolds*, Addison-Wesley, Reading, Mass., 1972.

[14] Lighthill, *Fourier analysis and generalised functions*, Cambridge Mathematical monographs on mechanics and applied mathematics, Cambridge University Press, Cambridge, 1959.

[15] B. Mendelson, *Introduction to topology*, 2nd ed., Allyn and Bacon, Boston, 1968.

[16] J. Milnor, *Morse theory* (based on lecture notes by M. Spivak and R. Wells), Annals of Mathematics Study No. 51, Princeton University Press, Princeton, 1963.

[17] J. Milnor, *Topology from the differentiable viewpoint*, The University Press of Virginia, 1965.

[18] F. J. Murray and K. S. Miller, *Existence theorems for ordinary differential equations,* New York University Press, New York, 1954.

[19] A. P. Robertson and Wendy Robertson, *Topological vector spaces*, Cambridge Tracts in Mathematical Physics, Cambridge University Press, Cambridge, 1973.

[20] A. Sommerfield, *Partial differential equations*, Academic Press, New York, 1969.

[21] A. Zygmund, *Trigonometrical series*, Dover Publications, New York, 1955.

Index

A CATALOG OF SELECTED

DOVER BOOKS

IN SCIENCE AND MATHEMATICS

Astronomy

CHARIOTS FOR APOLLO: The NASA History of Manned Lunar Spacecraft to 1969, Courtney G. Brooks, James M. Grimwood, and Loyd S. Swenson, Jr. This illustrated history by a trio of experts is the definitive reference on the Apollo spacecraft and lunar modules. It traces the vehicles' design, development, and operation in space. More than 100 photographs and illustrations. 576pp. 6 3/4 x 9 1/4. 0-486-46756-2

EXPLORING THE MOON THROUGH BINOCULARS AND SMALL TELESCOPES, Ernest H. Cherrington, Jr. Informative, profusely illustrated guide to locating and identifying craters, rills, seas, mountains, other lunar features. Newly revised and updated with special section of new photos. Over 100 photos and diagrams. 240pp. 8 1/4 x 11. 0-486-24491-1

WHERE NO MAN HAS GONE BEFORE: A History of NASA's Apollo Lunar Expeditions, William David Compton. Introduction by Paul Dickson. This official NASA history traces behind-the-scenes conflicts and cooperation between scientists and engineers. The first half concerns preparations for the Moon landings, and the second half documents the flights that followed Apollo 11. 1989 edition. 432pp. 7 x 10. 0-486-47888-2

APOLLO EXPEDITIONS TO THE MOON: The NASA History, Edited by Edgar M. Cortright. Official NASA publication marks the 40th anniversary of the first lunar landing and features essays by project participants recalling engineering and administrative challenges. Accessible, jargon-free accounts, highlighted by numerous illustrations. 336pp. 8 3/8 x 10 7/8. 0-486-47175-6

ON MARS: Exploration of the Red Planet, 1958-1978--The NASA History, Edward Clinton Ezell and Linda Neuman Ezell. NASA's official history chronicles the start of our explorations of our planetary neighbor. It recounts cooperation among government, industry, and academia, and it features dozens of photos from Viking cameras. 560pp. 6 3/4 x 9 1/4. 0-486-46757-0

ARISTARCHUS OF SAMOS: The Ancient Copernicus, Sir Thomas Heath. Heath's history of astronomy ranges from Homer and Hesiod to Aristarchus and includes quotes from numerous thinkers, compilers, and scholasticists from Thales and Anaximander through Pythagoras, Plato, Aristotle, and Heraclides. 34 figures. 448pp. 5 3/8 x 8 1/2. 0-486-43886-4

AN INTRODUCTION TO CELESTIAL MECHANICS, Forest Ray Moulton. Classic text still unsurpassed in presentation of fundamental principles. Covers rectilinear motion, central forces, problems of two and three bodies, much more. Includes over 200 problems, some with answers. 437pp. 5 3/8 x 8 1/2. 0-486-64687-4

BEYOND THE ATMOSPHERE: Early Years of Space Science, Homer E. Newell. This exciting survey is the work of a top NASA administrator who chronicles technological advances, the relationship of space science to general science, and the space program's social, political, and economic contexts. 528pp. 6 3/4 x 9 1/4. 0-486-47464-X

STAR LORE: Myths, Legends, and Facts, William Tyler Olcott. Captivating retellings of the origins and histories of ancient star groups include Pegasus, Ursa Major, Pleiades, signs of the zodiac, and other constellations. "Classic." – *Sky & Telescope*. 58 illustrations. 544pp. 5 3/8 x 8 1/2. 0-486-43581-4

A COMPLETE MANUAL OF AMATEUR ASTRONOMY: Tools and Techniques for Astronomical Observations, P. Clay Sherrod with Thomas L. Koed. Concise, highly readable book discusses the selection, set-up, and maintenance of a telescope; amateur studies of the sun; lunar topography and occultations; and more. 124 figures. 26 halftones. 37 tables. 335pp. 6 1/2 x 9 1/4. 0-486-42820-6

Chemistry

MOLECULAR COLLISION THEORY, M. S. Child. This high-level monograph offers an analytical treatment of classical scattering by a central force, quantum scattering by a central force, elastic scattering phase shifts, and semi-classical elastic scattering. 1974 edition. 310pp. 5 3/8 x 8 1/2. 0-486-69437-2

HANDBOOK OF COMPUTATIONAL QUANTUM CHEMISTRY, David B. Cook. This comprehensive text provides upper-level undergraduates and graduate students with an accessible introduction to the implementation of quantum ideas in molecular modeling, exploring practical applications alongside theoretical explanations. 1998 edition. 832pp. 5 3/8 x 8 1/2. 0-486-44307-8

RADIOACTIVE SUBSTANCES, Marie Curie. The celebrated scientist's thesis, which directly preceded her 1903 Nobel Prize, discusses establishing atomic character of radioactivity; extraction from pitchblende of polonium and radium; isolation of pure radium chloride; more. 96pp. 5 3/8 x 8 1/2. 0-486-42550-9

CHEMICAL MAGIC, Leonard A. Ford. Classic guide provides intriguing entertainment while elucidating sound scientific principles, with more than 100 unusual stunts: cold fire, dust explosions, a nylon rope trick, a disappearing beaker, much more. 128pp. 5 3/8 x 8 1/2. 0-486-67628-5

ALCHEMY, E. J. Holmyard. Classic study by noted authority covers 2,000 years of alchemical history: religious, mystical overtones; apparatus; signs, symbols, and secret terms; advent of scientific method, much more. Illustrated. 320pp. 5 3/8 x 8 1/2. 0-486-26298-7

CHEMICAL KINETICS AND REACTION DYNAMICS, Paul L. Houston. This text teaches the principles underlying modern chemical kinetics in a clear, direct fashion, using several examples to enhance basic understanding. Solutions to selected problems. 2001 edition. 352pp. 8 3/8 x 11. 0-486-45334-0

PROBLEMS AND SOLUTIONS IN QUANTUM CHEMISTRY AND PHYSICS, Charles S. Johnson and Lee G. Pedersen. Unusually varied problems, with detailed solutions, cover of quantum mechanics, wave mechanics, angular momentum, molecular spectroscopy, scattering theory, more. 280 problems, plus 139 supplementary exercises. 430pp. 6 1/2 x 9 1/4. 0-486-65236-X

ELEMENTS OF CHEMISTRY, Antoine Lavoisier. Monumental classic by the founder of modern chemistry features first explicit statement of law of conservation of matter in chemical change, and more. Facsimile reprint of original (1790) Kerr translation. 539pp. 5 3/8 x 8 1/2. 0-486-64624-6

MAGNETISM AND TRANSITION METAL COMPLEXES, F. E. Mabbs and D. J. Machin. A detailed view of the calculation methods involved in the magnetic properties of transition metal complexes, this volume offers sufficient background for original work in the field. 1973 edition. 240pp. 5 3/8 x 8 1/2. 0-486-46284-6

GENERAL CHEMISTRY, Linus Pauling. Revised third edition of classic first-year text by Nobel laureate. Atomic and molecular structure, quantum mechanics, statistical mechanics, thermodynamics correlated with descriptive chemistry. Problems. 992pp. 5 3/8 x 8 1/2. 0-486-65622-5

ELECTROLYTE SOLUTIONS: Second Revised Edition, R. A. Robinson and R. H. Stokes. Classic text deals primarily with measurement, interpretation of conductance, chemical potential, and diffusion in electrolyte solutions. Detailed theoretical interpretations, plus extensive tables of thermodynamic and transport properties. 1970 edition. 590pp. 5 3/8 x 8 1/2. 0-486-42225-9

Engineering

FUNDAMENTALS OF ASTRODYNAMICS, Roger R. Bate, Donald D. Mueller, and Jerry E. White. Teaching text developed by U.S. Air Force Academy develops the basic two-body and n-body equations of motion; orbit determination; classical orbital elements, coordinate transformations; differential correction; more. 1971 edition. 455pp. 5 3/8 x 8 1/2. 0-486-60061-0

INTRODUCTION TO CONTINUUM MECHANICS FOR ENGINEERS: Revised Edition, Ray M. Bowen. This self-contained text introduces classical continuum models within a modern framework. Its numerous exercises illustrate the governing principles, linearizations, and other approximations that constitute classical continuum models. 2007 edition. 320pp. 6 1/8 x 9 1/4. 0-486-47460-7

ENGINEERING MECHANICS FOR STRUCTURES, Louis L. Bucciarelli. This text explores the mechanics of solids and statics as well as the strength of materials and elasticity theory. Its many design exercises encourage creative initiative and systems thinking. 2009 edition. 320pp. 6 1/8 x 9 1/4. 0-486-46855-0

FEEDBACK CONTROL THEORY, John C. Doyle, Bruce A. Francis and Allen R. Tannenbaum. This excellent introduction to feedback control system design offers a theoretical approach that captures the essential issues and can be applied to a wide range of practical problems. 1992 edition. 224pp. 6 1/2 x 9 1/4. 0-486-46933-6

THE FORCES OF MATTER, Michael Faraday. These lectures by a famous inventor offer an easy-to-understand introduction to the interactions of the universe's physical forces. Six essays explore gravitation, cohesion, chemical affinity, heat, magnetism, and electricity. 1993 edition. 96pp. 5 3/8 x 8 1/2. 0-486-47482-8

DYNAMICS, Lawrence E. Goodman and William H. Warner. Beginning engineering text introduces calculus of vectors, particle motion, dynamics of particle systems and plane rigid bodies, technical applications in plane motions, and more. Exercises and answers in every chapter. 619pp. 5 3/8 x 8 1/2. 0-486-42006-X

ADAPTIVE FILTERING PREDICTION AND CONTROL, Graham C. Goodwin and Kwai Sang Sin. This unified survey focuses on linear discrete-time systems and explores natural extensions to nonlinear systems. It emphasizes discrete-time systems, summarizing theoretical and practical aspects of a large class of adaptive algorithms. 1984 edition. 560pp. 6 1/2 x 9 1/4. 0-486-46932-8

INDUCTANCE CALCULATIONS, Frederick W. Grover. This authoritative reference enables the design of virtually every type of inductor. It features a single simple formula for each type of inductor, together with tables containing essential numerical factors. 1946 edition. 304pp. 5 3/8 x 8 1/2. 0-486-47440-2

THERMODYNAMICS: Foundations and Applications, Elias P. Gyftopoulos and Gian Paolo Beretta. Designed by two MIT professors, this authoritative text discusses basic concepts and applications in detail, emphasizing generality, definitions, and logical consistency. More than 300 solved problems cover realistic energy systems and processes. 800pp. 6 1/8 x 9 1/4. 0-486-43932-1

THE FINITE ELEMENT METHOD: Linear Static and Dynamic Finite Element Analysis, Thomas J. R. Hughes. Text for students without in-depth mathematical training, this text includes a comprehensive presentation and analysis of algorithms of time-dependent phenomena plus beam, plate, and shell theories. Solution guide available upon request. 672pp. 6 1/2 x 9 1/4. 0-486-41181-8

Mathematics-Bestsellers

HANDBOOK OF MATHEMATICAL FUNCTIONS: with Formulas, Graphs, and Mathematical Tables, Edited by Milton Abramowitz and Irene A. Stegun. A classic resource for working with special functions, standard trig, and exponential logarithmic definitions and extensions, it features 29 sets of tables, some to as high as 20 places. 1046pp. 8 x 10 1/2. 0-486-61272-4

ABSTRACT AND CONCRETE CATEGORIES: The Joy of Cats, Jiri Adamek, Horst Herrlich, and George E. Strecker. This up-to-date introductory treatment employs category theory to explore the theory of structures. Its unique approach stresses concrete categories and presents a systematic view of factorization structures. Numerous examples. 1990 edition, updated 2004. 528pp. 6 1/8 x 9 1/4. 0-486-46934-4

MATHEMATICS: Its Content, Methods and Meaning, A. D. Aleksandrov, A. N. Kolmogorov, and M. A. Lavrent'ev. Major survey offers comprehensive, coherent discussions of analytic geometry, algebra, differential equations, calculus of variations, functions of a complex variable, prime numbers, linear and non-Euclidean geometry, topology, functional analysis, more. 1963 edition. 1120pp. 5 3/8 x 8 1/2. 0-486-40916-3

INTRODUCTION TO VECTORS AND TENSORS: Second Edition--Two Volumes Bound as One, Ray M. Bowen and C.-C. Wang. Convenient single-volume compilation of two texts offers both introduction and in-depth survey. Geared toward engineering and science students rather than mathematicians, it focuses on physics and engineering applications. 1976 edition. 560pp. 6 1/2 x 9 1/4. 0-486-46914-X

AN INTRODUCTION TO ORTHOGONAL POLYNOMIALS, Theodore S. Chihara. Concise introduction covers general elementary theory, including the representation theorem and distribution functions, continued fractions and chain sequences, the recurrence formula, special functions, and some specific systems. 1978 edition. 272pp. 5 3/8 x 8 1/2. 0-486-47929-3

ADVANCED MATHEMATICS FOR ENGINEERS AND SCIENTISTS, Paul DuChateau. This primary text and supplemental reference focuses on linear algebra, calculus, and ordinary differential equations. Additional topics include partial differential equations and approximation methods. Includes solved problems. 1992 edition. 400pp. 7 1/2 x 9 1/4. 0-486-47930-7

PARTIAL DIFFERENTIAL EQUATIONS FOR SCIENTISTS AND ENGINEERS, Stanley J. Farlow. Practical text shows how to formulate and solve partial differential equations. Coverage of diffusion-type problems, hyperbolic-type problems, elliptic-type problems, numerical and approximate methods. Solution guide available upon request. 1982 edition. 414pp. 6 1/8 x 9 1/4. 0-486-67620-X

VARIATIONAL PRINCIPLES AND FREE-BOUNDARY PROBLEMS, Avner Friedman. Advanced graduate-level text examines variational methods in partial differential equations and illustrates their applications to free-boundary problems. Features detailed statements of standard theory of elliptic and parabolic operators. 1982 edition. 720pp. 6 1/8 x 9 1/4. 0-486-47853-X

LINEAR ANALYSIS AND REPRESENTATION THEORY, Steven A. Gaal. Unified treatment covers topics from the theory of operators and operator algebras on Hilbert spaces; integration and representation theory for topological groups; and the theory of Lie algebras, Lie groups, and transform groups. 1973 edition. 704pp. 6 1/8 x 9 1/4. 0-486-47851-3

Browse over 9,000 books at www.doverpublications.com

Mathematics–Logic and Problem Solving

PERPLEXING PUZZLES AND TANTALIZING TEASERS, Martin Gardner. Ninety-three riddles, mazes, illusions, tricky questions, word and picture puzzles, and other challenges offer hours of entertainment for youngsters. Filled with rib-tickling drawings. Solutions. 224pp. 5 3/8 x 8 1/2. 0-486-25637-5

MY BEST MATHEMATICAL AND LOGIC PUZZLES, Martin Gardner. The noted expert selects 70 of his favorite "short" puzzles. Includes The Returning Explorer, The Mutilated Chessboard, Scrambled Box Tops, and dozens more. Complete solutions included. 96pp. 5 3/8 x 8 1/2. 0-486-28152-3

THE LADY OR THE TIGER?: and Other Logic Puzzles, Raymond M. Smullyan. Created by a renowned puzzle master, these whimsically themed challenges involve paradoxes about probability, time, and change; metapuzzles; and self-referentiality. Nineteen chapters advance in difficulty from relatively simple to highly complex. 1982 edition. 240pp. 5 3/8 x 8 1/2. 0-486-47027-X

SATAN, CANTOR AND INFINITY: Mind-Boggling Puzzles, Raymond M. Smullyan. A renowned mathematician tells stories of knights and knaves in an entertaining look at the logical precepts behind infinity, probability, time, and change. Requires a strong background in mathematics. Complete solutions. 288pp. 5 3/8 x 8 1/2. 0-486-47036-9

THE RED BOOK OF MATHEMATICAL PROBLEMS, Kenneth S. Williams and Kenneth Hardy. Handy compilation of 100 practice problems, hints and solutions indispensable for students preparing for the William Lowell Putnam and other mathematical competitions. Preface to the First Edition. Sources. 1988 edition. 192pp. 5 3/8 x 8 1/2. 0-486-69415-1

KING ARTHUR IN SEARCH OF HIS DOG AND OTHER CURIOUS PUZZLES, Raymond M. Smullyan. This fanciful, original collection for readers of all ages features arithmetic puzzles, logic problems related to crime detection, and logic and arithmetic puzzles involving King Arthur and his Dogs of the Round Table. 160pp. 5 3/8 x 8 1/2. 0-486-47435-6

UNDECIDABLE THEORIES: Studies in Logic and the Foundation of Mathematics, Alfred Tarski in collaboration with Andrzej Mostowski and Raphael M. Robinson. This well-known book by the famed logician consists of three treatises: "A General Method in Proofs of Undecidability," "Undecidability and Essential Undecidability in Mathematics," and "Undecidability of the Elementary Theory of Groups." 1953 edition. 112pp. 5 3/8 x 8 1/2. 0-486-47703-7

LOGIC FOR MATHEMATICIANS, J. Barkley Rosser. Examination of essential topics and theorems assumes no background in logic. "Undoubtedly a major addition to the literature of mathematical logic." – *Bulletin of the American Mathematical Society.* 1978 edition. 592pp. 6 1/8 x 9 1/4. 0-486-46898-4

INTRODUCTION TO PROOF IN ABSTRACT MATHEMATICS, Andrew Wohlgemuth. This undergraduate text teaches students what constitutes an acceptable proof, and it develops their ability to do proofs of routine problems as well as those requiring creative insights. 1990 edition. 384pp. 6 1/2 x 9 1/4. 0-486-47854-8

FIRST COURSE IN MATHEMATICAL LOGIC, Patrick Suppes and Shirley Hill. Rigorous introduction is simple enough in presentation and context for wide range of students. Symbolizing sentences; logical inference; truth and validity; truth tables; terms, predicates, universal quantifiers; universal specification and laws of identity; more. 288pp. 5 3/8 x 8 1/2. 0-486-42259-3

Mathematics–Algebra and Calculus

VECTOR CALCULUS, Peter Baxandall and Hans Liebeck. This introductory text offers a rigorous, comprehensive treatment. Classical theorems of vector calculus are amply illustrated with figures, worked examples, physical applications, and exercises with hints and answers. 1986 edition. 560pp. 5 3/8 x 8 1/2. 0-486-46620-5

ADVANCED CALCULUS: An Introduction to Classical Analysis, Louis Brand. A course in analysis that focuses on the functions of a real variable, this text introduces the basic concepts in their simplest setting and illustrates its teachings with numerous examples, theorems, and proofs. 1955 edition. 592pp. 5 3/8 x 8 1/2. 0-486-44548-8

ADVANCED CALCULUS, Avner Friedman. Intended for students who have already completed a one-year course in elementary calculus, this two-part treatment advances from functions of one variable to those of several variables. Solutions. 1971 edition. 432pp. 5 3/8 x 8 1/2. 0-486-45795-8

METHODS OF MATHEMATICS APPLIED TO CALCULUS, PROBABILITY, AND STATISTICS, Richard W. Hamming. This 4-part treatment begins with algebra and analytic geometry and proceeds to an exploration of the calculus of algebraic functions and transcendental functions and applications. 1985 edition. Includes 310 figures and 18 tables. 880pp. 6 1/2 x 9 1/4. 0-486-43945-3

BASIC ALGEBRA I: Second Edition, Nathan Jacobson. A classic text and standard reference for a generation, this volume covers all undergraduate algebra topics, including groups, rings, modules, Galois theory, polynomials, linear algebra, and associative algebra. 1985 edition. 528pp. 6 1/8 x 9 1/4. 0-486-47189-6

BASIC ALGEBRA II: Second Edition, Nathan Jacobson. This classic text and standard reference comprises all subjects of a first-year graduate-level course, including in-depth coverage of groups and polynomials and extensive use of categories and functors. 1989 edition. 704pp. 6 1/8 x 9 1/4. 0-486-47187-X

CALCULUS: An Intuitive and Physical Approach (Second Edition), Morris Kline. Application-oriented introduction relates the subject as closely as possible to science with explorations of the derivative; differentiation and integration of the powers of x; theorems on differentiation, antidifferentiation; the chain rule; trigonometric functions; more. Examples. 1967 edition. 960pp. 6 1/2 x 9 1/4. 0-486-40453-6

ABSTRACT ALGEBRA AND SOLUTION BY RADICALS, John E. Maxfield and Margaret W. Maxfield. Accessible advanced undergraduate-level text starts with groups, rings, fields, and polynomials and advances to Galois theory, radicals and roots of unity, and solution by radicals. Numerous examples, illustrations, exercises, appendixes. 1971 edition. 224pp. 6 1/8 x 9 1/4. 0-486-47723-1

AN INTRODUCTION TO THE THEORY OF LINEAR SPACES, Georgi E. Shilov. Translated by Richard A. Silverman. Introductory treatment offers a clear exposition of algebra, geometry, and analysis as parts of an integrated whole rather than separate subjects. Numerous examples illustrate many different fields, and problems include hints or answers. 1961 edition. 320pp. 5 3/8 x 8 1/2. 0-486-63070-6

LINEAR ALGEBRA, Georgi E. Shilov. Covers determinants, linear spaces, systems of linear equations, linear functions of a vector argument, coordinate transformations, the canonical form of the matrix of a linear operator, bilinear and quadratic forms, and more. 387pp. 5 3/8 x 8 1/2. 0-486-63518-X

Mathematics–Probability and Statistics

BASIC PROBABILITY THEORY, Robert B. Ash. This text emphasizes the probabilistic way of thinking, rather than measure-theoretic concepts. Geared toward advanced undergraduates and graduate students, it features solutions to some of the problems. 1970 edition. 352pp. 5 3/8 x 8 1/2. 0-486-46628-0

PRINCIPLES OF STATISTICS, M. G. Bulmer. Concise description of classical statistics, from basic dice probabilities to modern regression analysis. Equal stress on theory and applications. Moderate difficulty; only basic calculus required. Includes problems with answers. 252pp. 5 5/8 x 8 1/4. 0-486-63760-3

OUTLINE OF BASIC STATISTICS: Dictionary and Formulas, John E. Freund and Frank J. Williams. Handy guide includes a 70-page outline of essential statistical formulas covering grouped and ungrouped data, finite populations, probability, and more, plus over 1,000 clear, concise definitions of statistical terms. 1966 edition. 208pp. 5 3/8 x 8 1/2. 0-486-47769-X

GOOD THINKING: The Foundations of Probability and Its Applications, Irving J. Good. This in-depth treatment of probability theory by a famous British statistician explores Keynesian principles and surveys such topics as Bayesian rationality, corroboration, hypothesis testing, and mathematical tools for induction and simplicity. 1983 edition. 352pp. 5 3/8 x 8 1/2. 0-486-47438-0

INTRODUCTION TO PROBABILITY THEORY WITH CONTEMPORARY APPLICATIONS, Lester L. Helms. Extensive discussions and clear examples, written in plain language, expose students to the rules and methods of probability. Exercises foster problem-solving skills, and all problems feature step-by-step solutions. 1997 edition. 368pp. 6 1/2 x 9 1/4. 0-486-47418-6

CHANCE, LUCK, AND STATISTICS, Horace C. Levinson. In simple, non-technical language, this volume explores the fundamentals governing chance and applies them to sports, government, and business. "Clear and lively ... remarkably accurate." – *Scientific Monthly*. 384pp. 5 3/8 x 8 1/2. 0-486-41997-5

FIFTY CHALLENGING PROBLEMS IN PROBABILITY WITH SOLUTIONS, Frederick Mosteller. Remarkable puzzlers, graded in difficulty, illustrate elementary and advanced aspects of probability. These problems were selected for originality, general interest, or because they demonstrate valuable techniques. Also includes detailed solutions. 88pp. 5 3/8 x 8 1/2. 0-486-65355-2

EXPERIMENTAL STATISTICS, Mary Gibbons Natrella. A handbook for those seeking engineering information and quantitative data for designing, developing, constructing, and testing equipment. Covers the planning of experiments, the analyzing of extreme-value data; and more. 1966 edition. Index. Includes 52 figures and 76 tables. 560pp. 8 3/8 x 11. 0-486-43937-2

STOCHASTIC MODELING: Analysis and Simulation, Barry L. Nelson. Coherent introduction to techniques also offers a guide to the mathematical, numerical, and simulation tools of systems analysis. Includes formulation of models, analysis, and interpretation of results. 1995 edition. 336pp. 6 1/8 x 9 1/4. 0-486-47770-3

INTRODUCTION TO BIOSTATISTICS: Second Edition, Robert R. Sokal and F. James Rohlf. Suitable for undergraduates with a minimal background in mathematics, this introduction ranges from descriptive statistics to fundamental distributions and the testing of hypotheses. Includes numerous worked-out problems and examples. 1987 edition. 384pp. 6 1/8 x 9 1/4. 0-486-46961-1

Physics

MATHEMATICAL TOOLS FOR PHYSICS, James Nearing. Encouraging students' development of intuition, this original work begins with a review of basic mathematics and advances to infinite series, complex algebra, differential equations, Fourier series, and more. 2010 edition. 496pp. 6 1/8 x 9 1/4. 0-486-48212-X

TREATISE ON THERMODYNAMICS, Max Planck. Great classic, still one of the best introductions to thermodynamics. Fundamentals, first and second principles of thermodynamics, applications to special states of equilibrium, more. Numerous worked examples. 1917 edition. 297pp. 5 3/8 x 8. 0-486-66371-X

AN INTRODUCTION TO RELATIVISTIC QUANTUM FIELD THEORY, Silvan S. Schweber. Complete, systematic, and self-contained, this text introduces modern quantum field theory. "Combines thorough knowledge with a high degree of didactic ability and a delightful style." – *Mathematical Reviews.* 1961 edition. 928pp. 5 3/8 x 8 1/2. 0-486-44228-4

THE ELECTROMAGNETIC FIELD, Albert Shadowitz. Comprehensive undergraduate text covers basics of electric and magnetic fields, building up to electromagnetic theory. Related topics include relativity theory. Over 900 problems, some with solutions. 1975 edition. 768pp. 5 5/8 x 8 1/4. 0-486-65660-8

THE PRINCIPLES OF STATISTICAL MECHANICS, Richard C. Tolman. Definitive treatise offers a concise exposition of classical statistical mechanics and a thorough elucidation of quantum statistical mechanics, plus applications of statistical mechanics to thermodynamic behavior. 1930 edition. 704pp. 5 5/8 x 8 1/4. 0-486-63896-0

INTRODUCTION TO THE PHYSICS OF FLUIDS AND SOLIDS, James S. Trefil. This interesting, informative survey by a well-known science author ranges from classical physics and geophysical topics, from the rings of Saturn and the rotation of the galaxy to underground nuclear tests. 1975 edition. 320pp. 5 3/8 x 8 1/2. 0-486-47437-2

STATISTICAL PHYSICS, Gregory H. Wannier. Classic text combines thermodynamics, statistical mechanics, and kinetic theory in one unified presentation. Topics include equilibrium statistics of special systems, kinetic theory, transport coefficients, and fluctuations. Problems with solutions. 1966 edition. 532pp. 5 3/8 x 8 1/2. 0-486-65401-X

SPACE, TIME, MATTER, Hermann Weyl. Excellent introduction probes deeply into Euclidean space, Riemann's space, Einstein's general relativity, gravitational waves and energy, and laws of conservation. "A classic of physics." – *British Journal for Philosophy and Science.* 330pp. 5 3/8 x 8 1/2. 0-486-60267-2

RANDOM VIBRATIONS: Theory and Practice, Paul H. Wirsching, Thomas L. Paez and Keith Ortiz. Comprehensive text and reference covers topics in probability, statistics, and random processes, plus methods for analyzing and controlling random vibrations. Suitable for graduate students and mechanical, structural, and aerospace engineers. 1995 edition. 464pp. 5 3/8 x 8 1/2. 0-486-45015-5

PHYSICS OF SHOCK WAVES AND HIGH-TEMPERATURE HYDRO DYNAMIC PHENOMENA, Ya B. Zel'dovich and Yu P. Raizer. Physical, chemical processes in gases at high temperatures are focus of outstanding text, which combines material from gas dynamics, shock-wave theory, thermodynamics and statistical physics, other fields. 284 illustrations. 1966–1967 edition. 944pp. 6 1/8 x 9 1/4. 0-486-42002-7